Convergence of Science and Engineering: A Promising Window toward Improving the Public Health

Convergence of Science and Engineering: A Promising Window toward Improving the Public Health

Guest Editor

Ali Zarrabi

MDPI

Basel • Beijing • Wuhan • Barcelona • Belgrade • Novi Sad • Cluj • Manchester

Guest Editor
Ali Zarrabi
Biomedical Engineering
Department
Istinye University
Istanbul
Turkey

Editorial Office
MDPI AG
Grosspeteranlage 5
4052 Basel, Switzerland

This is a reprint of the Special Issue, published open access by the journal *Bioengineering* (ISSN 2306-5354), freely accessible at: www.mdpi.com/journal/bioengineering/special_issues/convergence_sci_eng.

For citation purposes, cite each article independently as indicated on the article page online and using the guide below:

Lastname, A.A.; Lastname, B.B. Article Title. *Journal Name* **Year**, *Volume Number*, Page Range.

ISBN 978-3-7258-2932-3 (Hbk)
ISBN 978-3-7258-2931-6 (PDF)
https://doi.org/10.3390/books978-3-7258-2931-6

Contents

About the Editor

Ali Zarrabi

Dr. Ali Zarrabi is an accomplished scientist and Associate Professor with extensive expertise in bioengineering and nanotechnology. He serves as the Principal Investigator at Istinye University in Istanbul, Turkey, and is the Director of Scientific Development at March Health Co. in San Francisco, California, USA. With a strong academic background and a passion for interdisciplinary research, Dr. Zarrabi has published extensively in prestigious journals and edited special issues that focus on innovative solutions to global health challenges. His work bridges the gap between science and engineering, fostering collaborations that drive impactful advancements in public health and biomedical technologies.

Preface

In recent decades, the interdisciplinary nature of scientific research has revolutionized our approach to solving the complex challenges that impact human health and well-being. Traditionally, the efforts of scientists and engineers have been fragmented across many specialized fields. However, the need to address multifaceted biological systems and public health challenges has highlighted the critical importance of convergence—melding diverse disciplines to create innovative solutions. This reprint, titled "Convergence of Science and Engineering: A Promising Window toward Improving Public Health", stems from the contributions to a Special Issue of *Bioengineering*. It reflects the synergies that arise from integrating advanced engineering techniques with foundational scientific principles.

The chapters within this reprint present groundbreaking research and comprehensive reviews at the intersection of fields such as bioengineering, nanotechnology, artificial intelligence, and healthcare technology. By bridging disciplines like biology, mechanical engineering, and cognitive sciences, this compendium showcases how such collaborations can lead to advancements in disease prevention, early diagnostics, and effective treatment modalities. From wearable devices and functional biomaterials to tissue regeneration and bioinspired robotics, the contents of this Book illuminate how convergent sciences are transforming the landscape of modern medicine.

This work represents a collective effort to envision and construct a future where interdisciplinary innovation translates to tangible improvements in public health. It serves not only as a repository of knowledge but also as a catalyst for further exploration and collaboration among researchers, engineers, and healthcare professionals.

I extend my deepest gratitude to the authors who contributed their cutting-edge research and insightful reviews and to the editorial and publishing teams whose dedication made this reprint a reality. It is my hope that the ideas and innovations presented herein will inspire readers to continue pushing the boundaries of what is possible in science and engineering, ultimately enhancing the quality of life for individuals and communities worldwide.

Ali Zarrabi
Guest Editor

bioengineering

Editorial

Could the Convergence of Science and Technology Guarantee Human Health in the Future?

Ali Zarrabi

Department of Biomedical Engineering, Faculty of Engineering & Natural Sciences, Istinye University, 34396 Istanbul, Türkiye; ali.zarrabi@istinye.edu.tr or alizarrabi@gmail.com

Due to the daily growth of the world population, there has been an increase in concerns regarding health, especially due to the increase in the number of aged people, the surge of pollution, and the appearance of new pandemic diseases such as COVID-19 and influenza H1N1. One exciting approaches that provides much hope is the convergence of science and technology, which can improve the performance of routinely used strategies (in both diagnosis and treatment) and even create new approaches for monitoring the healthcare of the global population. This could be put into practice by creating new therapeutic compounds against cancer, which could target the intracellular pathways [1], or by producing new nanomaterials that could carry therapeutic compounds [2]. The convergence of science and technology could also be used for the creation of low-cost highly sensitive biosensors used for point-of-care applications, such as monitoring heart rates [3], detecting therapeutic compounds, such as antibiotics in pharmaceutical and clinical samples [4], and as highly sensitive and selective aptasensors [5]. This convergence could also help overcome the problem of low efficiency in conventional in vitro and in vivo models in the development and assessment of new drug formulation using different types of organ-on-chip models [6]. In this context, Osouli Tabrizi and his coworkers represented a type of cells-on-a-chip platform for the modeling of complementary metal oxide semiconductors, as a low-cost sensor for the detection of living cells, which could help detect periodontal diseases early and with high accuracy. The results of this study showed the effectiveness of this new sensor in monitoring the cells in very small sample saliva volumes (1 μL) [7]. In another study, the application of hydrogels containing drug, cells, and growth factor for the treatment of osteoporosis showed promising signs of improvement in comparison to implants and metallic scaffolds [8]. On the other hand, coating the currently used stainless-steel (SS) implants with a polymeric shell of zein/Ag-Sr doped mesoporous bioactive glass nanoparticles has reduced the toxicity effects of the SS implants and exhibited beneficial effects in bone regeneration applications [9]. In another study, Sikder et al. evaluated the combination of ultrasound (US) and the 3D-printed bioactive piezoelectric scaffolds for the treatment of orthopedic defects. They have shown that the fabricated scaffold that contained bioactive PCL-$BaTiO_3$ piezoelectric composite formulations could improve osteogenesis. On the other hand, the application of US in combination with this scaffold could induce pre-osteoblast adhesion, proliferation, spreading (at 1 Hz), and osteoblast differentiation (at 3 HZ) [10]. In a recent study, Vieira et al. introduced a new type of theranostic platform based on gellan gum hydrogel functionalized with Mn molecules, which showed T1-weighted MRI features, and at the same time, the capability of delivering stem cells to the central nervous system (CNS) in a noninvasive manner. Therefore, the combination of hydrogel-based formulation, Mn, and MRI technology resulted in the real-time monitoring of cell delivery to the CNS system [11]. Another interesting direction that the simultaneous application of science and engineering can take is the utilization of microorganisms for removing pollutant from the environment, while simultaneously producing useful materials. We have shown that *Chlorella vulgaris* (*C. vulgaris*), as a type of microalgea, has the capability of removing antibiotic contamination from the water via a

Citation: Zarrabi, A. Could the Convergence of Science and Technology Guarantee Human Health in the Future?. *Bioengineering* **2023**, *10*, 589. https://doi.org/10.3390/bioengineering10050589

Received: 17 April 2023
Accepted: 12 May 2023
Published: 13 May 2023

3-steps bioreaction, bio-adsorption, bioaccumulation, and biodegradation. In addition, this study shows that the microalgae produce some useful byproducts, such as proteins, lipids, chlorophyll-a, and carbohydrates [12].

Recently, with the emergence of artificial intelligence (AI) and the rapid growth of its application in different fields of medicine, there is renewed hope for the fast detection of diseases and the prediction of the effectiveness of the performance of new drugs and therapeutic methods, which will help professionals make better and more informed decisions [13,14]. Computational modeling is one of the subcategories of AI that could be used, in combination with in vitro tests, to predict and evaluate the results of different treatments. This could reduce the time of treatment and improve its performance. We have used the in silico 3D and single cell ventricle simulations to evaluate the effects of mexiletine on cardiac cells, and the results showed that this drug could reduce the action potential duration (APD) in a long QT variant 3 (LQT3) patients with an A1656D mutation, shift the occurrence of alternants from a normal heart rate in the cells to a quicker one, and eliminate the possibility of a breakup of the spiral wave [15]. In another study, *Arippa* et al. used computational modeling for the evaluation of the kinematic parameters related to Parkinson's disease (PD). They selected "61 people with PD (aged 68.9 $\pm$ 9.3 years) and 47 unaffected individuals age- and sex-matched (66.0 $\pm$ 8.3 years)" to evaluate the differences between them in angular trends at hip, knee, and ankle joints by applying the "computerized 3D gait analysis performed using an optical motion-capture system". They presented a new approach for the early diagnosis of PD since, according to their results, the patients had significant alterations in interlimb coordination, which could be detected at the hip and ankle joints and had "a modified gait pattern particularly at the terminal stance/early swing phase of the gait cycle" [16].

According to the aforementioned research, this Special Issue aims to highlight the beneficial effects of the convergence of science and technology in different aspects of human health, from the daily monitoring of health to the introduction of novel or improved treatment and diagnosis methods for different types of disorders, which could be helpful in tackling the current challenges health management among nations.

Funding: This research received no external funding.

Institutional Review Board Statement: Not applicable.

Informed Consent Statement: Not applicable.

Data Availability Statement: No new data were created or analyzed in this study. Data sharing is not applicable to this article.

Conflicts of Interest: The authors declare no conflict of interest.

References

1. Ravish, A.; Shivakumar, R.; Xi, Z.; Yang, M.H.; Yang, J.-R.; Swamynayaka, A.; Nagaraja, O.; Madegowda, M.; Chinnathambi, A.; Alharbi, S.A.J.B. De Novo Design of Imidazopyridine-Tethered Pyrazolines That Target Phosphorylation of STAT3 in Human Breast Cancer Cells. *Bioengineering* **2023**, *10*, 159. [CrossRef] [PubMed]
2. Monfared, Y.K.; Mahmoudian, M.; Hoti, G.; Bisericaru, D.M.; Caldera, F.; Cavalli, R.; Zakeri-Milani, P.; Matencio, A.; Trotta, F.J.B. Hyper-Branched Cyclodextrin-Based Polymers as Anticoagulant Agents: In Vitro and In Vivo Studies. *Bioengineering* **2022**, *9*, 765. [CrossRef] [PubMed]
3. Salas-Monedero, M.; Lozano-Berrio, V.; Cazorla-Martínez, M.-J.; Ceruelo-Abajo, S.; Gil-Agudo, Á.; Hernández-Sánchez, S.; Jiménez-Díaz, J.-F.; DelosReyes-Guzmán, A.J.B. Is it Feasible to Use a Low-Cost Wearable Sensor for Heart Rate Monitoring within an Upper Limb Training in Spinal Cord Injured Patients?: A Pilot Study. *Bioengineering* **2022**, *9*, 763. [CrossRef] [PubMed]
4. Patil, V.B.; Ilager, D.; Tuwar, S.M.; Mondal, K.; Shetti, N.P.J.B. Nanostructured ZnO-Based Electrochemical Sensor with Anionic Surfactant for the Electroanalysis of Trimethoprim. *Bioengineering* **2022**, *9*, 521. [CrossRef] [PubMed]
5. Zhao, Z.; Das, S.; Zharnikov, M.J.B. Rational Design of Porous Poly (ethylene glycol) Films as a Matrix for ssDNA Immobilization and Hybridization. *Bioengineering* **2022**, *9*, 414. [CrossRef] [PubMed]
6. Gholizadeh, H.; Cheng, S.; Kourmatzis, A.; Xing, H.; Traini, D.; Young, P.M.; Ong, H.X.J.B. Application of Micro-Engineered Kidney, Liver, and Respiratory System Models to Accelerate Preclinical Drug Testing and Development. *Bioengineering* **2022**, *9*, 150. [CrossRef] [PubMed]

7. Osouli Tabrizi, H.; Panahi, A.; Forouhi, S.; Sadighbayan, D.; Soheili, F.; Haji Hosseini Khani, M.R.; Magierowski, S.; Ghafar-Zadeh, E.J.B. Oral Cells-On-Chip: Design, Modeling and Experimental Results. *Bioengineering* **2022**, *9*, 218. [CrossRef] [PubMed]
8. Alabdah, F.; Alshammari, A.; Hidalgo-Bastida, A.; Cooper, G.J.B. A Review of Conventional and Novel Treatments for Osteoporotic Hip Replacements. *Bioengineering* **2023**, *10*, 161. [CrossRef] [PubMed]
9. Batool, S.A.; Liaquat, U.; Channa, I.A.; Gilani, S.J.; Makhdoom, M.A.; Yasir, M.; Ashfaq, J.; Jumah, M.N.b.; Rehman, M.A.U.J.B. Development and Characterization of Zein/Ag-Sr Doped Mesoporous Bioactive Glass Nanoparticles Coatings for Biomedical Applications. *Bioengineering* **2022**, *9*, 367. [CrossRef] [PubMed]
10. Sikder, P.; Nagaraju, P.; Naganaboyina, H.P.J.B. 3D-Printed Piezoelectric Porous Bioactive Scaffolds and Clinical Ultrasonic Stimulation Can Help in Enhanced Bone Regeneration. *Bioengineering* **2022**, *9*, 679. [CrossRef] [PubMed]
11. Vieira, S.; Strymecka, P.; Stanaszek, L.; Silva-Correia, J.; Drela, K.; Fiedorowicz, M.; Malysz-Cymborska, I.; Janowski, M.; Reis, R.L.; Łukomska, B.J.B. Mn-Based Methacrylated Gellan Gum Hydrogels for MRI-Guided Cell Delivery and Imaging. *Bioengineering* **2023**, *10*, 427. [CrossRef] [PubMed]
12. Ricky, R.; Chiampo, F.; Shanthakumar, S.J.B. Efficacy of Ciprofloxacin and Amoxicillin Removal and the Effect on the Biochemical Composition of Chlorella vulgaris. *Bioengineering* **2022**, *9*, 134. [CrossRef] [PubMed]
13. Rong, G.; Mendez, A.; Assi, E.B.; Zhao, B.; Sawan, M.J.E. Artificial intelligence in healthcare: Review and prediction case studies. *Engineering* **2020**, *6*, 291–301. [CrossRef]
14. Cutillo, C.M.; Sharma, K.R.; Foschini, L.; Kundu, S.; Mackintosh, M.; Mandl, K.D.; MI in Healthcare Workshop Working Group. Machine intelligence in healthcare—Perspectives on trustworthiness, explainability, usability, and transparency. *NPJ Digit. Med.* **2020**, *3*, 47. [CrossRef] [PubMed]
15. Qauli, A.I.; Yoo, Y.; Marcellinus, A.; Lim, K.M.J.B. Verification of the Efficacy of Mexiletine Treatment for the A1656D Mutation on Downgrading Reentrant Tachycardia Using a 3D Cardiac Electrophysiological Model. *Bioengineering* **2022**, *9*, 531. [CrossRef] [PubMed]
16. Arippa, F.; Leban, B.; Monticone, M.; Cossu, G.; Casula, C.; Pau, M.J.B. A Study on Lower Limb Asymmetries in Parkinson's Disease during Gait Assessed through Kinematic-Derived Parameters. *Bioengineering* **2022**, *9*, 120. [CrossRef] [PubMed]

bioengineering

MDPI

Article

Is it Feasible to Use a Low-Cost Wearable Sensor for Heart Rate Monitoring within an Upper Limb Training in Spinal Cord Injured Patients?: A Pilot Study

Miriam Salas-Monedero [1,2], Vicente Lozano-Berrio [1], María-Jesús Cazorla-Martínez [3], Silvia Ceruelo-Abajo [4], Ángel Gil-Agudo [1,4], Sonsoles Hernández-Sánchez [5], José-Fernando Jiménez-Díaz [5] and Ana DelosReyes-Guzmán [1,*]

1 Biomechanics and Technical Aids Unit, Hospital Nacional de Parapléjicos (SESCAM), Finca La Peraleda s/n CP 45071, 14507 Toledo, Spain
2 International Doctoral School, Castilla La-Mancha University, 14507 Toledo, Spain
3 Nursing Department, Hospital Nacional de Parapléjicos (SESCAM), 14507 Toledo, Spain
4 Rehabilitation Department, Hospital Nacional de Parapléjicos (SESCAM), 14507 Toledo, Spain
5 Performance and Sport Rehabilitation Laboratory, Faculty of Sports Sciences, Castilla- La Mancha University, 14507 Toledo, Spain
* Correspondence: adlos@sescam.jccm.es

Citation: Salas-Monedero, M.; Lozano-Berrio, V.; Cazorla-Martínez, M.-J.; Ceruelo-Abajo, S.; Gil-Agudo, Á.; Hernández-Sánchez, S.; Jiménez-Díaz, J.-F.; DelosReyes-Guzmán, A. Is it Feasible to Use a Low-Cost Wearable Sensor for Heart Rate Monitoring within an Upper Limb Training in Spinal Cord Injured Patients?: A Pilot Study. *Bioengineering* **2022**, *9*, 763. https://doi.org/10.3390/bioengineering9120763

Academic Editor: Ali Zarrabi

Received: 28 October 2022
Accepted: 25 November 2022
Published: 3 December 2022

Abstract: (1) Background: Cervical spinal cord injury (SCI) patients have impairment in the autonomic nervous system, reflected in the cardiovascular adaption level during the performance of upper limb (UL) activities carried out in the rehabilitation process. This adaption level could be measured from the heart rate (HR) by means of wearable technologies. Therefore, the objective was to analyze the feasibility of using Xiaomi Mi Band 5 wristband (XMB5) for HR monitoring in these patients during the performance of UL activities; (2) Methods: The HR measurements obtained from XMB5 were compared to those obtained by the professional medical equipment Nonin LifeSense II capnograph and pulse oximeter (NLII) in static and dynamic conditions. Then, four healthy people and four cervical SCI patients performed a UL training based on six experimental sessions; (3) Results: the correlation between the HR measurements from XMB5 and NLII devices was strong and positive in healthy people (r = 0.921 and r = 0.941 ($p < 0.01$) in the static and dynamic conditions, respectively). Then, XMB5 was used within the experimental sessions, and the HR oscillation range measured was significantly higher in healthy individuals than in patients; (4) Conclusions: The XMB5 seems to be feasible for measuring the HR in this biomedical application in SCI patients.

Keywords: upper limb training; spinal cord injury; heart rate; wearable devices; biomedical application

1. Introduction

Recent epidemiological studies show that the prevalence of SCI in Spain is between 270 and 380 million per inhabitants. According to data from the Spanish National Institute of Statistics, 2.86% of the Spanish population was affected by SCI [1]. SCI of non-traumatic etiology has been the main cause of SCI (58%), followed by SCI of traumatic etiology (42%), thus breaking the tendency of the last ten years, during which the main origin was traumatic [2].

After a SCI, patients have an affectation in the autonomic nervous system in lesions above the T6 metameric level. For this reason, patients with cervical SCI present differences compared to the healthy condition in terms of the cardiovascular adaptation level during the execution of a therapy or effort [3]. Another consequence of cervical SCI is the impaired UL function, which affects the performance of activities of daily living (ADL). Thus, the main objective of the therapeutic programme is for patients to achieve the maximum level of independence in the performance of ADLs [4,5].

Therefore, some of the most widely used methods to evaluate training effectiveness in adults, children, and adolescents generally include the use of HR monitors or pulse oximeters [6,7] and the use of motion sensors such as pedometers or accelerometers [8,9], based on the movement (or acceleration) of the limbs, and the trunk is closely related to the global energy expenditure.

Longitudinal studies on HR in SCI have analyzed HR in the acute and chronic phases of the injury, paying attention to the PA level that the patient performs in order to improve the performance of ADL activities [10–12].

On the other hand, in relation to the medical equipment used for recording the HR, all these studies have used different equipment, such as Holters digital blood pressure systems to monitor pressures, which records for an approximate period of up to a week [11]; and a HR monitor with ECG signal, option of pulxioximetry, non-invasive pressure, with presentation of HR reader [12]. However, in other studies, wearable technology like watches with chest straps and a heart frequency reader has been used [10]. El-Amrawy et al. [13] showed that the new wireless technologies, such as blood pressure trackers and body sensor devices could have a great impact on health care systems and quality of life, and analyzed the accuracy and precision of 17 wearable devices for HR monitoring. Bent et al. [14] demonstrated that wearables also provide excess HR information during low-intensity physical activity, which could be a safety mechanism built into the devices to ensure that patients do not exceed their maximum HR during training. Other authors have analyzed different characteristics of sensors with the aim of optimizing a wider variety of applications [15]. This is especially important for clinicians, as they would be aware of HR measurements during the training when they perform workload intensity assessments in movement-based trainings.

There is evidence about the use of Xiaomi Mi band device, showing the best package and choice, taking into account the quality–price ratio [13]. Based on this previous study, the XMB5 was chosen for the present research. However, these devices, which are commercial, are being purchased for leisure purposes as a way to quantify daily physical activity and hours of sleep, without paying attention to the heart rate readings they provide. In this respect, special attention is needed when dealing with patients, especially neurological patients. From our point of view, the novelty of this study focuses on the intention to use this device, XMB5, in the context of a biomedical application following a methodology designed for patients with cervical SCI. The main objective is to analyze the feasibility of this wearable device for monitoring the HR in cervical SCI patients. To reach this objective, an experimental study was conducted with the aim of comparing the HR measured by means of XMB5 with those measured by a capnograph/pulse oximeter used in the clinical setting for the monitoring of patients, simultaneously. Then, the Xiaomi wearable device will be used in a biomedical application, within a proof-of-concept study, for monitoring the HR in healthy people and cervical SCI patients during the performance of UL trainings within the rehabilitation program.

The remainder of this paper consists of the following: Section 2 provides all the information about the methodology, describing the study design, the participants' characteristics, the equipment used, the experimental setup and methodology for data acquisition, and the statistical analysis; Section 3 is related to the results obtained in terms of variables at starting and ending conditions and variables related to each experimental session; Section 4 includes the discussion of the results, and Section 5 is the study's conclusion.

2. Materials and Methods

This section includes all the information regarding the experimental protocol applied in this research. For that purpose, subsections about the study design, the participants included in the study, the experimental setup and data acquisition, the outcomes variables, and the statistical analysis were included.

2.1. Study Design

The study performed was descriptive, responding to an observational design with an experimental phase.

2.2. Participants

In this study, 8 subjects participated; they were divided into two groups: a neurologically healthy group formed by 4 subjects and a group of 4 SCI patients, all of them with UL motor function impairments. All patients suffered a cervical SCI with a metameric level between C4 and C8, and AIS grade between A and D, as defined by the International Standards for the Neurological Classification of Spinal Cord injury [16], specifically the upper limb motor score of the right arm (UER). Patients who presented any vertebral deformity, joint restriction, surgery on any of the UL, balance disorders, dysmetria due to associated neurologic or orthopedic disorders, or visual acuity defects were excluded. The UER was obtained with the clinical staff assessment of the strength of five muscle groups of the dominant UL. Each muscle group can be evaluated between 0 (no function) and 5 (normal function), with a total of 25 points.

This study was carried out in the Biomechanics and Technical Aids Unit of the Hospital Nacional de Parapléjicos (Toledo, Spain). All patients signed an informed consent form before the study. The guidelines of the declaration of Helsinki were followed in every case, and the study design was approved by the local ethics committee. Subject demographics are provided in Table 1.

Table 1. Demographics and functional characteristics of the sample analyzed.

Variables	Sample Analyzed	
	Healthy (n = 4)	**Patients (n = 4)**
Sex(Male) *	1.00 ± 25.00	3.00 ± 75.00
Age(Years) +	31.50 ± 6.00	27.50 ± 13.20
Weight (kg)	62.00 ± 7.05	70.25 ± 7.25
Height (cm)	167.40 ± 10.24	172.50 ± 9.32
Etiology Injury (Traumatic)	-	4.00 ± 100.00
Time since injury (months)	-	5.50 ± 1.50
Injury Level		
C6	-	1.00 ± 25.00
C7	-	1.00 ± 25.00
C8	-	2.00 ± 50.00
AIS Classification		
A	-	2.00 ± 50
B	-	1.00 ± 25
C	-	-
D	-	1.00 ± 25
Right UE Motor score	25.00 ± 0.00 [a]	20.50 ± 3.50 [a]

[a] ($p < 0.01$); * categorical variables are expressed as frequency and percentage; + continuous variables are expressed as mean and standard deviation.

2.3. Experimental Setup and Data Acquisition

HR was monitored using the XMB5, which is considered a low cost-device among the wearable technology devices [13]. The characteristics of the XMB5 bracelet are as follows. The sensor placement is in the central hole, which is well adjusted for the comfortable use of the bracelet. Moreover, the HR sensor is kept in contact with the skin to allow for the taking of measurements. The bracelet is manufactured in silicon with several measures for adjusting the bracelet to different wrist sizes (Figure 1). The sensor has an ADI accelerometer for quantifying the physical activity by means of the steps taken or running performed during the day. A previous study analyzed the optoelectronic characteristics of a silicon light-emitting device [17].

Figure 1. All participants wore the Xiaomi Mi Band 5 on their right arm during the experiments.

In the present study, the application Mi Fit, running under Android, was used for monitoring the HR. Measurements obtained from XMB5 were compared to a positive control usually used in clinical setting for the monitoring of patients, the professional medical equipment Nonin LifeSense® II (NLII), capnograph and pulse oximeter, with CE marking as medical device [18–20]. This system was simultaneously used on the same hand wearing the XMB5.

Firstly, the HR measurements from XMB5 were compared to those measured by the NLII system on a sample of four healthy people. Both systems were used simultaneously, and Nonin was placed on the same hand wearing the XMB5, namely the left one. For each participant, 18 readings were recorded for each tracker simultaneously while the participant stayed in a resting position and then while the right arm performed the UL movements required by a UL activity. During the recording, the participant was seated in front of a table and the left hand remained at rest.

Following the completion of this first study, the UL training was performed in four healthy subjects and four subjects with cervical SCI. All participants performed the proposed UL training with the dominant arm, determined from the Edinburg dominance test [21]. The complete UL training in each participant lasted two weeks, with three training sessions per week of a duration of 30 min. Each session was based on UL activities by means of Leap Motion Controller (LMC) performing the "Robot Assembly" and "Petal picking" applications available in the Playground of LMC.

The duration of each session and the number of sessions per week were established while taking into account the recommendations of the resistance work of the American College of Sports Medicine (ACSM) for people with SCI [22].

The first step was to fit the XMB5 fitness wristband to the wrist of the dominant arm [21].

Each participant performed both therapeutic interventions in front of a table that was adjustable in height, with the aim of normalizing the starting position for all the participants. The arm was placed against the trunk; the elbow was flexed 90° in a neutral pronation-supination position and with the hand resting on the edge of the table, with the palm opened in medial position. Then, the LMC device was placed on the table at 75% of the maximum UL reaching movement of the participant to avoid compensatory movements with the trunk.

All the participants performed the experimental sessions in their own wheelchair or sitting in a conventional chair. The participants rested their back firmly against the back of the chair with an angle between the backrest and the seat of 90–100°, and their feet on the footrest at a 90° angle.

During the sessions, the participants wore the XMB5 used for the HR and the monitoring of the number of UL movements. The XMB5 was synchronized to the Smartphone of the researcher responsible of the experimental session. The mobile application used in this study was My Fit.

2.4. Outcome Measures

To compare both systems of HR monitoring, simultaneous measurements were taken in four healthy subjects. For each participant, 18 readings were taken in both experimental conditions: the static, in the absence of UL movement and during the performance of UL movements. In these situations, the variables analyzed were: the difference between the magnitude of the HR readings from both XMB5 and NLII medical equipment to analyze the agreement level of the simultaneous HR readings. Moreover, the trend of these readings was analyzed by means of a correlation analysis.

Therefore, the variables registered in the UL training previously described were as follows. As the training program was composed of several experimental sessions, two kinds of variables were registered: (a) variables within each experimental session and (b) variables measured at the beginning and at the end of the training over all the participants, both the healthy ones and the SCI patients.

Thus, within each experimental session, the HR was registered for analysis: the HR in baseline, the maximum, the minimum, the oscillation range, the HR at the end of the experimental session and the difference between the readings at the end and the baseline. In relation to the UL performance, the repetition number for each activity was registered. Moreover, the effort perception level by means of Borg Scale [23] was analyzed.

In relation to the UL training, all the subjects were evaluated at the beginning of the training and at the end by means of the Krupp scale, with the aim of determining the fatigue level associated to the neurological disease [24], and the amount of UL movements during a complete day was registered for all the subjects. To obtain this measurement, the subjects wore the XMB5 wristband over two complete days each time.

2.5. Statistical Analysis

The statistical analysis was performed by the SPSS 17.0 for Windows software (SPSS Inc., Chicago, IL, USA). The clinical and demographic characteristics of the participants were analyzed by descriptive statistics, showing the results as the mean and standard deviation.

In relation to the validation study between the XMB5 and NLII professional equipment, the paired t-test and a correlation analysis was conducted for the static and UL movement conditions.

In relation to the UL training program, the Mann–Whitney U non-parametric test was applied to analyze the differences between the two groups analyzed (healthy subjects and patients with SCI). The variables analyzed with this method were the HR reading at baseline and at the end of the experimental session, and the difference between both readings; the maximum, minimum, and the heart rate oscillation range during the performance of the experimental session; the number of UL movements; and the perception of effort according to the Borg scale. In addition, possible differences in the number of repetitions in each series were analyzed. Moreover, at the beginning and end of the complete training program, the punctuation in the Krupp scale and the number of total UL movements were analyzed for all participants.

3. Results

This section describes the results regarding the correlation between the HR measurements from both devices, XMB5 and NLII, in Section 3.1 and the analysis of the last experimental session in Section 3.2.

3.1. Feasibility of Using XMB5 Wristband

In this subsection, the results about the comparison between the HR readings from the XMB5 wristband and NLII medical equipment are shown.

The Pearson correlation coefficient between the readings from both systems suggests a strong and positive linear association between both variables: r = 0.921 ($p < 0.01$) in the static condition and r = 0.941 ($p < 0.01$) for the HR readings taken during the performance of UL movements. This trend is shown in the scatter graphics included in Figure 2.

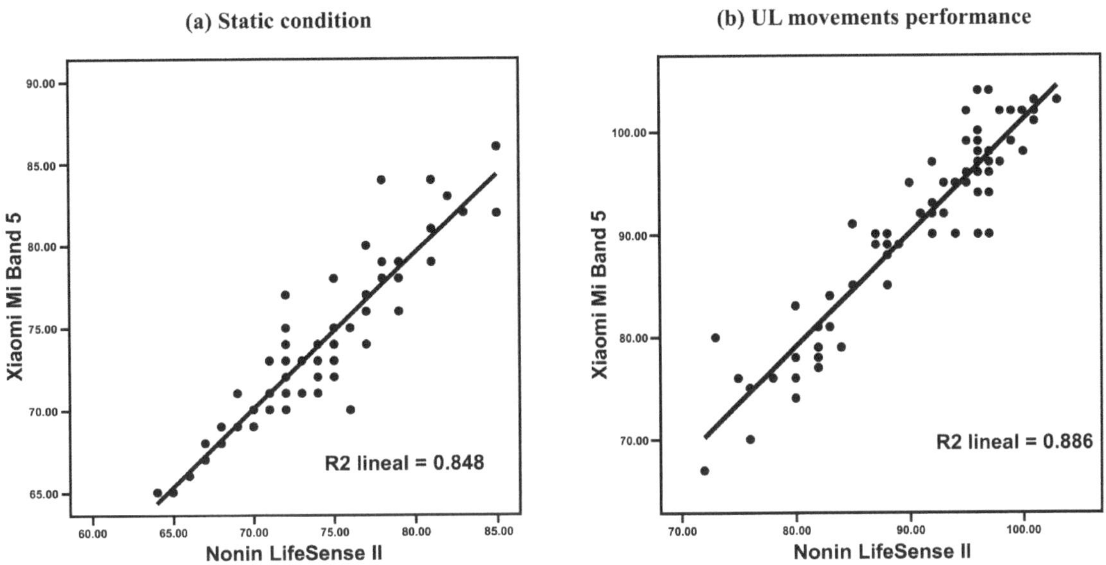

Figure 2. Scatter diagrams between the heart rate measured by Nonin Lifesense II and the heart rate measured by Xiaomi Mi Band 5 wristband during a static posture (**a**) and dynamic condition (**b**).

Both XMB5 and NLII systems registered similar HR readings in the static condition: 73.80 ± 4.79 and 73.97 ± 4.61, respectively. During the performance of UL movements, both systems also registered similar readings: 90.33 ± 9.38 for XMB5 and 90.20 ± 7.96 for Nonin medical equipment.

The differences observed between both XMB5 and NLII system were not statistically significant: t(71) = −0.754, p = 0.454 for the static condition, and t(71) = 0.323, p = 0.748 during the execution of UL movements.

3.2. Last Experimental Session within the UL Training

The sample analyzed was divided into two groups; neurologically healthy subjects and SCI subjects that were matched in age, weight, and height. Thus, no statistically significant differences were found for demographic characteristics between both groups analyzed (Table 1).

After performing the UL training based on the execution of six experimental sessions, the last session was chosen for analysis, because in that session, the healthy and SCI subjects all reached optimal performance.

Statistically significant differences were found between both groups (neurologically healthy and SCI) in variables related to the training performance: the number of repetitions in both activities within the training was lower in SCI subjects than in healthy people: in activity 1, 24.50 ± 12.25 and 11.00 ± 10.50 repetitions, respectively and in activity 2, 80.00 ± 34.25 and 26.50 ± 21.50, respectively, in each group. In relation to the HR, the statistical differences were found in the maximum value, in the oscillation range, and in the difference between the readings at baseline and at the end of the experimental session (Table 2).

Table 2. Outcome variables measured within the experimental session number 6 for healthy and SCI patients.

Variables	Healthy (n = 4)	SCI (n = 4)
Activities UL training (repetitions)		
Activity1 (Robot Assembly) *	24.50 ± 12.25 [a]	11.00 ± 10.50 [a]
Activity2 (Petal picking) *	80.00 ± 34.25 [a]	26.50 ± 21.50 [a]
Heart Rate (bpm)		
Baseline (before) +	63.50 ± 4.50	60,50±26.00
Ending session +	101.00 ± 38.25	78.00±36.00)
Difference (baseline-ending) +	38.50 ± 37.25 [a]	9.00 ± 15.50 [a]
Minimum +	53.00 ± 7.00	59.50 ± 9.50
Maximum +	149.50 ± 64.50 [a]	100.50 ± 20.75 [a]
Range +	92.50 ± 65.50 [a]	44.00 ± 14.25 [a]
Perceived effort		
Borg Scale (1–10)	3.00 ± 1.50	4.00 ± 1.50

a, $p < 0.05$. The results are expressed as mean and standard deviation, * this variable is expressed in repetitions achieved in the activity (rep.), + this variable is expressed in beats per minute (bpm).

Furthermore, it was observed that the HR before the session, in a baseline situation, was similar for both groups analyzed. However, after the experimental session, healthy subjects had a higher HR value than SCI subjects (101.00 ± 38.25 > 78.00 ± 36.00), but no statistically significant differences were found. In relation to the difference between the HR readings at the end and baseline conditions, statistically significant differences were found between healthy and SCI subjects (38.50 ± 37.25 > 9.00 ± 15.50, $p < 0.05$) (Table 2 and Figure 3).

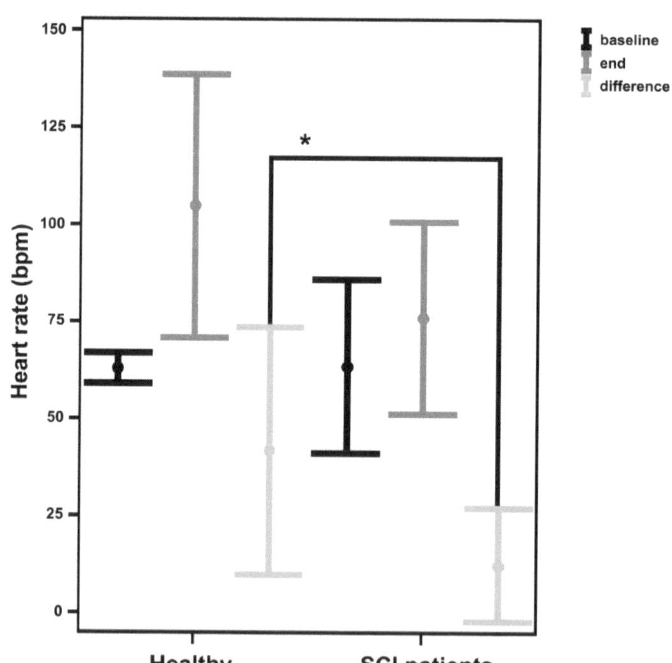

Figure 3. Mean heart rate measurements for neurologically healthy participants and SCI patients at baseline before the experimental session (black lines) and at the end of the experimental session (gray color), and the oscillation range (light gray color). * Significant statistically differences ($p < 0.05$).

Within the last experimental session, statistically significant differences were found between the two groups in the maximum value of heart rate and the oscillation range measured. The maximun value was higher in healthy people (149.50 ± 64.50) than in SCI patients (100.50 ± 20.75, $p < 0.05$) and, as a consequence, this difference was found in the oscillation range in healthy vs. SCI patients (92.50 ± 65.50 > 44.50 ± 14.25, $p < 0.05$).

The effort perception, analyzed through the Borg scale, was greater for the SCI patients group than for the healthy individuals, but no significant differences were found.

3.3. Variables at Baseline and at Ending the Complete UL Training

In this case, the variables related to the Krupp scale and the number of UL movements were analyzed, and the results are shown in Table 3.

Table 3. Results in the Krupp scale and the UL movements number at baseline (before starting the training) and at the end of all the experimental sessions.

	Baseline		Ending	
	Healthy (n = 4)	SCI (n = 4)	Healthy (n = 4)	SCI (n = 4)
Krupp Scale	2.33 ± 0.61	3.38 ± 1.28	1.99 ± 0.48	2.49 ± 0.69
UL Movement *	11.162 ± 5.055	1.418 ± 9.800	10.247 ± 6.739	12.409 ± 5.496

* The results are expressed as mean and standard deviation.

Fatigue decreased after the end of the complete training in both groups analyzed. However, no significant differences were found within the groups between the two assessments: in healthy people (2.33 ± 0.61 > 1.99 ± 0.48) and SCI patients (3.38 ± 1.28 > 2.49 ± 0.69). On the other hand, in relation to the UL movements, there was an important increase for this variable in SCI patients between both assessments (1.418 ± 9.800 < 12.409 ± 5.496), expressed as thousand units (Figure 4).

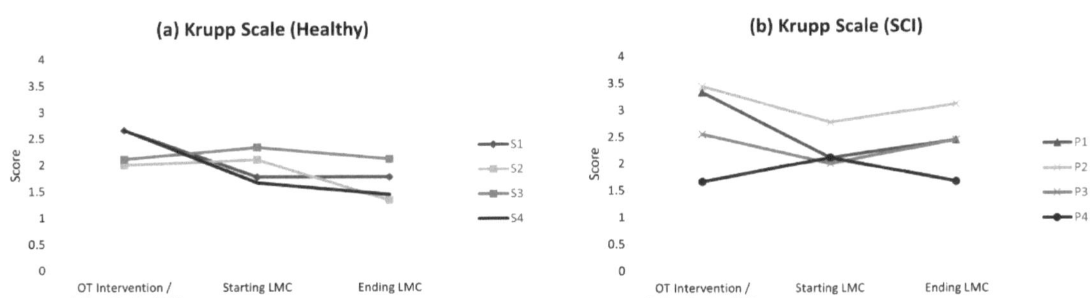

Figure 4. Evolution of the participants in the results obtained on the Krupp scale: (**a**) healthy participants and (**b**) SCI patients.

4. Discussion

The results obtained in this study as a proof of concept suggest that the XMB5 wristband seems to be a feasible way to monitor HR in SCI patients. This device was chosen for this study because other authors previously found that this device was the best in relation to quality and price [13]. In relation to the materials, the wristband was manufactured in silicon with several measures for adjusting the wristband to different wrist sizes. This characteristic was very important in the context of a biomedical application because it allowed for the wristband to be placed on different patients without difficulty.

The ADI accelerometer was used for measuring the number of UL movements in SCI patients from the number of directional changes in the arm, considering that the wristband is placement on the wrist. Other commercial wristbands based on GPS for quantifying physical activity were not suitable for measuring small and low-level UL activity in patients.

Obviously these measurements have an estimation error. For that reason, the first step was the feasibility analysis of the HR readings from XMB5 against those simultaneously measured by a clinical professional device used for HR monitoring in patients (NLII). In this study, a difference between both systems around the mean HR value of 0.166 bpm in absolute value was obtained with a standard deviation of 1.876 bpm in the absence of movement with both arms in a rest situation. However, during the execution of the UL movements, the standard deviation measured was higher (3.284 bpm). Nevertheless, the correlation analysis between the HR readings of both systems was high. There is no evidence from similar studies against which to compare the results obtained in this research, but there are those who suggest the importance of such studies in a clinical setting to monitor HR for any treatment [14]. It is very important for many people with acute CML to ensure that they do not exceed their maximal HR during exercise, which is a circumstance that can lead to adverse cardiac events [25].

The measurements obtained during the undertaking of activities by wearable sensors may contribute more accurate and objective insights into clinically meaningful changes in impairment, activity, and participation during the neurological rehabilitation process [26]. Furthermore, the XMB5 wearable device has been used in a biomedical application, within a small experimental study, for HR monitoring in healthy people and SCI patients during the performance of UL training. It has been proven that wearable sensors give us an excess of information about HR during low-intensity physical activity, such as that which is carried out in the different treatments and therapies in the neurorehabilitation process, which could be a safety mechanism incorporated into the patients' devices so that they do not exceed their maximum HR during the execution of therapeutic activities [14].

In this study, there was a decrease in the HR oscillation range between session 1 and session 6 in healthy subjects and people with SCI. This adaptation in healthy subjects is generally explained by an increase in the sympathetic pathway activity [27]. However, because people with cervical SCI have limited sympathetic efferent innervations to the cardiovascular system [28,29], studies suggest that changes in HR variability as soon as a the HR is low could be regulated by the parasympathetic branch [30]. This means that this adaptation to exercise should, in the case of SCI, lead to an improvement in the vagal modulation of the HR dynamics, related to the theory of HR variability "trainability" [31]. In fact, this study showed that people with cervical SCI are trained, since the HR readings at the end of the experimental session are higher than the readings at baseline condition. Although the sample size of the experimental groups is small and therefore the study has been conducted as a pilot study, for the sample analyzed, a different behavior was observed between healthy participants and SCI patients in the difference between the HR measured at the beginning of the session and at the end, showing a different cardiovascular behavior with training. This is reflected in Figure 3. Therefore, patients retain the ability to produce positive adaptations in autonomous cardiac regulation, coinciding with other studies focused on the health of people with cervical SCI, since it shows that moderate or vigorous physical activity can be scheduled in the short term [32].

Because the sympathetic pathway and the vagal modulation increase, the effort perception in patients with cervical SCI and healthy people decreases as the number of sessions progresses, leading to the performance of activities with lower energy expenditure. This lower energy expenditure results in an increase in the number of executions of the activities proposed. It should be noted that healthy subjects obtained a higher performance, but this is due to the fact that people with SCI have limitations in the execution of UL activities, in a greater extent depending on the injury level and the severity classified according the ASIA scale. Moreover, a practice effect has been observed across the experimental sessions focused on the UL training by means of reaching and grasping movements [33]. These movements are non-ballistic and in a closed loop; they allowed sensory information to arrive, obtaining a feedback in the performance of the activities, helping healthy subjects and patients with SCI to improve these movements through a planning of the movement, thus managing to store a "sensorial trace" [34] with which they could compare their future executions by reinforcing that movement, improving their effectiveness and increasing the UL movements number.

Finally, it is necessary to take into account the limitations of the present study. Firstly, there is a limitation in relation to the size of the sample analyzed because this is a feasibility study with the aim of obtaining preliminary results in relation to the use of a low-cost device for HR monitoring during the performance of UL trainings.

On the other hand, a comparison with other studies in the literature is difficult due to there being no evidence from similar studies. In other studies on SCI populations that aim to measure the HR, different seated conditions were analyzed [34]. Moreover, the lack of standardization in the experimental protocols and the functional tasks chosen complicate the comparison between studies. In fact, the HR oscillation ranges measured depend on the UL functional task chosen.

Furthermore, the patients sample analyzed was very heterogeneous. Thus, the patient less affected in relation to the UL function was classified as a metameric level injury C8 and AIS classification D (P3). This patient showed a general UL functionality similar to a healthy subject, whereas the functional deficit in relation to the dexterity and ability of the hand may appear in dexterity clinical tests. However, the results obtained in relation to the patient P4 were very different because the cervical injury was higher and more severe. In both patients, the HR oscillation range was lower than in healthy subjects because of the affectation in the autonomous nervous system.

Taking into account these aspects, a further study should be made with a greater sample of SCI patients. Therefore, an electromyographic registry should be made simultaneously with the aim of detecting the appearance of possible peripheral fatigue during the performance of UL training in SCI patients.

5. Conclusions

Low-cost technology, in this case XMB5, seems to be suitable for monitoring the HR during UL training by means of technology, LMC, in patients with cervical SCI. For this reason, this methodology will be incorporated within a biomedical application for registering the HR within an UL training based on immersive virtual reality applications for SCI patients with the aim of determining the optimal intensity and duration of the experimental sessions.

Author Contributions: Conceptualization, M.S.-M., S.C.-A. and A.D.-G.; methodology, M.-J.C.-M., V.L.-B., M.S.-M. and A.D.-G.; validation, M.S.-M., V.L.-B. and A.D.-G.; formal analysis, M.-J.C.-M., V.L.-B. and M.S.-M.; investigation, M.S.-M., S.H.-S., J.-F.J.-D. and A.D.-G.; resources, Á.G.-A., M.-J.C.-M. and S.C.-A.; data curation, M.-J.C.-M., S.C.-A. and Á.G.-A.; writing—original draft preparation, M.S.-M., V.L.-B., Á.G.-A., S.H.-S., J.-F.J.-D. and A.D.-G.; writing—review and editing, M.S.-M., M.-J.C.-M., S.C.-A., S.H.-S., J.-F.J.-D. and A.D.-G.; supervision, Á.G.-A. and A.D.-G.; project administration, A.D.-G.; funding acquisition, A.D.-G. All authors have read and agreed to the published version of the manuscript.

Funding: Grant PID2020-117361RB-C22 funded by MCIN/AEI/10.13039/501100011033.

Institutional Review Board Statement: The study was conducted in accordance with the Declaration of Helsinki and approved by the Institutional Review Board (or Ethics Committee) of Complejo Hospitalario de Toledo (protocol code 183 and date of approval 11 January 2018).

Informed Consent Statement: Informed consent was obtained from all subjects involved in the study.

Data Availability Statement: Data sharing is not applicable to this article.

Acknowledgments: Project RehabHand (Plataforma de bajo coste para rehabilitación del miembro superior basado en Realidad Virtual, ref. DPI2016-77167-R).

Conflicts of Interest: The authors declare no conflict of interest.

References

1. Statistical Portal Health, Social Services and Equality Spain Ministery. Available online: https://www.msssi.gob.es/estadEstudios/portada/home.htm (accessed on 10 September 2022).
2. Torres Alaminos, M.A. Epidemiological aspects of spinal cord injury at the National Paraplegic Hospital. *Ene 12(2) ago*, 2018.
3. Wecht, J.M.; Harel, N.Y.; Guest, J.; Kirshblum, S.C.; Forrest, G.F.; Bloom, O.; Ovechkin, A.V.; Harkema, S. Cardiovascular Autonomic Dysfunction in Spinal Cord Injury: Epidemiology, Diagnosis, and Management. *Semin Neurol.* **2020**, *40*, 550–559. [CrossRef]
4. Fulk, G.D.; Bowden, M.; Behrman, A.L. Traumatic Spinal Cord Injury. In *Physical Rehabilitation*, 7th ed.; O'Sullivan, S.B., Schmitz, T.J., Fulk, G., Eds.; McGraw Hill: New York, NY, USA, 2019; Available online: https://fadavispt.mhmedical.com/content.aspx?bookid=2603&sectionid=214793570 (accessed on 27 October 2022).
5. Beninato, M.; O'Kane, K.S.; Sullivan, P.E. Relationship between motor FIM and muscle strength in lower cervical-level spinal cord injuries. *Spinal Cord.* **2004**, *42*, 533–540. [CrossRef] [PubMed]
6. Treiber, F.A.; Musante, L.; Hartdagan, S.; Davis, H.; Levy, M.; Strong, W.B. Validation of a heart rate monitor with children in laboratory and field settings. *Med. Sci. Sports Exerc.* **1989**, *21*, 338–342. [CrossRef] [PubMed]
7. DuRant, R.H.; Baranowski, T.; Davis, H.; Rhodes, T.; Thompson, W.O.; Greaves, K.A.; Puhl, J. Reliability and variability of indicators of heart-rate monitoring in children. *Med. Sci. Sports Exerc.* **1993**, *25*, 389–395. [CrossRef] [PubMed]
8. Janz, K.F.; Witt, J.; Mahoney, L.T. The stability of children's physical activity as measured by accelerometry and self-report. *Med. Sci. Sports Exerc.* **1995**, *27*, 1326–1332. [CrossRef]
9. Epstein, L.H.; Paluch, R.A.; Coleman, K.J.; Vito, D.; Anderson, K. Determinants of physical activity in obese children assessed by accelerometer and self-report. *Med. Sci. Sports Exerc.* **1996**, *28*, 1157–1164. [CrossRef]
10. Ravensbergen, H.J.; De Groot, S.; Post, M.W.; Slootman, H.J.; Van Der Woude, L.H.; Claydon, V.E. Cardiovascular function after spinal cord injury: Prevalence and progression of dysfunction during inpatient rehabilitation and 5 years following discharge. *Neurorehabil. Neural Repair.* **2014**, *28*, 219–229. [CrossRef]
11. Malmqvist, L.; Biering-Sørensen, T.; Bartholdy, K.; Krassioukov, A.; Welling, K.L.; Svendsen, J.H.; Kruse, A.; Hansen, B.; Biering-Sørensen, F. Assessment of autonomic function after acute spinal cord injury using heart rate variability analyses. *Spinal Cord.* **2015**, *53*, 54–58. [CrossRef]
12. Thayer, J.F.; Sollers, J.J., 3rd; Clamor, A.; Koenig, J.; Hagglund, K.J. The association of resting state heart rate variability and 24-hour blood pressure variability in spinal cord injury. *J. Neurol Sci.* **2016**, *361*, 52–59. [CrossRef]
13. El-Amrawy, F.; Nounou, M.I. Are Currently Available Wearable Devices for Activity Tracking and Heart Rate Monitoring Accurate, Precise, and Medically Beneficial? *Healthc. Inform. Res.* **2015**, *21*, 315–320. [CrossRef]
14. Bent, B.; Goldstein, B.A.; Kibbe, W.A.; Dunn, J.P. Investigating sources of inaccuracy in wearable optical heart rate sensors. *NPJ Digit. Med.* **2020**, *3*, 18. [CrossRef]
15. Xu, K.; Chen, Y.; Okhai, T.A.; Snyman, L.W. Micro optical sensors based on avalanching silicon light-emitting devices monolithically integrated on chips. *Opt. Mater. Express* **2019**, *9*, 3985–3997. [CrossRef]
16. Kirshblum, S.; Snider, B.; Rupp, R.; Read, M.S. International Standards Committee of ASIA and ISCoS. Updates of the International Standards for Neurologic Classification of Spinal Cord Injury: 2015 and 2019. *Phys. Med. Rehabil. Clin. N. Am.* **2020**, *31*, 319–330. [CrossRef]
17. Xu, K. Silicon electro-optic micro-modulator fabricated in standard CMOS technology as components for all silicon monolithic integrated optoelectronic systems. *J. Micromech. Microeng.* **2021**, *31*, 054001. [CrossRef]
18. Kaditis, A.G.; Polytarchou, A.; Moudaki, A.; Panaghiotopoulou-Gartagani, P.; Kanaka-Gantenbein, C. Measures of nocturnal oxyhemoglobin desaturation in children with neuromuscular disease or Prader-Willi syndrome. *Pediatric Pulmonol.* **2020**, *55*, 2089–2096. [CrossRef]
19. Paiva, R.; Krivec, U.; Aubertin, G.; Cohen, E.; Clément, A.; Fauroux, B. Carbon dioxide monitoring during long-term noninvasive respiratory support in children. *Intensive Care Med.* **2009**, *35*, 1068–1074. [CrossRef]
20. Albino, P.; Pavone, M.; Chiarini Testa, M.B.; Petreschi, F.; Bertini, E.S.; Cutrera, R. Noninvasive Ventilation in Children with Spinal Muscular Atrophy Types 1 and 2. *Am. J. Phys. Med. Rehabil.* **2007**, *86*, 216–221.
21. Oldfield, R.C. The assessment and analysis of handedness: The Edinburgh inventory. *Neuropsychologia* **1971**, *9*, 97–113. [CrossRef]
22. Martin Ginis, K.A.; Jörgensen, S.; Stapleton, J. Exercise and sport for persons with spinal cord Injury. *PM R* **2012**, *4*, 894–900. [CrossRef]
23. Chen, M.J.; Fan, X.; Moe, S.T. Criterion-related validity of the Borg ratings of perceived exertion scale in healthy individuals: A meta-analysis. *J. Sports Sci.* **2002**, *20*, 873–899. [CrossRef]
24. Krupp, L.B.; LaRocca, N.G.; Muir-Nash, J.; Steinberg, A.D. The Fatigue Severity Scale: Application to Patients With Multiple Sclerosis and Systemic Lupus Erythematosus. *Arch. Neurol.* **1989**, *46*, 1121–1123. [CrossRef] [PubMed]
25. Prawiro, E.A.P.J.; Yeh, C.I.; Chou, N.K.; Lee, M.W.; Lin, Y.H. Integrated wearable system for monitoring heart rate and step during physical activity. *Mob. Inf. Syst* **2016**, *2016*. [CrossRef]
26. Dobkin, B.H.; Martinez, C. Wearable Sensors to Monitor, Enable Feedback, and Measure Outcomes of Activity and Practice. *Curr. Neurol. Neurosci. Rep.* **2018**, *18*, 87. [CrossRef] [PubMed]
27. Marek, M. Task Force of the European Society of Cardiology and the North American Society of Pacing and Electrophysiology. Heart rate variability: Standards of measurement, physiological interpretation and clinical use. *Eur. Heart J.* **1996**, *17*, 354–381.

28. Bunten, D.C.; Warner, A.L.; Brunnemann, S.R.; Segal, J.L. Heart rate variability is altered following spinal cord injury. *Clin Auton. Res.* **1998**, *8*, 329–334. [CrossRef] [PubMed]
29. Grimm, D.R.; De Meersman, R.E.; Almenoff, P.L.; Spungen, A.M.; Bauman, W.A. Sympathovagal balance of the heart in subjects with spinal cord injury. *Am. J. Physiol.* **1997**, *272 Pt 2*, H835–H842. [CrossRef]
30. Takahashi, M.; Matsukawa, K.; Nakamoto, T.; Tsuchimochi, H.; Sakaguchi, A.; Kawaguchi, K.; Onar, K. Control of heart rate variability by cardiac parasympathetic nerve activity during voluntary static exercise in humans with tetraplegia. *J. Appl. Physiol.* **2007**, *103*, 1669–1677. [CrossRef]
31. Brizuela Costa, G.; Sinz, S.; Aranda Malaves, R.; Martinez Navarro, I. Effect of armcrank pedaling on the cardiorespiratory system of the people with tetraplegia. *Rev. Int. Cienc. Deporte* **2010**, *6*, 297–310. [CrossRef]
32. Palazón-García, R.; Alcobendas-Maestro, M.; Esclarín-de Ruz, A.; Benavente-Valdepeñas, A.M. Treatment of spasticity in spinal cord injury with botulinum toxin. *J. Spinal Cord. Med.* **2019**, *42*, 281–287. [CrossRef]
33. Fernandez del Olmo, M.A. (Ed.) *Neurophysiology Applied to Physical Activity*, 2nd ed.; Sintesís: Madrid, Spain, 2012; Volume 1, pp. 20–25.
34. West, C.R.; Mills, P.; Krassioukov, A.V. Influence of the neurological level of spinal cord injury on cardiovascular outcomes in humans: A meta-analysis. *Spinal Cord.* **2012**, *50*, 484–492. [CrossRef]

bioengineering

MDPI

Article

Oral Cells-On-Chip: Design, Modeling and Experimental Results

Hamed Osouli Tabrizi [1,2], Abbas Panahi [1,2], Saghi Forouhi [1,2], Deniz Sadighbayan [1,3], Fatemeh Soheili [1,3], Mohammad Reza Haji Hosseini Khani [1], Sebastian Magierowski [2] and Ebrahim Ghafar-Zadeh [1,2,3,*]

1 Biologically Inspired Sensors and Actuators (BioSA) Laboratory, Lassonde School of Engineering, York University, Toronto, ON M3J 1P3, Canada; htabrizi@cse.yorku.ca (H.O.T.); panahiyu@yorku.ca (A.P.); sforouhi@yorku.ca (S.F.); denizsdg@yorku.ca (D.S.); fsoheili@yorku.ca (F.S.); hosseinkhani.farhad@gmail.com (M.R.H.H.K.)

2 Department of Electrical Engineering and Computer Science, Lassonde School of Engineering, York University, Toronto, ON M3J 1P3, Canada; magiero@eecs.yorku.ca

3 Department of Biology, Faculty of Science, York University, Toronto, ON M3J 1P3, Canada

* Correspondence: egz@cse.yorku.ca

Abstract: Recent advances in periodontal studies have attracted the attention of researchers to the relation between oral cells and gum diseases, which is a real threat to overall human health. Among various microfabrication technologies, Complementary Metal Oxide Semiconductors (CMOSs) enable the development of low-cost integrated sensors and circuits for rapid and accurate assessment of living cells that can be employed for the early detection and control of periodontal diseases. This paper presents a CMOS capacitive sensing platform that can be considered as an alternative for the analysis of salivatory cells such as oral neutrophils. This platform consists of two sensing electrodes connected to a read-out capacitive circuitry designed and fabricated on the same chip using Austria Mikro Systeme (AMS) 0.35 μm CMOS process. A graphical user interface (GUI) was also developed to interact with the capacitive read-out system and the computer to monitor the capacitance changes due to the presence of saliva cells on top of the chip. Thanks to the wide input dynamic range (IDR) of more than 400 femto farad (fF) and high resolution of 416 atto farad (aF), the experimental and simulation results demonstrate the functionality and applicability of the proposed sensor for monitoring cells in a small volume of 1 μL saliva samples. As per these results, the hydrophilic adhesion of oral cells on the chip varies the capacitance of interdigitated electrodes (IDEs). These capacitance changes then give an assessment of the oral cells existing in the sample. In this paper, the simulation and experimental results set a new stage for emerging sensing platforms for testing oral samples.

Keywords: capacitive sensor; complementary metal-oxide-semiconductor; oral epithelial cells; oral neutrophils; saliva

Citation: Osouli Tabrizi, H.; Panahi, A.; Forouhi, S.; Sadighbayan, D.; Soheili, F.; Haji Hosseini Khani, M.R.; Magierowski, S.; Ghafar-Zadeh, E. Oral Cells-On-Chip: Design, Modeling and Experimental Results. *Bioengineering* **2022**, *9*, 218. https://doi.org/10.3390/bioengineering9050218

Academic Editor: Ali Zarrabi

Received: 30 April 2022
Accepted: 17 May 2022
Published: 19 May 2022

Publisher's Note: MDPI stays neutral with regard to jurisdictional claims in published maps and institutional affiliations.

1. Introduction

Recent periodontal research studies have found a correlation between gum diseases and other conditions such as diabetes, osteoporosis, and human immunodeficiency virus (HIV) [1–3]. It is also well documented that patients with severe osteoporosis will likely experience periodontal breakdown [4]. There are also reports regarding the increased populations of periodontal-related bacteria in patients with HIV [5,6]. Moreover, there is a high risk of heart diseases in people with unhealthy gums, which is a result of bacterial infection in the bloodstream [7,8]. In addition, recent studies show a link between the salivatory oral cells such as Oral Polymorphonuclear Neutrophils (oPMNs) and gum diseases [9–11]. Indeed, the quantification and analysis of oral cells can improve our understanding of the cellular activities in saliva and might deliver a clear picture of oral health. For example, it has been verified that the number of oPMNs in patients with periodontitis is 4 times higher in comparison to healthy individuals [12]. The presence of an excessive number of oPMNs might have a negative impact on the oral tissue's integrity. The secretion of effective active mediators into the oral cavity may endanger oral homeostasis [13]. As a result of periodontitis and chronic activation of oral neutrophils, there might be damage to the periodontal connective tissue, which leads to the loss of attachment, alveolar bone, and tooth loss [14].

Another example is a link between the epithelial cells derived from oral cavities and oral health. As per recent studies, epithelial cells contain biomarkers for oral diseases such as oral cancer. These research findings further show the importance of oral cell analysis using advanced sensing technologies to improve our understanding of the role of salivary cells in the gum and other inflammatory diseases [15,16].

Until now, there has been no established sensing platform for analyzing oral cellular activities. Researchers employ standard methods such as fluorescence microscopy and flow cytometry [17]. These methods need specialized equipment with tens of thousands of dollars, a costly service contract, and trained staff to operate it. Therefore, their use is restricted to well-funded laboratories [18]. In this paper, we address the challenge of developing a sensing platform for analyzing oral cells using Complementary Metal-Oxide-Semiconductor (CMOS) technology. Among various sensing technologies, CMOS has offered great advantages for integrating a single small chip featuring millions of sensors with high signal-to-noise ratios (SNRs), low cost, and high speed [19]. To date, many types of CMOS biosensors have been developed by researchers for cell monitoring applications using various approaches such as optical [20], capacitive [21–23], impedimetric and multi-modal [24] techniques. In these works, cells are cultured on the chip surface, and the sensors work based on the adhesion of cells to the substrate due to the adsorption of cell surface protein on the substrate. They are mostly reported for cancer cells like human ovarian cancer cells [21], human lung carcinoma cells [22], breast cancer cells [20], and also bacterial cells such as *E. coli* [23]. In this paper, we investigate the advantages of the CMOS capacitive sensor for studying oral cells.

Capacitive sensors are one of the popular CMOS biosensors reported for precise measuring of the capacitance variations at the electrode-sample interface in different cellular assays. They can offer significant advantages for studying growth and other cellular activities [22,25]. In these applications, the cells are firmly attached to the surface through integrin-containing multi-protein structures [26]. However, there are some challenges to studying oral cells and doing research on them [27]. Unlike the other cells, oral cells have very low culture ability. Thus, they have less affinity to the surface even if they are in a culture medium [28–30]. Although researchers have made some efforts to culture oral cells in saliva-like medium, they have not been completely successful [27]. In many experiments for culturing neutrophils, the cells underwent apoptosis, died after less than an hour and their morphology had changed [27]. In most tests on periodontal diseases, normal saline was used to take samples and study saliva cells without culturing them. Oral cells should be studied in the material most similar to the saliva-like medium in order to keep them alive during the experiments [31]. These cells cannot be cultured in an in vitro environment because they have a short life span [32]. It is worth mentioning that since oral neutrophils are activated during their migration from blood circulation to the oral cavity, they cannot be divided anymore. Therefore, they are in the last stage of their life cycle, and their nucleus cannot make cytokines as well [32,33]. There are different types of cluster of differentiation (CD) markers on the surface of neutrophil cells that might be responsible for cell attachment. According to the literature, the essential CD markers accountable for cell attachment are Mac-1 (CD 11b and CD 18) [33–36]. These CD markers may increase the affinity of cells to the hydrophilic surface of the silicon oxide or aluminum oxide above the CMOS chip.

This paper addresses the challenge of monitoring the hydrophilic adhesion using a CMOS capacitive sensing platform. As seen in Figure 1, the proposed platform includes a CMOS capacitive sensor featuring sensing electrodes and an interface circuit. The chip is encapsulated with epoxy with an opening for introducing the sample. This chip is connected to a computer through a printed circuit board (PCB) populated with off-chip components. This board provides the chip's controlling signals and makes its output signal ready to send to the computer. A graphical user interface (GUI) is also used to help capture the output data in a user-friendly manner by employing interactive visual components of computer software.

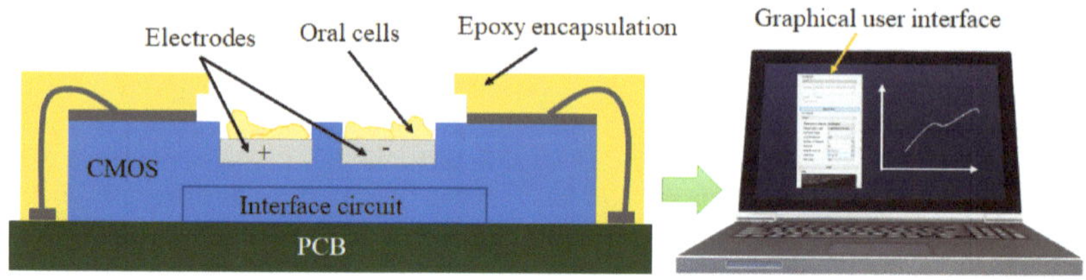

Figure 1. A CMOS-based capacitive sensor encapsulated with epoxy for oral cell monitoring.

The rest of the paper is organized as follows. Section 2 gives a brief overview of the related CMOS capacitive sensors. In Section 3, the materials and methods are presented. Section 4 demonstrates the implementation and experimental results. In Section 5, we discuss the practical considerations, and future works. Finally, our conclusions are drawn in Section 6.

2. Related Works

To date, many papers have reported the advantages of CMOS capacitive sensors for monitoring the growth or proliferation of living cells such as breast cancer cells [1], human lung carcinoma cells [2,3], or *E. coli* [4]. As seen in Table 1, for tracking the proliferation of MDA-MB-231 breast cancer cells, a capacitive sensor, with a resolution of 5 fF and passivated aluminum (Al) electrodes, was employed [34]. In another work, Senevirathna et al. reported the development of a capacitive sensor with a higher resolution of 14.4 aF for analyzing human ovarian cancer cells with passivated aluminum electrodes [21,35]. In another effort, Widdershoven et al. [25,36] used a capacitive sensor with gold nanoelectrodes with a radius of 90 nm and a resolution of 1 aF for imaging living cells such as BEAS, THP1, and MCF7 cancer cell lines. In addition to cancer cells, CMOS capacitive sensors have successfully been employed for bacteria detection. For instance, Couniot et al. proposed a CMOS capacitive sensor with Al/Al_2O_3 electrodes for the detection of *S. epidermidis* [37,38]. In another effort, Ghafar-Zadeh et al. used a technique called charge-based capacitance measurement (CBCM) for the development of a $\sum\Delta$ capacitance-to-digital converter (CDC) [23]. This sensor was employed for monitoring the growth of *E. coli* [23] on a passivated aluminum interdigitated electrode (IDE) with a window in between the fingers (Al/wPass). Although this sensor showed a high resolution of about 10 aF, its input dynamic range (IDR) was limited to less than 3 fF. Indeed, other reported capacitive sensors [21,25,34–36] have also demonstrated a low resolution or high sensitivity but low IDR. Thus they suffer from parasitic capacitances created due to the debris and other remnants above the capacitive sensor. Our group has addressed this challenge by developing a new core-CBCM interface circuitry with a wide IDR of 400 fF [39,40] and a resolution of about 450 aF. This circuit works based on a current-mode technique using an extended charge-based capacitance measurement (ECBCM) structure and a current-controlled oscillator (CCO), which gains the advantage of a calibration-free approach utilizing a bank of reference capacitors as briefly described in Section 3.4.1. In this paper, we employ this CMOS sensor for oral cell monitoring for the first time. For this purpose, we proposed a new electronic circuit board and graphical interface to read out the sensing data from the capacitive sensor chip and demonstrate it on the computer screen. This sensing platform enables the measurement of capacitance changes caused by oral cells. The experimental results agree with the proposed equivalent circuit and the Multiphysics simulation results discussed in Section 3.2. Furthermore, the sensing electrodes in this paper and other reported papers include metal or metal oxides demonstrating hydrophilic properties suitable for measuring the hydrophilic materials such as oral neutrophils, which can attach to the surface of these electrodes [41].

Table 1. Comparison of the capacitive sensors reported for cellular applications.

CMOS Technology	Type of Cell	Electrode Material	Resolution	IDR (fF)	Reference
0.5 µm	hBC	Al/1pass	5 fF	NA	[34]
0.35 µm	hOC	$Al/SiO_2/Si_3N_4$	14.4 aF	12	[21,35,42]
0.25 µm	*S. epidermidis*	Al/Al_2O_3	10 fF	NA	[37]
0.25 µm	*S. epidermidis*	Al/Al_2O_3	450 aF	57	[38]
0.18 µm	*E.Coli*	Al/wPass	10 aF	2.7	[23]
0.35 µm	hLC	Al/PEM	10 aF	10	[22,43]
90 nm	hBC	AuCu	<10 aF	NA	[25,36]
0.35 µm	Oral cells	Al/A_2O_3	416 aF	400	This work

IDR: Input Dynamic Range, hBC: Human Breast Cancer cell, hOC: Human Ovarian Cancer cell, hLC: Human Lung Carcinoma cell, Al/1pass: Al electrode with One passivation layer, Al/wPass: Passivated Aluminum interdigitated electrode with a window in between the fingers, PEM: Polyelectrolyte Multilayer.

3. Materials and Methods

In this section, new modeling is presented for the analysis of the electrodes–analyte interface and the biological and chemical protocols used for the experiments are introduced. Section 3.1 discusses the electrical equivalent modeling of the interface, and Section 3.2 is dedicated to the simulations of the electrodes' response using COMSOL software. The biological and chemical protocols are described in Section 3.3. Section 3.4 is dedicated to introducing the capacitive sensing platform.

3.1. Electrical Equivalent Modeling

In this section, the electrical equivalent modeling is explained for three states: (1) When there is no sample on the electrode; (2) When the sample droplet exists on the electrode; (3) After the evaporation of the droplet. Then, the modeling of the oral samples on the chip is discussed based on these three states.

3.1.1. Electrical Equivalent Modeling without Sample

Figure 2 illustrates three distinct phases in the sample–electrode interface. A cross-sectional schematic of the electrodes implemented on the topmost metal layer of the CMOS chip is given on the left side of each row, whereas the proposed electrical equivalent model is shown on the right side of that row. One plate is always grounded in the implemented IDE topology, while the other plate is exposed to a pulsated voltage. In our topology, the voltage of the positive electrode reaches about one threshold voltage (0.6 V) less than V_{dd} (3.3 V) [39,40]. This voltage difference creates an electric field between the two plates of the IDEs with a fringe electric field passing above the electrodes, shown as C_{sens}. As depicted in Figure 2a, other parasitic capacitances exist due to the underlying layers. C_{sub} represents the offset to the top metal and the substrate. C_{sub} can be calculated using the given value of the parasitic capacitance per area from the CMOS technology datasheets (1).

$$C_{sub} = \alpha_1 \times A + \alpha_2 \times P \tag{1}$$

where A and P are the area and the perimeter of the implemented electrode, and α_1 and α_2 are process-dependent coefficients that are known for each fabrication technology. Another parasitic capacitance that is due to the direct electric field between the grounded plate and the voltage plate is shown as C_{dir} and is a function of the area of the side of the plates that coincide with each other, as well as the electrode pitch, d, with the equation given in (2).

$$C_{dir} = \varepsilon_0 \frac{\varepsilon_{Sio_2} \times P_{coincide} \times W_{th}}{d} \tag{2}$$

where ε_0 is the vacuum permittivity, ε_{SiO_2} denotes the relative permittivity of silicon oxide as the dielectric between the electrodes, $P_{coincide}$ stands for the perimeter of the plates that coincide with each other, W_{th} represents the thickness of the electrodes, and d is the pitch of the electrodes. Considering the symmetry in the geometry of the IDEs, it can be assumed that C_{fringe} is ε_{SiO_2} times larger than C_{sens}, as expressed in (3), since the dielectric material for C_{fringe} is silicon oxide while it is air for C_{sens} when there is no sample placed on the electrodes.

$$C_{fringe} = \varepsilon_{Sio_2} \times C_{sens} \quad (3)$$

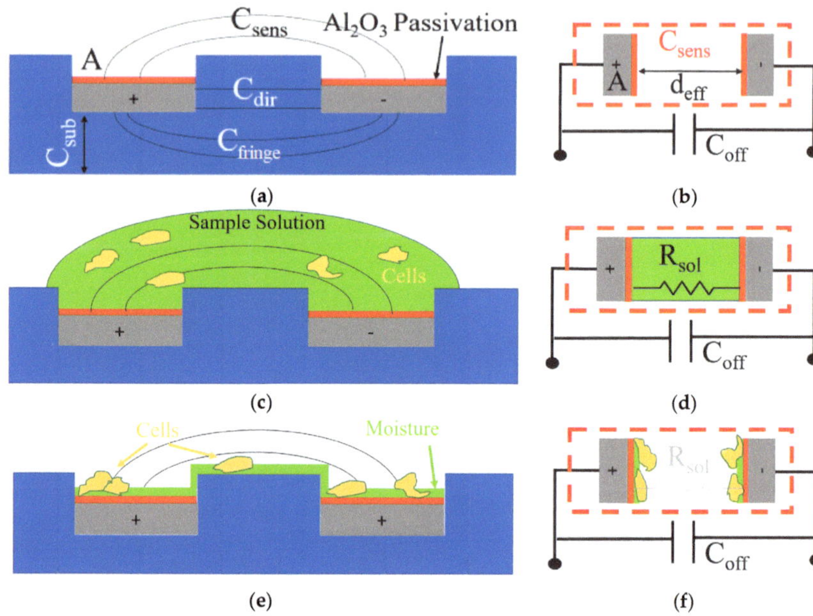

Figure 2. (**a**) Cross-section schematic of the implemented IDE electrode without sample; (**b**) Equivalent capacitance model without sample; (**c**) Cross-section schematic of the IDE electrode with the sample before the evaporation of the sample solution; (**d**) Equivalent capacitance model of the sample before the evaporation of the sample solution; (**e**) Cross-section schematic of the IDE electrode with the sample after the evaporation of the sample solution; (**f**) Equivalent capacitance model of the sample after the evaporation of the sample solution.

The value of the equivalent offset capacitance, C_{off}, is thus equal to the addition of C_{sub}, C_{dir}, and C_{fringe}, as given in (4).

$$C_{off} = C_{sub} + C_{dir} + C_{fringe} \quad (4)$$

Figure 2b illustrates an equivalent electric model for the electrodes shown in Figure 2a, where A is the surface area of the electrodes and d_{eff} is an adequate equivalent direct distance for the fringe distance. The strips shown in red represent the equal capacitance that forms due to the native aluminum oxide layer created on the aluminum electrodes because of its exposure to air. The thickness of the aluminum oxide layer naturally formed above the electrodes can reach less than 10 nm [44]. Since the relative dielectric constant of aluminum oxide is 9.5, this layer can create approximately 8.4 fF/µm^2, which results in a high capacitance of about 59 pF considering that the dimensions of the implemented IDE with a surface area of 7056 µm^2. The total equivalent capacitance for the electrodes without being exposed to any sample can be calculated using (5).

$$C_{equivalant} = \varepsilon_0 \frac{\varepsilon_{PL} \times \varepsilon_{Air} \times A}{\varepsilon_{PL} \times d_{eff-Air} + \varepsilon_{Air} \times (2d_{PL})} + C_{off} \quad (5)$$

where ε_{PL} stands for the relative permittivity of the aluminum oxide passivation layer, ε_{Air} is the relative permittivity of air, and d_{PL} represents the thickness of the passivation layer.

3.1.2. Electrical Equivalent Modeling When the Sample Droplet Is Present

Figure 3c demonstrates the electrodes when the sample solution is placed on top of the electrodes. As seen in the electric equivalent model given in Figure 2d, due to the conductivity of the sample solution that contains ions, a resistive path forms between the electrodes, R_{sol}. In addition, a sizeable double-layer capacitance forms in the interface of the electrodes and the sample. The overall effect is saturating the sensor readings.

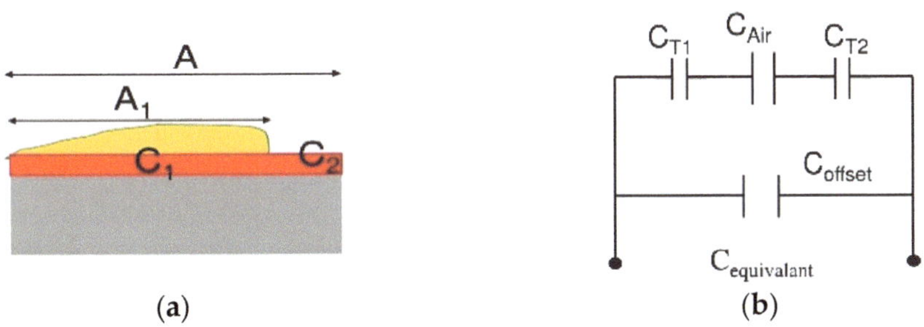

Figure 3. (**a**) Simplified model for the attachment of the cells on the electrodes after complete evaporation of the sample; (**b**) Equivalent circuit model.

3.1.3. Electrical Equivalent Modeling after Evaporation of the Droplet

Figure 2e demonstrates the phase in which the sample has evaporated while some moisture remains on the surface and the cells are deposited. In this phase, cell remnants play the role of a third dielectric in the equivalent capacitance of the passivation layer. Since the sample solution has evaporated, the value of resistance, R_{sol}, is negligible. The amount of change in the total capacitance of the electrodes in the presence of cells compared to the capacitance of the electrodes with no sample is related to the percentage of the area of the electrode that is covered by cells, referred to as confluence, $conf_n = A_1/A$ (see Figure 3a). Figure 3b illustrates the equivalent electric circuit model for this phase. The deposition of cells on the electrode's surface affects the passivation layer's capacitance, shown as C_{T1} and C_{T2}. Equations (6) and (7) describe C_{Tn} and $C_{equivalant}$, respectively.

$$C_{Tn} = \varepsilon_0 \times A\left(\frac{\varepsilon_{PL} \times \varepsilon_{cell} \times conf_n}{\varepsilon_{PL} \times d_{eff-cell} + \varepsilon_{cell} \times d_{PL}} + \frac{\varepsilon_{PL}\,(1 - conf_n)}{d_{PL}}\right) n = 1,\, 2 \quad (6)$$

$$C_{equivalant} = C_{offset} + C_{T1}||\, C_{Air}||\, C_{T2} \quad (7)$$

where ε_{cell} stands for the relative permittivity of the moist cells, and $d_{eff-cell}$ represents the average thickness of the sedimented cells. Equation (6) shows that the value of C_{Tn} depends on the percentage of the coverage of cells on the electrodes, $conf_n$. Thus, an increase in the cell coverage leads to a rise in C_{Tn}. From (7), equivalent capacitance is equal to the combination of C_{T1}, C_{T2}, and C_{Air}. As a result, an increase in the measured $C_{equivalent}$ will be the representative of an increase in C_{Tn}. Therefore, the number of cells were present in the sample and were deposited on the electrode. This increase was observed in the simulation results discussed in Section 3.2.

3.1.4. Oral Samples on the Chip

The addition of oral neutrophil samples on the chip mimics the three distinct states that were explained in Sections 3.1.1–3.1.3. In the beginning, to have a baseline, we perform a full reading of the sensor in phase one without introducing the sample. The readings reveal the C_{off} + C_{sens} mentioned in Figure 2b. Immediately after introducing the sample, the outputs become saturated due to the presence of the water droplet, as explained in detail in Section 3.1.2. After the water droplet evaporates, the model enters the third phase discussed in Section 3.1.3. In this phase, the sensor outputs reflect the amount of the change in the C_{sens} due to the presence of wet oral cells on the electrodes. The experimental results are explained in detail in Section 4.2.

3.2. COMSOL Simulation

In this section, the capacitive response of the electrodes is simulated for two modes: (1) In the dry mode; and (2) when electrodes are exposed to cells in the water. In the wet mode, we also study the effect of the thick and thin layers of liquid. These simulations cover all three phases.

3.2.1. Dry Mode

A COMSOL Multiphysics simulation was carried out to model the CMOS capacitive sensor's static electrical response in the dry mode and when the sensor is subjected to a specific cell concentration. For this purpose, the sensor was modeled in the electrostatic module of COMSOL following the size and physical boundary conditions maintained in the fabricated sensor. In this module, the following Formulas (8) and (9), are solved numerically, which determine the sensor's electrical potential for measuring the capacitance.

$$E = -\nabla V \tag{8}$$

$$\nabla .\, (\varepsilon_0 \varepsilon_r) = \rho_v \tag{9}$$

where E is the electric field; ε_0 denotes the vacuum permittivity; ε_r is the relative permittivity; ρ stands for the electric charge density, and V is the electric voltage. To perform the simulation, the sensor was designed in the CAD module of COMSOL and then meshed to perform numerical simulation. A mesh analysis was performed to avoid the meshing effect on the result; consequently, the extra-fine mesh was utilized. The relative permittivity was used as the dielectric model. A schematic of the boundary conditions and the meshed structure are demonstrated in Figure 4.

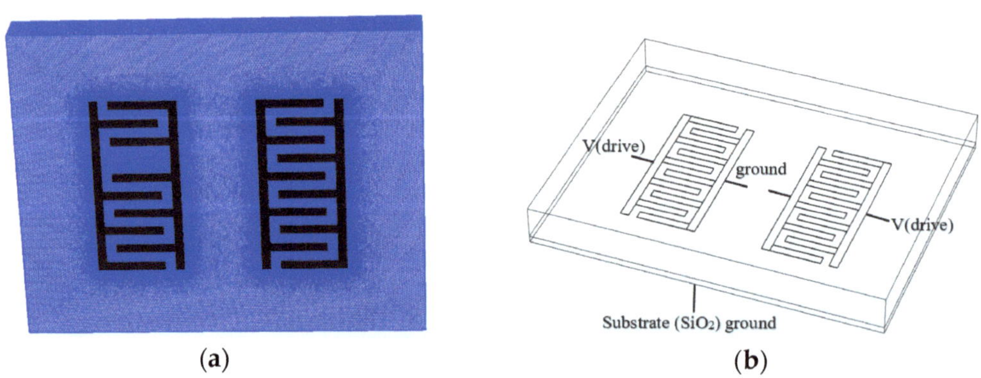

Figure 4. (**a**,**b**) Boundary conditions and meshing structure of the simulated sensor in COMSOL software.

Figure 5 demonstrates the electrical potential contours of the CMOS capacitive sensor when it is subjected to 2.7 V bias. The sensor was meshed with an extremely fine approximation to yield the best convergence. The sensor surrounding material is a thin layer of SiO_2 identical to the CMOS fabrication materials. The simulation considers the capacitance due to the thin layer of SiO_2 as the native material between the layers and the P-type silicon substrate of the sensor, which is electrically grounded. According to the simulation results, the equivalent capacitance for the sensor in the dry mode is around 117.415 fF. The capacitance is calculated by assuming that the sensor is surrounded by air.

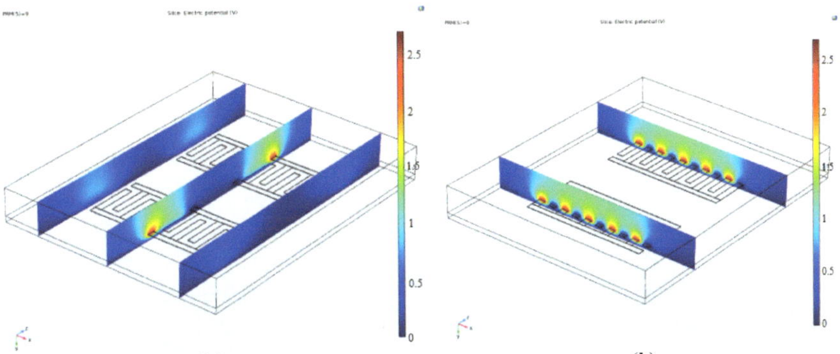

(a) (b)

Figure 5. (**a**,**b**) The electrical potential counters of the sensor for a bias voltage of 2.7 Volts. The substrate is grounded.

3.2.2. Exposing the Electrodes to Cellular Sample

According to the experimental evidence, the capacitance upsurges when exposed to a liquid volume. After complete evaporation of liquid on top of the sensor, the response gets back to the baseline considering the presence of cells. This also occurs when the medium encompasses cells or other components. Due to a change in the medium's dielectric, the sensor response changes, and the cycle of rise-fall of capacitance would be different from the previous cycle with a different dielectric. To prove this observation and validate it, COMSOL Multiphysics was employed to simulate the cycle. Following the experiments, in modeling, we filled the top of the sensor with a thick liquid layer (here water) below the air layer as can be seen in Figure 6. Afterward, we changed the thickness of the water layer from 400 µm to 5 µm atop the sensor. This gradual decrease of the water layer thickness resembles the evaporation of the droplet. At the same time, the capacitance of each step was recorded. As seen in Figure 6, the sensor initially gets saturated, and then after complete evaporation of the droplet, it returns to the baseline. However, we expect different levels in the return cycle due to the existence of remnants of cells on top.

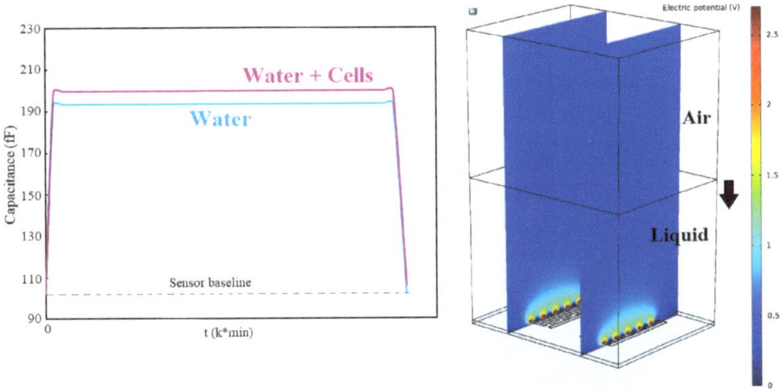

(a) (b)

Figure 6. (**a**,**b**) COMSOL simulation showing the saturation of the sensor in liquids and returning to baseline.

As shown in Figure 6, the electric potential on top of the sensor is decaying with the distance from the sensor surface. If the sensor environment is filled with a liquid (with a specific dielectric constant), the sensor response reaches a saturation level and constant level. When the liquid is evaporating, the cells' concentration becomes more visible to the sensor by increasing the dielectric constant. Therefore, we can sense the mass of cells on the sensor surface. COMSOL simulation validates this idea that, with the capacitive sensor, we can measure the concentration of biological entities in the droplet by measuring the capacitance change during the sedimentation of cells on the sensor in a loop shown in Figure 6 and experiments.

3.3. *Biological and Chemical Protocol*

After introducing the employed materials and instruments in Section 3.3.1, the cleaning procedure is explained in Section 3.3.2. Then, the sample preparation protocol and control measurement techniques are described in Sections 3.3.3 and 3.3.4, respectively.

3.3.1. Materials and Instruments

In this work, various materials were used. These materials are 50 polypropylene conical tubes (Baxter, Toronto, ON, Canada); Sterile cell strainer: 40 μm nylon mesh (Millipore, Burlington, MA, USA); 11 μm nylon filters (Millipore, Burlington, MA, USA); 1.5 mL Eppendorf tubes (Fisherscience, Saint-Laurent, QC, Canada); 0.9% irrigation-grade sodium chloride solution (Baxter, Toronto, ON, Canada). We also employed several instruments, including an incubator (Heracell 150i, Thermo Fisher Scientific, Waltham, MA, USA); Lab Centrifuge (Sorvall ST 8, Thermo Scientific, Waltham, MA, USA); Hemocytometer (BLAUBRAND® Neubauer Millipore Sigma, Burlington, MA, USA); Inverted and phase contract Microscope (isherbrand™ Inverted Infinity, Phase contrast 10× and 20×, light splitter (100% or 20/80%), Fisher Scientific, Hampton, NH, USA).

3.3.2. Cleaning Procedure

Although the goal in the field of CMOS-based biosensors is mass-production and designing single-use devices, it is important to wash and reuse them while they are still in the research stage. Because their cost is high when fabricated in limited quantity despite their cost-effectiveness in batch-production [45]. To test the functionality of the device, we gradually increase the cell concentration in each step to observe the differences between the device's response to an increased number of cells. However, to have a precise measurement, it is crucial first to clean the surface and reuse it for the subsequent trial. We tried drop-casting and dipping in washing materials such as acetone, isopropanol alcohol (IPA), and ethanol; however, this method did not completely clean the surface. Therefore, we came up with using mechanical force to remove the cell residuals. Using a micrometric brush and droplets of Ethanol or IPA approximately solved the problem, and the surface was cleared from biological remnants.

3.3.3. Sample Preparation

To prepare neutrophil samples, participants were advised to consume nothing but the water within 30 min before sample collection, rinse the oral cavity with 10 mL of tap water for 15 s, and discharge the rinsed fluid. The participants were asked to wait for 2 min before rinsing the oral cavity with 10 mL of a 0.9 percent *w/v* saltwater solution (normal saline) measured using the 15 mL falcon tube and eject the sample into the 50 mL falcon tube. After one wash, there was a 2-min wait time, and the same process was repeated five times. Then, the final sample was filtered gently via 40 μm filters to isolate the cells from the debris larger than 40 μm sizes. The tubes were centrifuged at 4 °C for 5 min with 2600 RPM and the supernatant was discarded until 1 mL of the sample remained in the falcon tube. Then, 1 mL of ultrapure water was added to the falcon tube and mixed with the cells. The related ethical approval for this research, including the human participants, was provided by York University.

3.3.4. Control Measurement Technique

The hemocytometry cell counting technique was used to control the measurement results. At this step, the sample was thoroughly mixed. The hemocytometer was washed and cleaned with 70% (*v/v*) ethanol and allowed to dry. The coverslip was washed with 70% ethanol, allowed to dry, and placed on the hemocytometer counting chamber. A pipette was used to mix 10 μm of cells and 90 μm of Trypan Blue in a 1.5 mL Eppendorf tube. Then, 10 μL of cell suspension was added to Trypan Blue under the coverslip on a hemocytometer. After 2 min, the cells were counted. We estimated that the number of neutrophil cells in each 10 μL of saliva sample was about 7. The number of epithelial cells was around 13 in each microliter of the saliva sample. Figure 7 shows the microscopic images of the hemocytometer showing neutrophil and epithelial cells.

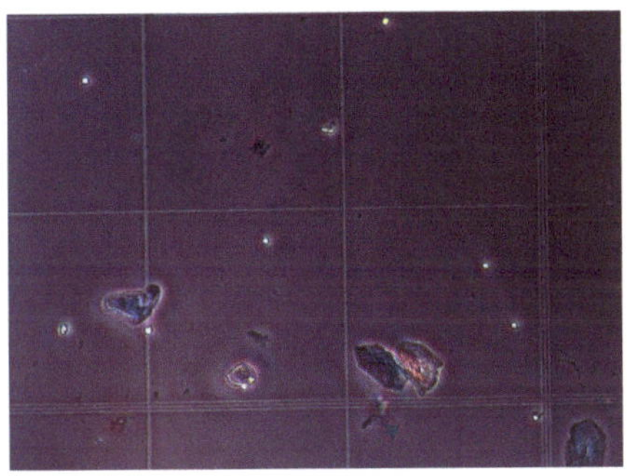

Figure 7. The microscopic images of the hemocytometer showing neutrophil and epithelial cells.

3.4. Capacitive Sensing Platform

The capacitance sensing platform is composed of on-chip and off-chip parts. A block diagram of the sensing platform is shown in Figure 8. On the CMOS chip, which was discussed in Section 3.4.1, the conversion of differential capacitance to digital output takes place in three modules. The chip requires external clock signals that were provided by utilizing a microcontroller circuit (based on an Arduino DUE board). The same microcontroller was used to read the digital output data of the chip through its serial peripheral interface (SPI) port. A PCB board was designed that hosts the chip and voltage supply (circuitry including voltage regulators) and allows the Arduino DUE board to be connected. This was explained in Section 3.4.2. A GUI was designed to fulfill the requirements for user preference settings and data visualization, which was explained in Section 3.4.3. The capacitance calculation algorithm was also described in Section 3.4.4.

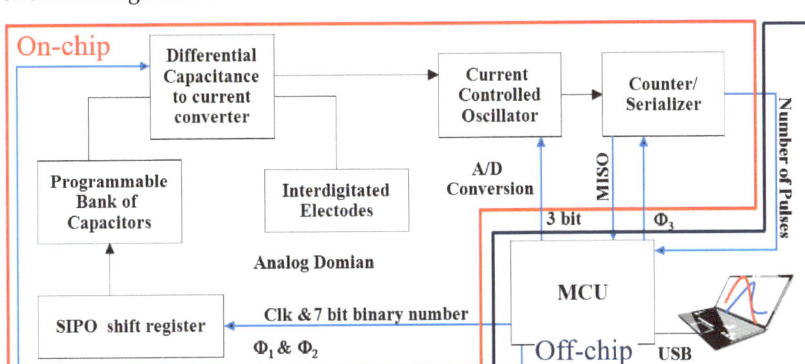

Figure 8. The block diagram of the proposed circuit and system.

3.4.1. CMOS Capacitive Sensor

The implemented on-chip circuit is composed of a differential capacitance-to-current converter. The reference capacitance in this topology is a digitally programmable bank of capacitors that makes it possible to sweep the reference capacitance value. The output differential current of this block transfers to the CCO block. The oscillation frequency of the CCO is linearly related to the magnitude of the input current.

A counter is used to generate the digital output of the chip. The output of the chip is, as a result, the number of pulses, which is linearly related to the differential capacitance at the input stage of the capacitance of the IDE and the reference electrode ($C_{interdigitated}$-$C_{reference}$).

The value of this digitally programmable bank of reference capacitors can change from 200 fF up to 1270 fF with a step of 10 fF. This wide range of reference capacitors, instead of a single capacitor, enables measurement of a wide IDR. As a result, an additional axis, C_R, was introduced to the measurements, changing the data from a single point to a 2D curve. The addition of a timestamp to the 2D curves results in the creation of 3D plots as a footprint of the sample (see Section 3.4.3).

3.4.2. Off-Chip Circuit and System

An Arduino DUE microcontroller board together with a custom-made electronic board that holds the chip and the required circuitry for voltages and the clock pulses are the testbench hardware, which is shown in Figure 9. Pulse Width Modulation (PWM) clocks with a frequency of 66.6 kHz are generated in channels 0 and 1 of the microcontroller with a duty cycle of 4/15 and 2/15 where they form in a non-overlapping fashion, and a 1 MHz clock with a duty cycle of 1/2 is generated in channel 3, which is used as the clock for the SPI port. A 2-to-1 multiplexer was also used to switch between the outputs of the right and left electrodes. The details of the circuitry can be seen in Appendix A.

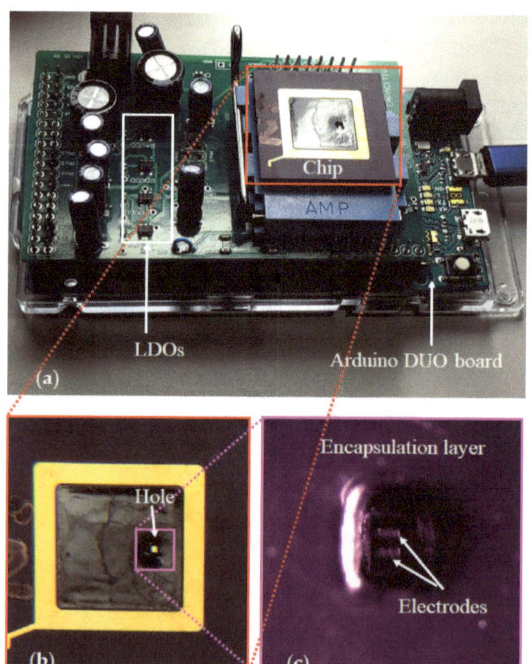

Figure 9. The testbench hardware: (**a**) An Arduino DUE microcontroller board together with a custom-made electronic board holding the chip; (**b**) The chip; (**c**) The microscopic image of the two electrodes and the surrounding encapsulation layer with a hole for the non-passivated sensing electrodes.

3.4.3. GUI Development

To practically utilize the chip, a dedicated testbench including both hardware and software was developed. The chip requires two sets of supply voltages for digital and analog parts and a reference voltage. The supply voltages are fed through Texas instruments low dropout voltage regulators TPS75901KTTR while adjusting the output to 3.3 V. Considering that we separated the left and right circuits, four voltage regulators in total were used. The CCO requires a 1.85 V reference that was supplied through the NCP705MTADJTCG adjustable output regulator from the ON Semiconductor. Required clocking signals were generated using an Arduino DUE board, and the SPI port of the board was used to capture the digital output data.

The use of a testbench without a GUI is not very practical and user-friendly. As a result, a GUI was implemented in Python, enabling the users to configure their required set of experiments as well as perform data visualization. Figure 10 illustrates a snapshot of the GUI. Users can select capturing data from right, left, or both channels as well as select obtaining curves when the reference bank of the capacitor's values are swept in the whole range or measured at a single point by determining the reference point value. In addition, by determining the number of samples, multiple measurements will be performed at a single reference point that can be averaged for better noise immunity. Users can set the GUI to obtain as many curves as desired or set the time-based measurement settings for the time interval of data acquisition and the experiment time. The visualized data is a three-dimensional (3D) curve with the chip's output (number of pulses) on the z-axis, the values of the reference capacitance on the x-axis, and time on the y-axis. The GUI also extracts the value of capacitance from the captured data (see Section 3.4.4). A snapshot of the visualized data for both right and left channels is shown in Figure 11a,b, respectively. The variations of the extracted value of capacitance during the time are depicted in Figure 12.

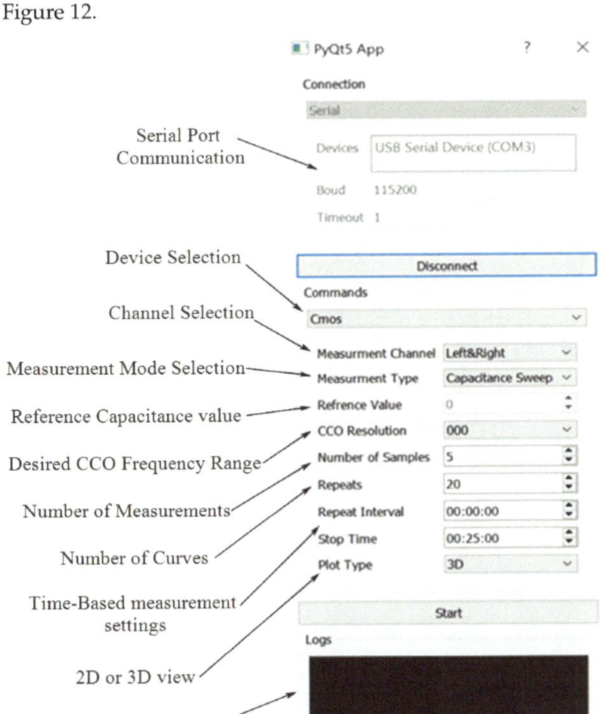

Figure 10. A snapshot of the GUI.

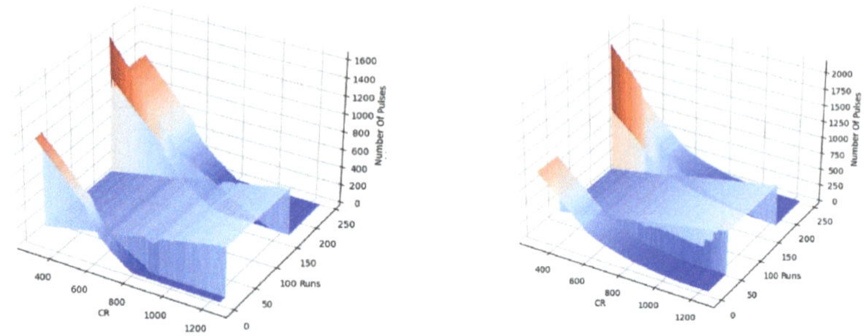

Figure 11. The visualized data captured by the GUI in real-time: (**a**) for the left IDE, (**b**) for the right IDE.

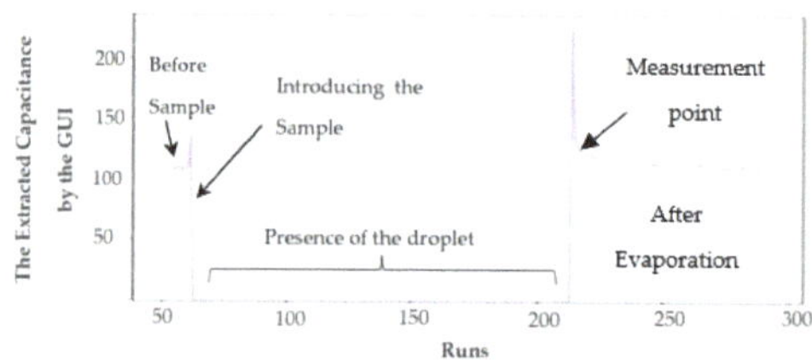

Figure 12. The extracted capacitance variations versus time.

The real-time visualization and signal processing capability that the GUI provides enables monitoring of the experiment and detection of the time that the sample was introduced as well as the time that the sample evaporates, as shown in Figure 12. Figure 13 illustrates the flowchart of the operation of the GUI that communicates with the microcontroller (Arduino DUE board).

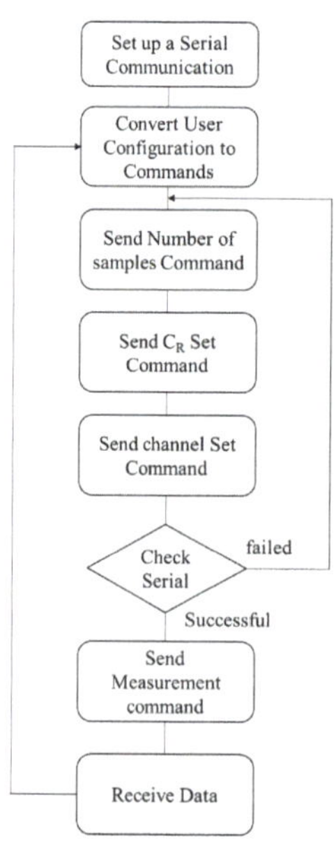

Figure 13. The flowchart of the GUI data acquisition from the Microcontroller: After a serial connection is set up between the GUI and the microcontroller, the user can input all the settings shown in

Figure 12a. The GUI generates three commands based on the user settings and sends them to the microcontroller. Then, the serial port is checked again before establishing the measurement command on the serial port. Next, the data is received from the chip and read from the microcontroller's SPI port buffer.

After a serial communication channel has been established between the GUI and the microcontroller unit (MCU), the GUI converts the user input settings to four different types of command, namely the number of samples, and the value of the reference capacitance, and the channel of interest. If writing the commands is successful, the GUI initiates the measurement command. After receiving this command, the MCU transmits the measurement data through the SPI port to the GUI.

3.4.4. Capacitance Calculation

Figure 14 illustrates the 3D output of the chip for the first run of the experiments with oral cells. The chip's output is the number of pulses related to the difference in the capacitance seen for the IDEs and the reference capacitance. For a given IDE capacitance, an increase in the reference capacitance results in a decrease in the output number of pulses within the dynamic range of the chip. The sensor's resolution is about 0.5 fF, which is enough to detect the presence of small oral cells such as neutrophils based on the COMSOL simulation results explained in Section 3.2. Sweeping the bank of capacitors provides a calibration-free capacitance measurement technique, which we previously reported in [40]. The capacitance can be extracted from the 3D curves based on the principle of operation of the sensor. In the differential mode, the sensor's output will be the same when the sensing and the reference side are in equilibrium. Internal capacitances of 400 fF were added in parallel with the IDEs. A transmission gate on the right circuit enables connecting and disconnecting the IDE to the circuit. The pseudocode given in Algorithm 1 demonstrates the capacitance extraction steps to evaluate the value of the offset capacitance of the IDEs as well as the capacitance change due to the presence of samples.

Algorithm 1. Capacitance extraction algorithm from the 3D footprint.

1- Switch off the transmission gate and obtain the output curve versus the value of sweeping reference capacitance (number of pulses versus C_R).
2- Calculate the output of the chip for C_R = 400 fF from the curve obtained in step 1.
3- Turn on the switch and obtain the number of pulses versus the C_R curve.
4- For the calculated output in step 2, calculate the amount of shift to right. The shifted value is the offset capacitance of the IDE, C_{IDE}.
5- Calculate the number of pulses for C_R = 400 + C_{IDE} that was obtained in step 4.
6- For the calculated output in step 5, calculate the amount of shift to right after putting the samples. The shifted value is the capacitance increase due to the presence of samples on the electrodes.
7- Repeat the steps for all the next obtained curves to achieve a time-resolved capacitance plot (as shown in Figure 12).

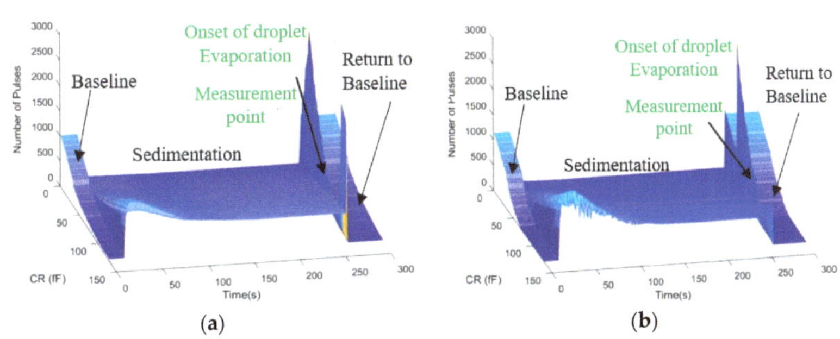

Figure 14. The 3D output of the chip for the first run of the oral samples experiments for (**a**) Left IDE, (**b**) Right IDE.

In Figure 14, The sedimentation time refers to when the sample solution still exists on the chip, and as a result, a huge capacitance is created. This capacitance saturates the sensors. Here, saturation is shown by zero in order to differentiate from other measurements.

4. Results

In this section, after explaining the fabrication and measurement setup in Section 4.1, the experimental results are demonstrated in Section 4.2.

4.1. Fabrication and Measurement Setup

Chip fabrication and its encapsulation are outlined in Section 4.1.1, and next, the whole measurement setup for real-time data acquisition is demonstrated in Section 4.1.2.

4.1.1. Chip Fabrication

Our CMOS capacitive chips were fabricated in AMS 0.35 μm high voltage CMOS technology. The chips are composed of two IDEs with five fingers, 12 μm finger width, and pitch, forming an active sensing surface area of 228 μm by 108 μm. Each IDE is connected to a separate differential CDC read-out circuit. Figure 15a represents a die micrograph of the chip. The chip was packaged on a commercial CPGA85 ceramic package with a cavity size of 8.9 mm by 8.9 mm, where the chip occupies a 1 mm × 2 mm area. After packaging, a partial encapsulation was performed on the chip, using a dam and fill technique to cover all the pads on the chip as well as on the package and the bond wires with a non-conductive resin, Hysol CB064/FP4653. After partial encapsulation, a chamber with a rectangular-shaped cross-section with an approximate size of 350 μm by 600 μm was created to serve as the container of the liquid sample, allowing the samples to have direct contact with the two underlying electrodes as depicted in Figure 15b. To be able to distinguish between the electrodes, we call them right and left as they are shown in Figure 15b. There are separate pins for the output of each side.

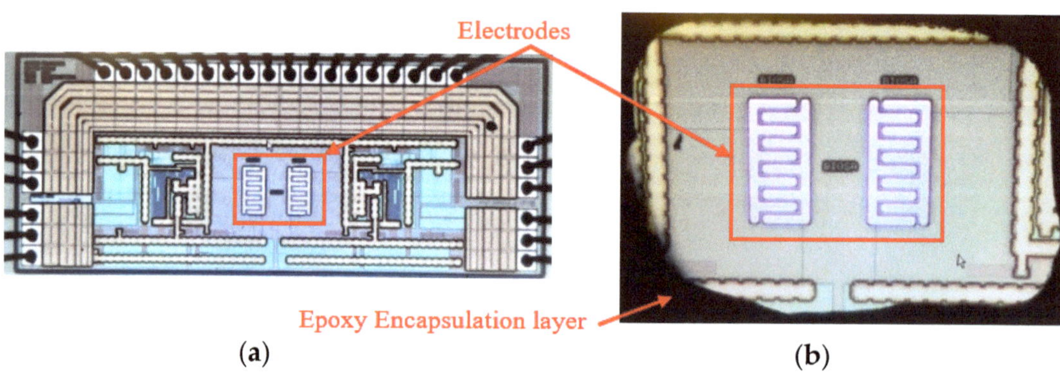

Figure 15. (**a**) Die Micrograph without encapsulation, (**b**) the chamber created after partial encapsulation of the chip.

4.1.2. Measurement Setup

Figure 16 illustrates the testbench prepared for the real-time data acquisition using the GUI and capturing images before and after cell experiments by employing the high-resolution reflective camera. Imaging the surface was challenging for many reflective microscopes due to the partial encapsulation of the chip. We were able to utilize the reflective high-resolution camera of the prob station, TS200-SE. A micropipette was used to manually put 1 μL of oral cell samples on the chip.

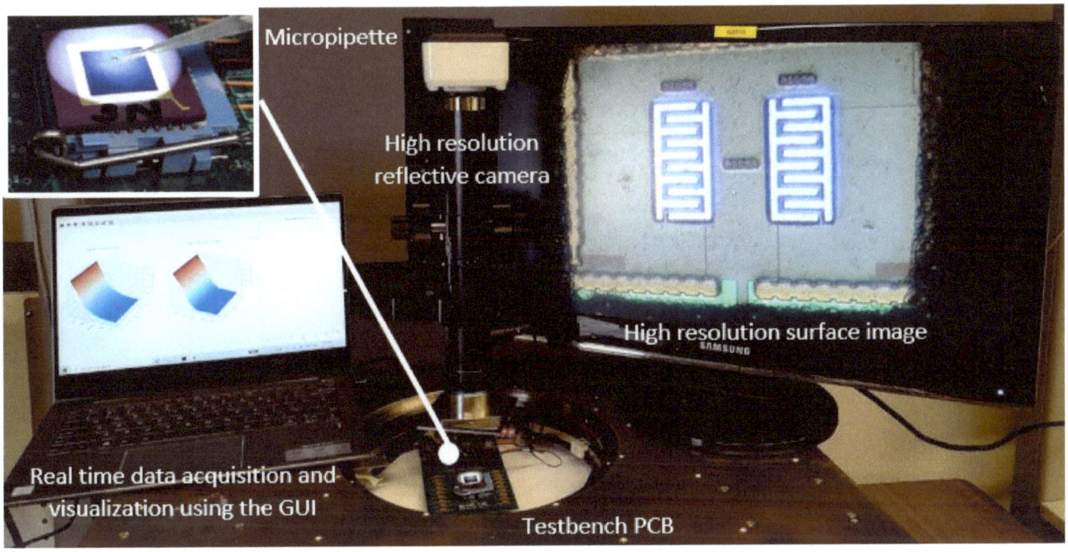

Figure 16. The prepared testbench for real-time data acquisition using the GUI and capturing images before and after cell experiments using the high-resolution reflective camera.

4.2. Oral Cell-Surface Interaction Results

In this section, the capacitance measurement results are demonstrated in two phases: (1) wet phase and (2) dry phase.

4.2.1. Capacitance Measurement in the Wet Phase

In the beginning, the baseline of each electrode, which is a result of the inherent offset capacitance of the IDE, was evaluated by obtaining the chip's output without introducing the samples. The output of the chip was obtained for the whole range of values of the bank of capacitors to have a reference point at all possible reference capacitance values. One microliter of each oral sample was placed on the electrodes. The presence of water on the non-passivated chips could increase the capacitance to a saturation level even for the highest value of reference capacitance. As a result, the data obtained before introducing the sample and before water evaporation was considered to determine the cell coverage on the electrodes in the wet mode (see measurement point, Figure 12).

Figure 17a illustrates the microscopic image of the surface of the electrodes before the deposition of samples. The deposition of cells, including epithelial and neutrophil cells (shown in Figure 17b) on the electrodes depends upon the hydrodynamic of the evaporation and the tendency of the cells to attach to the surface. Figure 17b shows the microscopic images after the first trial. We put 1 μL of oral sample on the chip and waited until the preliminary water evaporation occurred to eliminate the effect of significant capacitance change (see the Zero value, Figure 12) due to the high volume of water. In Figure 17b, two different types of cells are recognized in the microscopic images. The smaller spherical cells are neutrophils, and the larger cells with nuclei are epithelial cells. Figure 17c–e demonstrate the surface after the second, third, and fourth trials, putting 1 μL of oral sample on the chip and after partial evaporation. It is noteworthy, that the onset of evaporation is where the capacitance sharply drops, as shown in Figure 12. However, this is not the condition that the cells are completely dried.

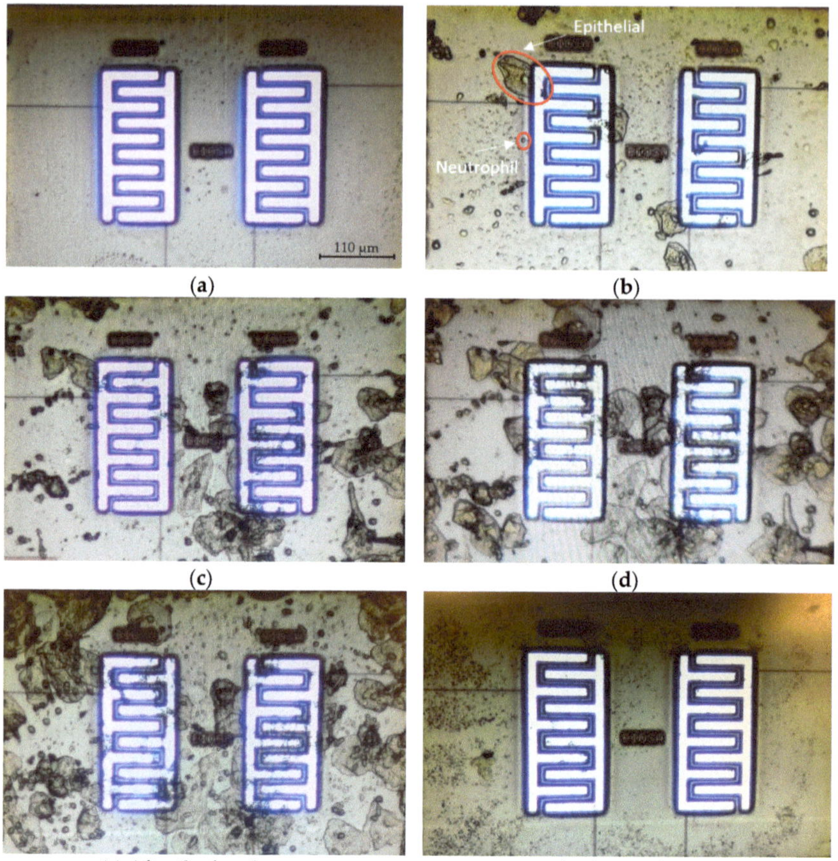

(**a**) (**b**) (**c**) (**d**)

(**e**) After the fourth run (**f**) After washing

Figure 17. The microscopic image of the surface of the chip for the first experiment: (**a**) before putting samples; (**b**) after the first sample; (**c**) after the second sample; (**d**) after the third sample; (**e**) after the fourth sample; (**f**) after washing (the size of the electrodes is 220 µm × 110 µm, which shows the scale of the images).

Figure 18 demonstrates the magnitude of capacitance changes after each run with respect to the baseline before the first run ($C_{AfterRun\#}$-$C_{BeforeRun1}$). The results were obtained by performing the capacitance extraction procedure explained in Algorithm 1. By comparing the values of differential capacitance ($C_{AfterRun\#}$-$C_{BeforeRun\#}$) and the images shown in Figure 17, there is a correlation between the number of cells that cover the electrode surface and the change in the capacitance. After the first run, there is some coverage of cells on both electrodes, with higher coverage on the left side electrode. The capacitance changes after the first run ($C_{AfterRun1}$-$C_{BeforeRun1}$) are 26.1 fF and 8.8 fF for the left electrode and the right one, respectively. After the second run, there is more coverage of cells on the right side, and the capacitance change ($C_{AfterRun2}$-$C_{BeforeRun2}$) is 21.8 fF for the left side, while it is 31.2 fF for the right side. After the third run, cells deposited on the left side have increased as the capacitance change also shows a higher value for the left electrode after the third run. Figure 17f illustrates the microscopic image of the chip after washing, and it shows that the cells were successfully removed with the procedure explained in Section 3.3.2.

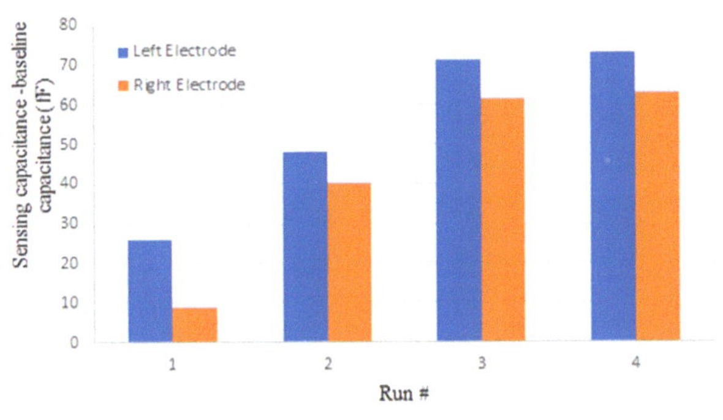

Figure 18. The difference between the capacitance after each run and the capacitance before the first run (fF) ($C_{AfterRun\#}$-$C_{BeforeRun1}$) for four consecutive runs of the first experiment for the two electrodes.

4.2.2. Capacitance Measurement in the Dry Phase

In the second experiment, the output of the sensor was monitored until the surface completely dried and the output capacitance reached a steady-state value. In the first step, the sensor surface was cleaned with the procedure explained in Section 3.3.2 as shown in Figure 19a. Then, a new prepared sample was introduced to the sensor. The outputs were measured consecutively during two runs without any washing steps in between. Figure 19b,c show the microscopic images of the chip surface after the first and the second runs, respectively. As seen in these figures, the coverage of the chip surface in Figure 19c is more than in Figure 19b. Figure 19d depicts the cleaned chip surface after the second run.

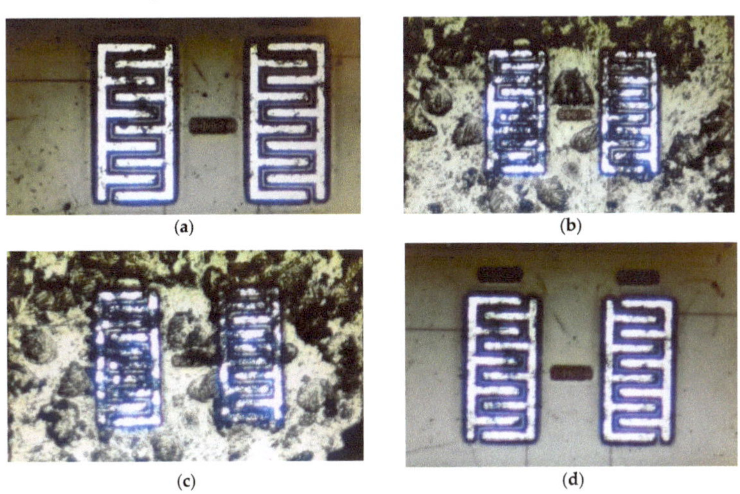

Figure 19. The microscopic images of the surface of the chip for the second experiment: (**a**) After washing and before Run 1; (**b**) After putting the first sample (Run 1); (**c**) After putting the second sample (Run 2); (**d**) The cleaned surface after Run 2.

The experimental results almost agree with the COMSOL simulation results presented in Section 3.2. The experimental equivalent capacitance for the sensor in the dry mode when the sensing area is empty and surrounded by air is around 109 fF while the simulation showed a capacitance of around 117 fF for this condition. The small difference between these values can be because of the environmental issues in the real experiment, as well as the differences in the models of CMOS layers used in COMSOL in comparison to the fabricated circuit.

The capacitance was extracted according to the algorithm explained in Section 3.4.4 Figure 20a,b show the capacitance variations versus time. Real-time measurement of the output during these two trials and after the washing procedure results in three groups of 3D profiles, as shown in Figure 21. Each group includes two 3D profiles for the two electrodes on the left and right.

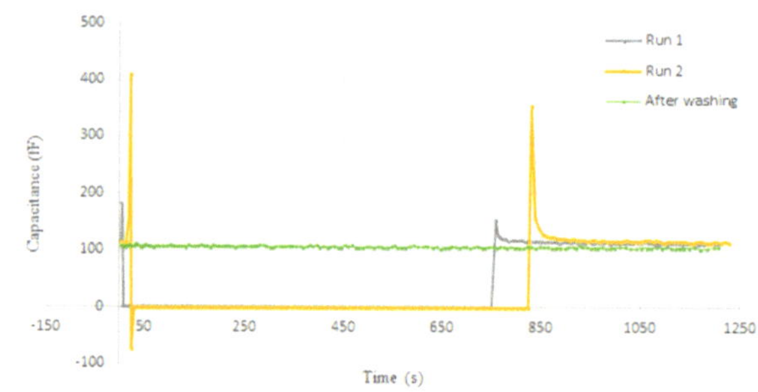

(**a**)

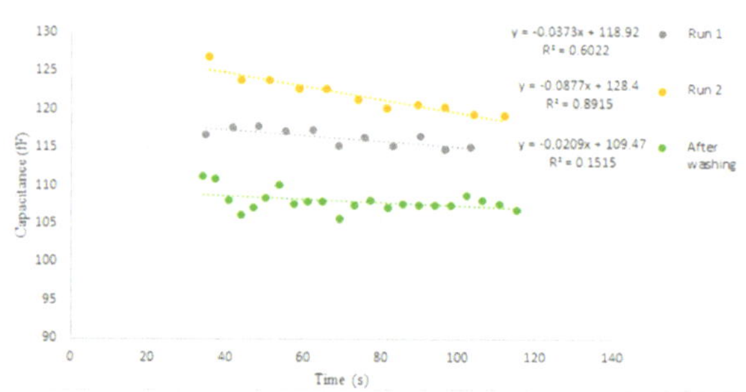

(**b**)

Figure 20. Real-time capacitance measurement result from the right electrode for the second experiment showing: (**a**) the baseline, the saturation state, the transient from the saturation to steady state, as well as the steady state (the saturation state is shown by zero in the figure); (**b**) Steady state (from about 30 s after the peak of the curves depicted in (**a**) for each run).

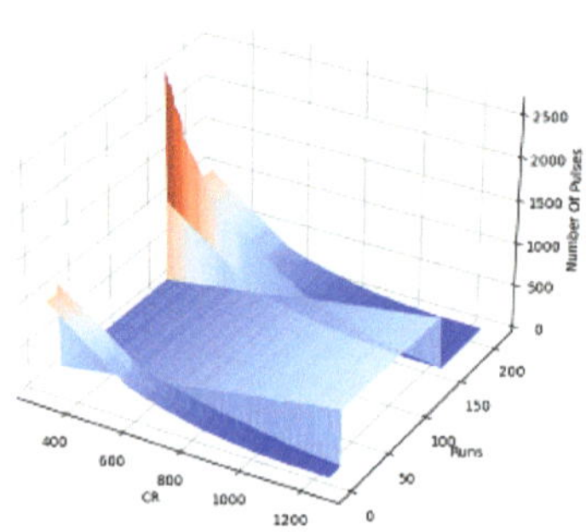

(**a**)

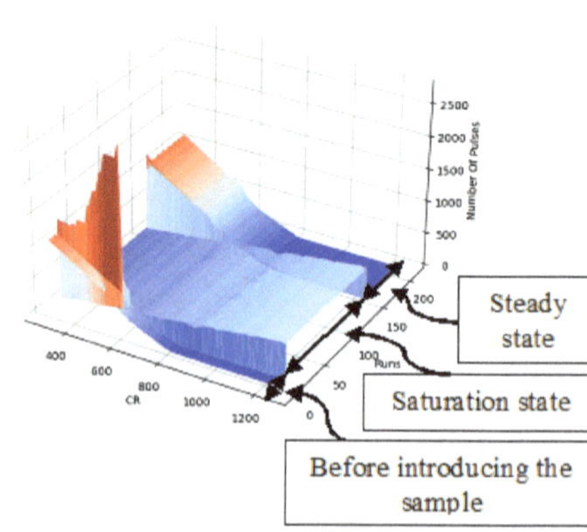

(**b**)

Group 1: The first run

Figure 21. *Cont.*

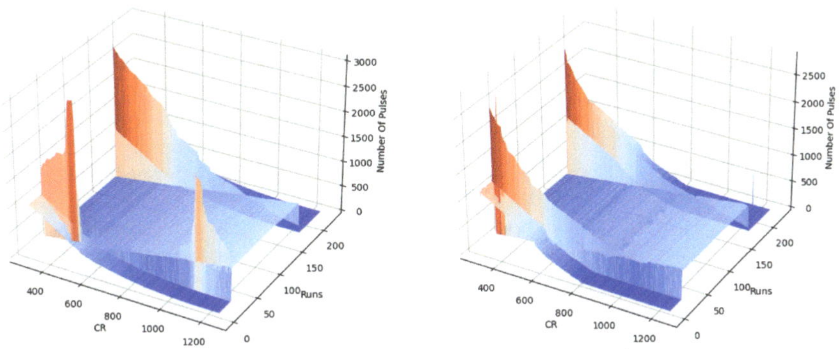

(**c**) (**d**)

Group 2: The second run

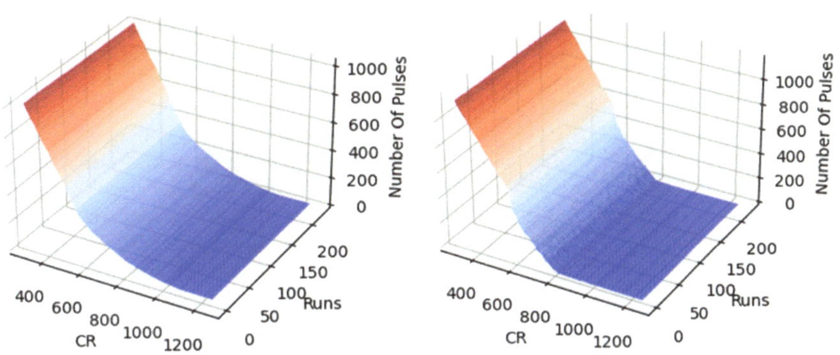

(**e**) (**f**)

Group 3: After washing

Figure 21. The 3D output of the chip for the second experiment for: (**a**) Run 1, left electrode; (**b**) Run 1, right electrode; (**c**) Run 2, left electrode; (**d**) Run 2, right electrode; (**e**) After washing, left electrode; (**f**) After washing, right electrode.

According to the simulations, the sensor initially gets saturated by introducing the sample and then drops down during the evaporation of the liquid. As shown in Figures 20a and 21, the extracted capacitance and the digital output follow the same trend over time. As aforementioned, the saturation state due to the large capacitance of the liquid is shown by zero in Figures 20a and 21. During the evaporation of the liquid, the output decreases until it reaches a steady-state level.

Figure 20b illustrates the steady states of the two trials and after washing the chip surface. As seen in this figure, adding the new sample to the surface in run 2 without washing after run 1 results in an accumulative output that is higher than the steady-state level of run 1. More coverage of the electrodes by the cells, shown in Figure 19, led to this increasing trend for steady-state capacitance. Washing the surface causes the output to return to the baseline, which is around 109 fF.

5. Discussion

In this paper, we introduced a capacitive sensing platform for monitoring oral cells and likely, in the future, analyzing them to understand their links to inflammatory diseases. Herein, we demonstrated the presence of epithelial and neutrophil cells and their effects on capacitive sensors. This takes us one step closer to developing the capacitive sensing PoC device for monitoring and counting the cells. In this direction, there are several practical considerations to be addressed in the future, as discussed below.

5.1. Isolation of Cells from Saliva

Saliva is composed of 99% water and only 1% immunoglobulins, protein, mucus, enzymes, salts, and electrolytes such as sodium, potassium, calcium, magnesium, bicarbonate, and phosphates as well as different types of cells like epithelial cells or intact and lysed inflammatory cells, specifically oPMNs [45,46]. Herein, the standard filtering technique is used to isolate oral cells from saliva. Neutrophils can only be kept alive within a short time frame, and physical techniques are used to purify the sample. To develop a handheld sensing device for monitoring oral neutrophils, which are small, microfluidic techniques are also required to separate them from epithelial cells and the debris in the saliva. The greater the purity of the sample, the greater the accuracy of the measurement result.

5.2. The Effects of the Evaporation of the Sample

As explained in Section 4.2, the measurement results depend on the evaporation of the sample. For a reliable measurement, a microfluidic device can be designed to direct the sample toward the sensing site and prevent or control the evaporation of the liquid.

5.3. Bubble Creation

Since pipetting the sample on the chip surface is done manually, some errors are inevitable. For instance, one of the practical problems during the experimental results was the creation of bubbles on top of the sensing area. This phenomenon results in a meaningless output, including fluctuations depending on the bubbles' presence or absence, as shown in Figure 22. Providing a microfluidic device to control and even automate putting the sample on the chip surface can help avoid bubble formation. In addition, microfluidics makes it possible to put the sample on top of the sensing electrodes more precisely, leading to more accurate measurement results.

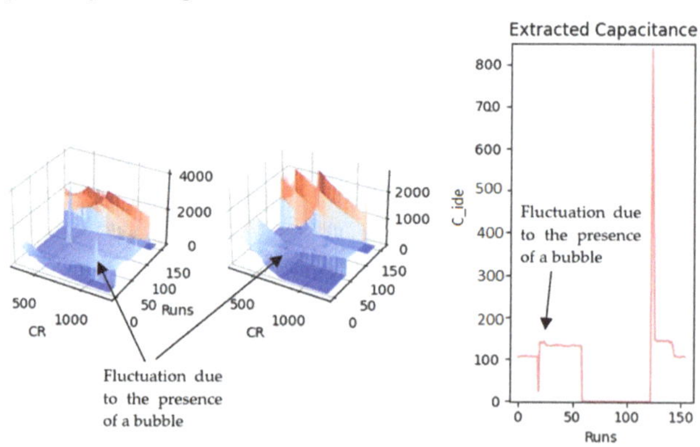

Figure 22. The fluctuated pattern due to the presence of a bubble created during sample placement.

5.4. Sensing Electrodes

The appropriate location of the cells on top of the sensing area and the amount of cell coverage can affect the measurement results. Since the electrodes do not cover the whole area of the substrate, the cells might be placed over the non-sensitive or less sensitive area, for example in between the electrodes. As a result, the sensor cannot sense them perfectly. To avoid such errors, a new integrated circuit is required whose electrodes cover the whole area of the substrate.

The size of the electrode can also affect its sensitivity. Although we almost cancel the offset in the proposed circuit, the value of the offset and parasitic capacitances might be higher than the sensing capacitance for electrodes that are too large, resulting in a low sensitive measurement. To quantify this, consider that the electrode capacitive ratio (ECR) is defined as $\alpha C = (\Delta C/C_0) \times 100$, where ΔC is the variations of the sensing capacitance variations due to the presence of the cells and C_0 is the offset and parasitic shares of the electrode–analyte interface. If the size of the electrode is much bigger than the size of the cells, ΔC will be much smaller than C_0 leading to a small ECR. On the other hand, although the smaller sizes of the electrodes help increase the ECR, we will need a large number of them to cover the whole area of the substrate so that the circuit can read their outputs in parallel. Herein, we tested a two-electrode sensor with oral cells as a proof of concept. To further our research, we intend to present a new sensor composed of many electrodes for this application.

5.5. Read-Out Circuit Specifications

Various factors such as parasitic effects, systematic errors, environmental factors, and experiment-time offset variation due to the remnants of the cells can create a time-variant offset capacitance that might saturate the sensor's output. Therefore, the sensor should have a wide IDR to show the target cells' concentration. According to Table 1, the proposed sensor in this work offers a wider IDR than the other reported capacitive sensors. In addition, the IDR of the sensor is programmable, and the employed calibration-free technique is based on sweeping the reference capacitor [40]. This has paved the way to mitigate the effects of undesired time-variant offsets significantly.

On the other hand, a higher sensor resolution can lead to more accuracy. If the cell is not completely located on the electrode, the sensor can detect the part of the cell covering the electrode. There is a trade-off between IDR and the resolution of a circuit.

Raising the gain of current amplifiers and the sensitivity of the CCO can help to improve the sensitivity of the sensor. Increasing the IDR of the CCO and the size of the counter can also result in a wider IDR. These topics are deferred to future work.

6. Conclusions

This paper demonstrated the applicability of a calibration-free and wide-IDR CMOS capacitive sensor for monitoring oral cells in saliva samples. The capacitance of the on-chip IDEs was mathematically modeled by considering the existing parasitic and fringe capacitances. A COMSOL simulation was performed to qualitatively gain insight from the sensor response to the biological cells. Simulations were qualitatively in agreement with the experiment and confirmed the functionality of devices in terms of sensitivity to change in the dielectric of added material on its sensing surface. The difference between the capacitance values before and after water evaporation could represent the amount of electrode surface coverage by oral cells. By taking advantage of the wide IDR of the sensor and the programmable bank of reference capacitors, the accumulation of the cells could be assessed over four runs without washing. To avoid the electrodes' conditions being manipulated after each run, no washing steps were applied to the electrodes exposed to the oral cells sample. The experiments showed that the accumulation of the cells after each run decreases the sensor's sensitivity. Therefore, a cleaning procedure was employed to remove the cells from the chip surface. The results were encouraging regarding developing reusable, integrated sensing devices that can set the stage for quantifying and analyzing oral cells, including neutrophils and epithelial cells, in the future.

Author Contributions: Conceptualization, H.O.T. and E.G.-Z.; methodology, H.O.T., A.P. and E.G.-Z.; software, H.O.T., A.P. and M.R.H.H.K.; validation, H.O.T., A.P. and E.G.-Z.; Investigation, H.O.T., D.S., F.S. and S.F.; writing—original draft preparation, H.O.T., D.S., A.P., F.S., S.F. and E.G.-Z.; writing—review and editing, H.O.T., S.F., D.S., A.P., F.S., E.G.-Z. and S.M.; supervision, E.G.-Z. and S.M.; funding acquisition, E.G.-Z. All authors have read and agreed to the published version of the manuscript.

Funding: This research was funded by Natural Sciences and Engineering Research Council (NSERC) Canada (RGPIN-2021-03618).

Institutional Review Board Statement: Human subject participation approval received from York University.

Informed Consent Statement: Informed consent was obtained from all subjects involved in the study.

Data Availability Statement: The data are not publicly available.

Acknowledgments: The authors would like to acknowledge the support of NSERC Canada, the Canadian Microsystem Corporation (CMC), and Mitacs.

Conflicts of Interest: The authors declare no conflict of interest.

Appendix A

Figure A1 illustrates the schematic of the PCB of the sensor shown in Figure 9. In this figure, the block shown with U0 represents the chip while U1 represents the 2:1 multiplexer chip that is implemented to switch between right and left channels. Vref is a voltage regulator block of 1.85 V output that is utilized for the CCO block. RAVDD, RDVDD, LAVDD, and LDVDD blocks are 3.3 V-output voltage regulators that provide the VDD voltage to the right analog block, right digital block, left analog block, and left digital block, respectively. The input is an external 5 V DC source shown as X1. Connectors are for the Arduino DUE board, making the digital interface between the board and Arduino. Multiple decoupling capacitors are used between voltage and ground rails.

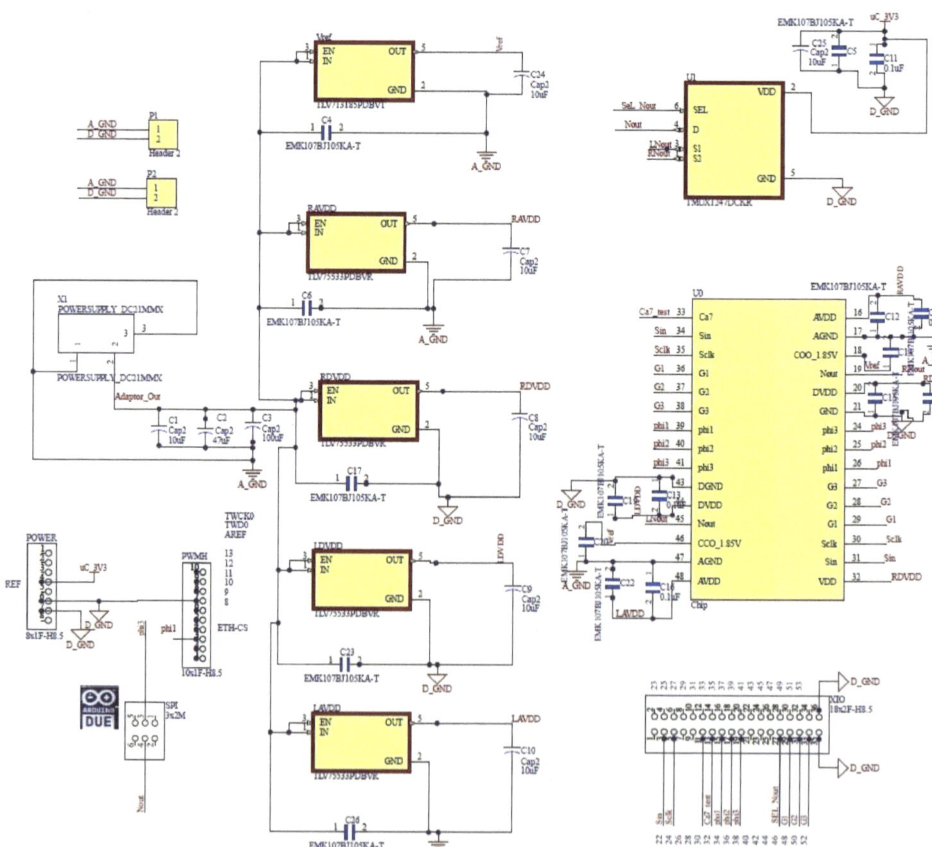

Figure A1. The schematic of the PCB of the sensor shown in Figure 9.

References

1. Llambés, F.; Arias-Herrera, S.; Caffesse, R. Relationship between Diabetes and Periodontal Infection. *World J. Diabetes* **2015**, *6*, 927. [CrossRef] [PubMed]
2. Wang, C.W.; McCauley, L.K. Osteoporosis and Periodontitis. *Curr. Osteoporos. Rep.* **2016**, *14*, 284–291. [CrossRef] [PubMed]
3. Pólvora, T.L.S.; Nobre, Á.V.V.; Tirapelli, C.; Taba, M.; de Macedo, L.D.; Santana, R.C.; Pozzetto, B.; Lourenço, A.G.; Motta, A.C.F. Relationship between Human Immunodeficiency Virus (HIV-1) Infection and Chronic Periodontitis. *Expert Rev. Clin. Immunol.* **2018**, *14*, 315–327. [CrossRef] [PubMed]
4. Hajishengallis, G.; Moutsopoulos, N.M.; Hajishengallis, E.; Chavakis, T. Immune and Regulatory Functions of Neutrophils in Inflammatory Bone Loss. *Semin. Immunol.* **2016**, *28*, 146–158. [CrossRef] [PubMed]
5. Perez Rosero, E.; Heron, S.; Jovel, J.; O'Neil, C.R.; Turvey, S.L.; Parashar, P.; Elahi, S. Differential Signature of the Microbiome and Neutrophils in the Oral Cavity of HIV-Infected Individuals. *Front. Immunol.* **2021**, *12*, 4716. [CrossRef]
6. Levine, A.M.; Karim, R.; Mack, W.; Gravink, D.J.; Anastos, K.; Young, M.; Cohen, M.; Newman, M.; Augenbraun, M.; Gange, S.; et al. Neutropenia in Human Immunodeficiency Virus Infection: Data From the Women's Interagency HIV Study. *Arch. Intern. Med.* **2006**, *166*, 405–410. [CrossRef]
7. Sanz, M.; del Castillo, A.M.; Jepsen, S.; Gonzalez-Juanatey, J.R.; D'Aiuto, F.; Bouchard, P.; Chapple, I.; Dietrich, T.; Gotsman, I.; Graziani, F.; et al. Periodontitis and Cardiovascular Diseases: Consensus Report. *J. Clin. Periodontol.* **2020**, *47*, 268–288. [CrossRef]
8. Liccardo, D.; Cannavo, A.; Spagnuolo, G.; Ferrara, N.; Cittadini, A.; Rengo, C.; Rengo, G. Periodontal Disease: A Risk Factor for Diabetes and Cardiovascular Disease. *Int. J. Mol. Sci.* **2019**, *20*, 1414. [CrossRef]
9. The Neutrophil in Oral Health and Disease: The New Diagnostic Biomarker—Oral Health Group. Available online: https://www.oralhealthgroup.com/features/the-neutrophil-in-oral-health-and-disease-the-new-diagnostic-biomarker/ (accessed on 22 April 2022).
10. Fine, N.; Hassanpour, S.; Borenstein, A.; Sima, C.; Oveisi, M.; Scholey, J.; Cherney, D.; Glogauer, M. Distinct Oral Neutrophil Subsets Define Health and Periodontal Disease States. *J. Dent. Res.* **2016**, *95*, 931–938. [CrossRef]
11. Çetin Özdemir, E.; Bilen, E.; Yazar, F.M. Can the Delta Neutrophil Index Be Used as a Preliminary Biomarker in the Evaluation of Periodontal Disease: A Pilot Study. *J. Appl. Oral Sci.* **2022**, *30*, e20210555. [CrossRef]
12. Nicu, E.A.; Rijkschroeff, P.; Wartewig, E.; Nazmi, K.; Loos, B.G. Characterization of Oral Polymorphonuclear Neutrophils in Periodontitis Patients: A Case-Control Study. *BMC Oral Health* **2018**, *18*, 149. [CrossRef]
13. Hirschfeld, J. Neutrophil Subsets in Periodontal Health and Disease: A Mini Review. *Front. Immunol.* **2020**, *10*, 3001. [CrossRef] [PubMed]
14. Khoury, W.; Glogauer, J.; Tenenbaum, H.C.; Glogauer, M. Oral Inflammatory Load: Neutrophils as Oral Health Biomarkers. *J. Periodontal Res.* **2020**, *55*, 594–601. [CrossRef] [PubMed]
15. Rijkschroeff, P.; Loos, B.G.; Nicu, E.A. Oral Polymorphonuclear Neutrophil Contributes to Oral Health. *Curr. Oral Health Rep.* **2018**, *5*, 211–220. [CrossRef] [PubMed]
16. Rijkschroeff, P.; Jansen, I.D.C.; Van Der Weijden, F.A.; Keijser, B.J.F.; Loos, B.G.; Nicu, E.A. Oral Polymorphonuclear Neutrophil Characteristics in Relation to Oral Health: A Cross-Sectional, Observational Clinical Study. *Int. J. Oral Sci.* **2016**, *8*, 191–198. [CrossRef] [PubMed]
17. McKinnon, K.M. Flow Cytometry: An Overview. *Curr. Protoc. Immunol.* **2018**, *120*, 5.1.1–5.1.11. [CrossRef]
18. Vembadi, A.; Menachery, A.; Qasaimeh, M.A. Cell Cytometry: Review and Perspective on Biotechnological Advances. *Front. Bioeng. Biotechnol.* **2019**, *7*, 147. [CrossRef]
19. Forouhi, S.; Dehghani, R.; Ghafar-Zadeh, E. CMOS Based Capacitive Sensors for Life Science Applications: A Review. *Sens. Actuators A Phys.* **2019**, *297*, 111531. [CrossRef]
20. Musayev, J.; Altiner, C.; Adiguzel, Y.; Kulah, H.; Eminoglu, S.; Akin, T. Capturing and Detection of MCF-7 Breast Cancer Cells with a CMOS Image Sensor. *Sens. Actuators A Phys.* **2014**, *215*, 105–114. [CrossRef]
21. Senevirathna, B.P.; Lu, S.; Dandin, M.P.; Basile, J.; Smela, E.; Abshire, P.A. Real-Time Measurements of Cell Proliferation Using a Lab-on-CMOS Capacitance Sensor Array. *IEEE Trans. Biomed. Circuits Syst.* **2018**, *12*, 510–520. [CrossRef]
22. Nabovati, G.; Ghafar-Zadeh, E.; Letourneau, A.; Sawan, M. Smart Cell Culture Monitoring and Drug Test Platform Using CMOS Capacitive Sensor Array. *IEEE Trans. Biomed. Eng.* **2018**, *66*, 1094–1104. [CrossRef] [PubMed]
23. Ghafar-Zadeh, E.; Sawan, M.; Chodavarapu, V.P.; Hosseini-Nia, T. Bacteria Growth Monitoring through a Differential CMOS Capacitive Sensor. *IEEE Trans. Biomed. Circuits Syst.* **2010**, *4*, 232–238. [CrossRef] [PubMed]
24. Park, J.S.; Aziz, M.K.; Li, S.; Chi, T.; Grijalva, S.I.; Sung, J.H.; Cho, H.C.; Wang, H. 1024-Pixel CMOS Multimodality Joint Cellular Sensor/Stimulator Array for Real-Time Holistic Cellular Characterization and Cell-Based Drug Screening. *IEEE Trans. Biomed. Circuits Syst.* **2017**, *12*, 80–94. [CrossRef]
25. Widdershoven, F.; Cossettini, A.; Laborde, C.; Bandiziol, A.; Van Swinderen, P.P.; Lemay, S.G.; Selmi, L. A CMOS Pixelated Nanocapacitor Biosensor Platform for High-Frequency Impedance Spectroscopy and Imaging. *IEEE Trans. Biomed. Circuits Syst.* **2018**, *12*, 1369–1382. [CrossRef]
26. Burridge, K.; Burridge, C.K. Focal Adhesions: A Personal Perspective on a Half Century of Progress. *FEBS J.* **2017**, *284*, 3355–3361. [CrossRef] [PubMed]
27. Li, R.; Wang, Z.; Huang, J.; He, S.; Peng, Y.; Wan, Y.; Ma, Q. Neutrophils Culture in Collagen Gel System. *Front. Immunol.* **2022**, *13*, 816037. [CrossRef] [PubMed]

28. Primary Culture of Human Oral Epithelial Cells. Growth Requirements and Expression of Differentiated Characteristics—PubMed. Available online: https://pubmed.ncbi.nlm.nih.gov/2433501/ (accessed on 25 April 2022).
29. Oda, D.; Watson, E. Human Oral Epithelial Cell Culture I. Improved Conditions for Reproducible Culture in Serum-Free Medium. *Vitr. Cell. Dev. Biol.* **1990**, *26*, 589–595. [CrossRef]
30. Sheth, R.; Neale, M.H.; Shortt, A.J.; Massie, I.; Vernon, A.J.; Daniels, J.T. Culture and Characterization of Oral Mucosal Epithelial Cells on a Fibrin Gel for Ocular Surface Reconstruction. *Mol. Vis.* **2014**, *40*, 1077–1087. [CrossRef]
31. Yang, Y.; Knight, R.; Stephens, P.; Zhang, Y. Three-Dimensional Culture of Oral Progenitor Cells: Effects on Small Extracellular Vesicles Production and Proliferative Function. *J. Oral Pathol. Med.* **2020**, *49*, 342–349. [CrossRef]
32. Amulic, B.; Cazalet, C.; Hayes, G.L.; Metzler, K.D.; Zychlinsky, A. Neutrophil Function: From Mechanisms to Disease. *Annu. Rev. Immunol.* **2012**, *30*, 459–489. [CrossRef]
33. Nussbaum, G.; Shapira, L. How Has Neutrophil Research Improved Our Understanding of Periodontal Pathogenesis? *J. Clin. Periodontol.* **2011**, *38*, 49–59. [CrossRef] [PubMed]
34. Prakash, S.B.; Abshire, P. Tracking Cancer Cell Proliferation on a CMOS Capacitance Sensor Chip. *Biosens. Bioelectron.* **2008**, *23*, 1449–1457. [CrossRef] [PubMed]
35. Senevirathna, B.; Lu, S.; Dandin, M.; Basile, J.; Smela, E.; Abshire, P. High Resolution Monitoring of Chemotherapeutic Agent Potency in Cancer Cells Using a CMOS Capacitance Biosensor. *Biosens. Bioelectron.* **2019**, *142*, 111501. [CrossRef] [PubMed]
36. Laborde, C.; Pittino, F.; Verhoeven, H.A.; Lemay, S.G.; Selmi, L.; Jongsma, M.A.; Widdershoven, F.P. Real-Time Imaging of Microparticles and Living Cells with CMOS Nanocapacitor Arrays. *Nat. Nanotechnol.* **2015**, *10*, 791. [CrossRef] [PubMed]
37. Couniot, N.; Bol, D.; Poncelet, O.; Francis, L.A.; Flandre, D. A Capacitance-to-Frequency Converter with on-Chip Passivated Microelectrodes for Bacteria Detection in Saline Buffers up to 575 MHz. *IEEE Trans. Circuits Syst. II Express Briefs* **2014**, *62*, 159–163. [CrossRef]
38. Couniot, N.; Francis, L.A.; Flandre, D. A 16×16 CMOS Capacitive Biosensor Array towards Detection of Single Bacterial Cell. *IEEE Trans. Biomed. Circuits Syst.* **2016**, *10*, 364–374. [CrossRef]
39. Tabrizi, H.O.; Farhanieh, O.; Owen, Q.; Magierowski, S.; Ghafar-Zadeh, E. Wide Input Dynamic Range Fully Integrated Capacitive Sensor for Life Science Applications. *IEEE Trans. Biomed. Circuits Syst.* **2021**, *15*, 339–350. [CrossRef]
40. Tabrizi, H.O.; Forouhi, S.; Farhanieh, O.; Bozkurt, A.; Magierowski, S.; Ghafar-Zadeh, E. Calibration-Free CMOS Capacitive Sensor for Life Science Applications. *IEEE Trans. Instrum. Meas.* **2021**, *70*, 2006512. [CrossRef]
41. Bae, J.; Samek, I.A.; Stair, P.C.; Snurr, R.Q. Investigation of the Hydrophobic Nature of Metal Oxide Surfaces Created by Atomic Layer Deposition. *Langmuir* **2019**, *35*, 5762–5769. [CrossRef]
42. Bunnfors, K.; Abrikossova, N.; Kilpijärvi, J.; Eriksson, P.; Juuti, J.; Halonen, N.; Brommesson, C.; Lloyd Spetz, A.; Uvdal, K. Nanoparticle Activated Neutrophils-on-a-Chip: A Label-Free Capacitive Sensor to Monitor Cells at Work. *Sens. Actuators B Chem.* **2020**, *313*, 128020. [CrossRef]
43. Nabovati, G.; Ghafar-Zadeh, E.; Letourneau, A.; Sawan, M. Towards High Throughput Cell Growth Screening: A New CMOS 8 × 8 Biosensor Array for Life Science Applications. *IEEE Trans. Biomed. Circuits Syst.* **2017**, *11*, 380–391. [CrossRef] [PubMed]
44. Vargel, C. *Corrosion of Aluminium*; Elsevier: Amsterdam, The Netherlands, 2020.
45. Tiwari, M. Science behind Human Saliva. *J. Nat. Sci. Biol. Med.* **2011**, *2*, 53. [CrossRef] [PubMed]
46. Klinkhamer, J.M.; Zimmerman, S. The Function and Reliability of the Orogranulocytic Migratory Rate as a Measure of Oral Health. *J. Dent. Res.* **1969**, *48*, 709–715. [CrossRef] [PubMed]

bioengineering

Review

Application of Micro-Engineered Kidney, Liver, and Respiratory System Models to Accelerate Preclinical Drug Testing and Development

Hanieh Gholizadeh [1,2,3], Shaokoon Cheng [3], Agisilaos Kourmatzis [4], Hanwen Xing [5], Daniela Traini [1,2], Paul M. Young [2,6] and Hui Xin Ong [1,2,*]

1 Macquarie Medical School, Faculty of Medicine, Health, and Human Sciences, Macquarie University, Ryde, NSW 2109, Australia; hanieh.mohammad-gholizadeh-@hdr.mq.edu.au (H.G.); daniela.traini@mq.edu.au (D.T.)
2 Respiratory Technology, The Woolcock Institute of Medical Research, The University of Sydney, Sydney, NSW 2037, Australia; p.young@mq.edu.au
3 School of Engineering, Faculty of Science and Engineering, Macquarie University, Ryde, NSW 2113, Australia; shaokoon.cheng@mq.edu.au
4 School of Aerospace, Mechanical and Mechatronic Engineering, The University of Sydney, Sydney, NSW 2006, Australia; agisilaos.kourmatzis@sydney.edu.au
5 Faculty of Engineering and Information Technology, University of Technology Sydney, Sydney, NSW 2007, Australia; xinghanwen9@gmail.com
6 Department of Marketing, Macquarie Business School, Macquarie University, Ryde, NSW 2109, Australia
* Correspondence: huixin.ong@mq.edu.au; Tel.: +61-2-9114-0373

Abstract: Developing novel drug formulations and progressing them to the clinical environment relies on preclinical in vitro studies and animal tests to evaluate efficacy and toxicity. However, these current techniques have failed to accurately predict the clinical success of new therapies with a high degree of certainty. The main reason for this failure is that conventional in vitro tissue models lack numerous physiological characteristics of human organs, such as biomechanical forces and biofluid flow. Moreover, animal models often fail to recapitulate the physiology, anatomy, and mechanisms of disease development in human. These shortfalls often lead to failure in drug development, with substantial time and money spent. To tackle this issue, organ-on-chip technology offers realistic in vitro human organ models that mimic the physiology of tissues, including biomechanical forces, stress, strain, cellular heterogeneity, and the interaction between multiple tissues and their simultaneous responses to a therapy. For the latter, complex networks of multiple-organ models are constructed together, known as multiple-organs-on-chip. Numerous studies have demonstrated successful application of organ-on-chips for drug testing, with results comparable to clinical outcomes. This review will summarize and critically evaluate these studies, with a focus on kidney, liver, and respiratory system-on-chip models, and will discuss their progress in their application as a preclinical drug-testing platform to determine in vitro drug toxicology, metabolism, and transport. Further, the advances in the design of these models for improving preclinical drug testing as well as the opportunities for future work will be discussed.

Keywords: organ-on-chip; metabolism; toxicology; drug transport; body-on-chip; disease-on-chip; drug discovery

Citation: Gholizadeh, H.; Cheng, S.; Kourmatzis, A.; Xing, H.; Traini, D.; Young, P.M.; Ong, H.X. Application of Micro-Engineered Kidney, Liver, and Respiratory System Models to Accelerate Preclinical Drug Testing and Development. *Bioengineering* **2022**, *9*, 150. https://doi.org/10.3390/bioengineering9040150

Academic Editor: Ali Zarrabi

Received: 28 February 2022
Accepted: 28 March 2022
Published: 2 April 2022

1. Introduction

Intensive and costly preclinical tests on novel therapeutic agents often involve animal studies prior to human trials. In addition to being ethically controversial, animal models have been increasingly criticized for their limited ability to predict the efficacy, safety, and toxicity of numerous drugs in humans [1,2]. Animal testing is predisposed—by its xenogeneic nature and failure to represent the complicated anatomical and physiological systems in humans—to lead to discordant results and, consequently, to failures in translating the results to clinical trials. Conventional in vitro models that utilize two-dimensional (2D) or three-dimensional (3D) cell cultures, on the other hand, are limited in their predictive capabilities of in vivo conditions due to the absence of critical physiological factors, such as fluid flow and biomechanical forces [3]. Consequently, the development of novel methodologies that could accurately represent the physiological conditions of the human body in an in vitro platform to perform drug testing is of critical importance to accelerate the development of novel therapeutics that can be used in clinical settings.

The development of micro-structured functional human organ models, known as organ-on-chip technology, is a potential solution for providing a physiologically relevant in vitro platform and has attracted increasing interest [3]. Organ-on-chips aim to provide a translational model for human organs to predict human responses to therapeutic agents. As a result, they can deliver reliable and accurate outcome measures and also likely have the potential to predict clinical trial results. Since its advent, organ-on-chip technology has been widely studied and found to offer numerous advantages in recapitulating organ physiology to study human diseases and its potential for drug testing. Current reviews on the application of organ-on-chips for in vitro drug tests mainly focus on models of certain organs, their microfluidic designs and structures, and their relevance to healthy or diseased tissues whilst providing examples of their potential application for testing drugs [4–8]. In addition, these reviews elaborate on the advances in organ-on-chips and their future opportunities for commercialization of these platforms for drug testing [3,9–11]. Table 1 provides an overview of the different relevant reviews and their comparison with the current review on the application of organ-on-chips for drug testing. This review will provide a pharmaceutical perspective by focusing on the application of organ-on-chips for different preclinical drug tests with improved relevance. Hence, the studies on organ-on-chips will be discussed based on the assays used to evaluate drug efficacy and toxicity in vitro, which are important for drug development and commercialization. Further, the advantages of organ-on-chips in drug testing will be elaborated based on their physiologically relevant dynamic design [12–14], emulation of the inter-organ crosstalk in the human body [15,16], enabling patient-specific drug testing [17], and high throughput time-efficient assays by integrated miniaturised analytical tools [18,19]. Liver, kidney, and respiratory system models will be the focus of this review due to their broad applicability in toxicology [20], drug metabolism [21], and drug delivery/transport [22] studies.

Table 1. Overview of the primary focus of relevant reviews in the application of OOCs for preclinical drug testing compared with the current review.

Refs.	Focus of the Review
[3–5]	• Application of organ-on-chip models of certain organs in providing a relevant platform for drug testing in terms of the design, structure, and cell culture techniques • The different assays used to evaluate drug efficacy in those organ-on-chip models
[3,9–11]	• Progress, challenges, and opportunities for the application of organ-on-chip technology in preclinical drug discovery • The commercialization outlook of organ-on-chip technology for drug testing
[6,7]	• Drug toxicity studies performed on organ-on-chips with improved physiological relevance
[8]	• Organ-on-chips as potential platforms for screening nanocarrier drug delivery with improved physiological relevance
This review	• Improvements of in vitro drug efficacy assays when conducted in organ-on-chips to ensure outcomes are clinically relevant • Advantages of organ-on-chip technology in providing a translational model for physiologically relevant in vitro drug testing • Discussion of the various drug compounds that have been tested on organ-on-chips • Relevance of drug testing outcomes from organ-on-chips to clinical observations

2. Drug Testing Capabilities Using Liver-, Kidney-, and Lung-on-Chip Models

Liver-, kidney-, and lung-on-chip models have recreated some key functions of their respective organs for drug metabolism, expression of cytotoxic response to drugs, and barrier function against drug permeation. Figure 1 summarizes how these organ-on-chips have been used for studies on drug metabolism, toxicology, and transport.

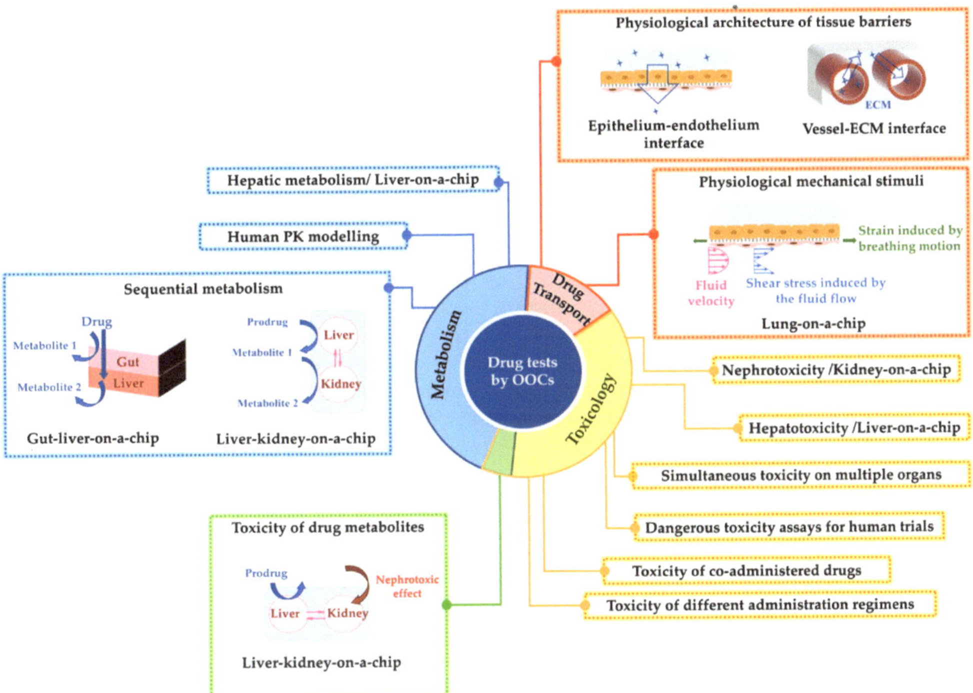

Figure 1. A summary of the investigations on three important drug assessments: drug metabolism, toxicology, and drug transport using liver-, kidney-, and lung-on-chip models. The figure summarizes the advantages of these organ-on-chip models to enhance drug testing.

2.1. Drug Metabolism Studies

The principal, although not the sole, site for metabolism of nearly all ingested drugs is the liver [23]. Consequently, first-pass hepatic metabolism has a paramount influence on the efficacy and side effects of drugs, which highlights the importance of simulating this biological process during drug testing. Organ-on-chip technology has been utilized for screening drug metabolism, where recapitulation conditions of the liver tissue have been the main focus [24,25]. Table 2 summarizes the studies on the metabolism of different drugs using liver-on-chip models and the interconnection of the liver with other organs-on-chip.

The utilization of organ-on-chip technology to construct the network of multiple organs is beneficial to mimic the complex interaction between organs in metabolizing drugs, prodrugs, and micronutrients in vivo. For instance, a liver–kidney-on-chip with interconnected chambers for liver and kidney cell culture was developed by Tehobald et al. [26] to imitate sequential hepatic and renal metabolism of vitamin D. The device could mimic hepatic metabolism of vitamin D to 25-hydroxyvitamin D and its further metabolism to 1,25-dihydroxyvitamin D by the kidney analog. The latter bioactive metabolite is known for its anti-tumor effect and for inducing differentiation in multiple tumor cell types, such as acute myeloid leukemia cells.

Studying the drug metabolism process in multi-organs-on-chip models and their comparison with single-organ-on-chip models can shed insights into understanding the tissue–tissue crosstalk and individual contribution to the metabolic process [27]. In another study, Choe et al. reproduced the first-pass-metabolism process in a microfluidic gut–

liver chip with two chambers separated by a membrane to culture gut epithelial and liver cells [28]. The gut cells' culture chamber was located on top of the liver cells' chamber, so that the drug molecules passing across the gut epithelial barrier could reach the liver cells. This enables emulation of the simultaneous drug absorption in the gut and metabolic reaction in the liver that the drug goes through after oral intake. The metabolic activity of the gut–liver chip was compared to the gut monoculture system, and the gut epithelial cells were highlighted to have significantly higher contribution in the metabolism of apigenin.

Another important factor that needs to be considered when designing in vitro drug-metabolism models is that the rates of absorption and metabolism are influenced by the volumes and dimensions of the designed organ model [29]. Hence, it is important to replicate the relative sizes of organ analogs and the circulating fluid flow connecting them based on the physiology of the human body in order to achieve accurate pharmacokinetic (PK) modelling using multi-organs-on-chip. Some studies on the application of multi-organs-on-chips to replicate human drug metabolism have focused on the design of devices based on mathematically modelled PK profiles [30,31]. These microfluidic chips are promising tools to emulate human-relevant PK in vitro. For instance, the first-pass metabolism of orally taken paracetamol was replicated in a gut–liver-on-chip [31]. The design parameters of the chip were such that the surface area of the gut and the volume of the liver compartment was optimized based on a mathematical PK model close to the human PK.

Taken together, organ-on-chips have been able to successfully model the metabolism of numerous drugs, either by a single tissue or multiple tissues interconnected. As shown in Table 2, the majority of these devices are evaluated based on the testing of only one drug molecule. Future research is needed to test a wide variety of drug candidates and formulations, with their excipients using these devices to validate them against existing in vivo observations and to study any drug–drug interactions that may occur as a result of concomitant drug uptake.

Table 2. Summary of the drug metabolism and drug toxicity studies that include liver and/or kidney tissue models on organ-on-chips (OOCs) and multi-organ-on-chips (MOCs).

Drug	Toxicology	Metabolism	Tissue(s)	Reference
diclofenac acetaminophen	✓		liver	[18]
troglitazone	✓		liver	[19]
acetaminophen	✓		liver	[32–36]
acetaminophen	✓	✓	liver	[37,38]
acetaminophen isoniazid rifampicin	✓	✓	liver	[39]
rifampin ketoconzazole acetaminophen	✓	✓	liver	[40]
bupropion tolbutamide omeprazole testosterone		✓	liver	[24]

Table 2. *Cont.*

Drug	Toxicology	Metabolism	Tissue(s)	Reference
7-ethoxy-4-trifluoromethyl coumarin		✓	liver	[25]
acetaminophen chlorpromazine tacrine	✓		liver	[41]
ccetaminophen fialuridine	✓	✓	liver	[42]
diclofenac	✓	✓	liver	[43]
cadmium aspirin caffein troglitazone rosiglitazone pioglitazone acetaminophen	✓		liver	[2]
cisplatin	✓		kidney	[14]
adriamycin	✓		kidney	[44]
gentamicin	✓		kidney	[45]
polymyxin B	✓		kidney	[46]
carboxylated polystyrene nanoparticles	✓		GI tract–liver	[47]
troglitazone	✓	✓	liver–intestine liver–skin	[48]
apigenin		✓	gut–liver	[28]
epirubicine irinotecan cyclophosphamide		✓	small intestine–liver–lung	[29]
ifosfamide verapamil	✓	✓	liver–kidney	[49]
paracetamol		✓	liver–gut	[31]
mannitol propranolol caffeine		✓	GI–liver	[50]
combination of genistein and dacarbazine		✓	intestine–liver	[51]
5-fluorouracil	✓		liver–tumor–marrow	[52]
paracetamol	✓	✓	liver–kidney	[53]
diclofenac ketoconazole hydrocortisone acetaminophen	✓		liver–heart–skin	[54]
luteolin		✓	liver–tumor	[30]

Table 2. *Cont.*

Drug	Toxicology	Metabolism	Tissue(s)	Reference
capecitabine tegafur		✓	liver–cancer intestine–liver–cancer–connective tissue	[27]
digoxin	✓		intestine–kidney	[55]
ifosfamide	✓	✓	liver–kidney	[56]
vitamin D		✓	liver–kidney	[26]

2.2. *Toxicology*

Drug-induced toxicity is one of the major reasons for the failure of drug candidates and the withdrawal of approved drugs from the market [57]. The main reason for drug toxicity is the undesirable off-target activity of drug molecules or their reactive metabolites. Determining toxicity early in the drug discovery pipeline remains a challenge, as animal studies do not efficiently predict toxicity responses in human [3,58]. Therefore, accurate high-throughput assays for toxicity prediction are highly valuable in the pharmaceutical industry to reject potentially toxic drug candidates at the early stages. In this regard, the application of new technologies that could enable toxicity studies to be done in the context of organotypic biology as highly predictive models has attracted significant interest [57,59]. Table 2 summarizes the studies utilizing kidney and liver tissues-on-chip for drug toxicity assays (since nephrotoxicity [14,44–46,49,55,56] and hepatotoxicity [32,33,38,40–43,51,60] are the most common adverse effects reported in drug development).

2.2.1. Toxicology Studies by the Kidney- or Liver-on-Chip Models

The liver- and kidney-on-chip models were developed to have in vivo-like tissue functionalities, and they have demonstrated superiority over the conventional cell culture platforms in predicting toxic drug responses [14]. For instance, a kidney-glomerulus-on-chip was developed using pluripotent stem cells differentiated into kidney podocytes [44]. The cells in this device expressed morphological, molecular, and functional characteristics similar to mature human podocytes. Indeed, the renal toxicity associated with albuminuria induced by a cancer drug, Adriamycin, was successfully simulated by this device. Furthermore, the hepatotoxic effect of acetaminophen was also modelled by liver-on-chip models, where the toxic response of the tissue model to the drug treatment was evaluated based on the disruption in basic liver function or secretion of biomarkers similar to in vivo observations [34,35].

The application of organ-on-chips in toxicology has also been expanded to simulate specific drug administration schedules. Previous studies have shown that organ-on-chips can differentiate dosage regimens, where the modality of drug administration or the interaction of drugs may influence the toxicity response of organ tissues. A kidney-on-chip model simulated the nephrotoxicity of gentamicin in two different administration regimens: bolus injection and continuous infusion [45]. These different treatment regimens, given in the same dose, led to significantly different nephrotoxicity outcomes, with increased cytotoxicity detected for continuous infusion. Organ-on-chips also enabled the simulation of toxicity response to drug–drug interactions in patients undergoing treatment of multiple diseases with coadministration of medicines. A study by Ma et al. reported an in vitro, 3D, liver-lobule-like microtissue [39] that was used to simulate adverse drug reactions caused by the interaction of acetaminophen and omeprazole, rifampicin, ciprofloxacin, or probenecid. The consequent hepatotoxic effects reported in vivo could be simulated in vitro by this chip. It was observed that the pretreatment of the chip with omeprazole or probenecid resulted in an increased hepatotoxic effect of acetaminophen. However, the toxicity effect was alleviated by pretreatment of the cells with rifampicin or ciprofloxacin.

Another benefit of performing toxicology studies in organ-on-chips is that they can be used to assess poisonous molecules that are not ethically acceptable to subject healthy humans to. One example of this application are studies on radiation-induced injuries and testing the efficacy of radioprotective drugs or assessing radiotherapies [61]. Since the experimental exposure of healthy people to radiation is unethical, the organ-on-chips can offer valuable input in this area to conduct these assessments safely and rapidly. Another study that supports this application is the coculture of mammary epithelial cells and hepatic carcinoma cells on a chip [62]. The chip was exposed to gamma radiation to mimic the space-like environment, where the radioprotection effect of amifostine prodrug on human mammary epithelial cells after metabolism by the liver could be modelled in vitro.

2.2.2. Toxicology Studies by the Kidney- and Liver-on-Chip Models Interconnected with Other Organs

Simulating a network of organs based on physiological PK has been promising in toxicology studies [63]. An example of this network is a lung–liver-on-chip that simulated the decreased toxicity of inhaled toxicants in the lung tissue because of the detoxification process in the liver tissue [64]. Such multi-organs-on-chip platforms also enable the simulation of drug metabolism in one organ and the consequent toxic effect of the formed metabolites in another organ [47,53,55]. Similarly, severe nephrotoxicity of ifosfamide anticancer prodrug observed in the clinic could be emulated by a liver–kidney-on-chip that mimicked the interaction of kidney and liver [56]. Multi-organs-on-chips also allow simultaneous evaluation of the cytotoxic effects on the targeted and untargeted organs, where a study by Theobald et al. was able to model both the hepatotoxicity and nephrotoxicity of Aflatoxin B1 in a kidney–liver-on-chip [65].

The multi-organ-on-chips—in their advanced form involving more organ analogs—could eventually mimic the drug toxicity in the whole human body, significantly helping the prediction of side effects of drug candidates at a very early stage in vitro.

2.3. *Drug Delivery/Transport*

Transport of therapeutics across tissue barriers is one of the major challenges in drug delivery that can influence the bioavailability of drugs [66] and, hence, requires considerations and optimizations during drug discovery. Skin, epithelium, intestine, and blood-brain barrier (BBB) are examples of tissue in the human body that provide effective barriers against the delivery of therapeutics via transdermal, respiratory, oral, and intravenous drug delivery routes, respectively. Drug transport across in vitro tissue barrier models is another assessment where the application of the organ-on-chip models can be beneficial.

2.3.1. Simulation of In Vivo-Level Barrier Functions On-Chip

Organ-on-chips can model tissue-specific barrier functions with reliable physiological relevance to study drug permeation. The organ-on-chip models used for drug permeation studies are mainly designed as dual-chamber structures with a donor and an acceptor chamber. The donor chamber is mainly located on top of the acceptor chamber, and the tissue barrier is represented by the culture of cells on a permeable membrane that separates the two chambers. The quantity of drug transported from the donor to the acceptor chamber is measured to evaluate the efficient permeation of the developed drugs across the tissue barriers [12,50,54]. A study by Chen et al. modelled gastrointestinal (GI) barrier functionality in a GI–liver system [50]. This coculture cell model was able to express the desired physiological relevance, as assessed by the transepithelial electrical resistance (TEER) similar to the human native gut, and allowed emulation of the in vivo absorption of drugs across the gut wall. TEER measures the electrical resistance of the developed epithelium model and is a quantitative representation of the tissue barrier function and formation of tight junctions (TJs) between the cells [67]. Additionally, the GI barrier functionality was demonstrated in terms of the permeability of mannitol, propranolol, and caffeine across the primary intestinal monolayer in this chip. Organ-on-chips can also recreate heterogenous cellular structures, where the transport of substances occurs across adjacent tissues. The goal is to simulate the neighbouring tissues-on-chip with the coculture of different cell types to create structures similar to the native organs in the human body. This was proven feasible by a lung-on-chip developed by Huh et al. that could mimic the interaction between the pulmonary alveoli and the neighboring vascular endothelium in the human lung [12]. Therefore, this model enables the consideration of regulatory functions of both the epithelial and endothelial tissues as a more reliable barrier model for pulmonary drug-delivery tests.

2.3.2. Simulation of Multiple Drug-Delivery Routes On-Chip

Multi-organs-on-chip structured by the coculture of different organs also enabled the evaluation of different drug delivery routes and resultant transport of drug to the target tissue. This was demonstrated in a study by de Melo et al., where the skin barrier and dermal drug absorption were modelled by a Strat-M® membrane incorporated in a heart–liver-body-on-chip system [54]. The chip system mimicked both transdermal and systemic drug delivery routes and, hence, could predict both hepatotoxicity and cardiotoxicity of four model drugs administered to the chip: diclofenac, ketoconazole, hydrocortisone, and acetaminophen. In another study, Ozkan et al. conducted drug permeation studies using a two-vessel structure-on-chip mimicking the vascularized microenvironments of the liver and breast tumors [68]. The two vessels were surrounded by collagen-based extracellular matrix (ECM) with breast cancer cells or liver cells cultured in each vessel. This chip allowed evaluation of particle diffusion from a vessel into its surrounding ECM and back into the vessel, as well as transportation between the two vessels. For this purpose, particles of different sizes were perfused through these vascularized microenvironments to replicate chemotherapy drugs and drugs conjugated with nanoparticles. The resultant permeability of the tumor microenvironment and the accumulation of particles observed in ECM from this chip were consistent with in vivo findings [68–72].

2.3.3. Drug Delivery Tests under In Vivo-Inspired Dynamic Conditions On-Chip

Another added benefit of organ-on-chips is the capability to perform drug transport studies under mechanically dynamic conditions. The physiological mechanical cues in native organs could be emulated in the engineered design of organ-on-chips. This can enable drug-transport studies across tissue barriers under more physiologically realistic conditions. This feature was prominent and was highlighted in the lung-on-chip model that had capabilities to emulate both the strain exerted on the alveolar–capillary barriers by breathing motions and the shear stress on the capillary endothelium induced by the blood flow [12]. The lung-on-chip could mimic the physiological breathing motion via cyclic vacuuming of two channels on the sides. This resulted in cyclic stretching motions of a membrane inside the cell-culture channel, inducing physiological levels of cyclic strain. This dynamic microenvironment was eventually found to influence the permeability of the alveolar–endothelium interface. Higher permeability of nanoparticles across the alveolar–capillary interface was observed when the physiological strain was emulated compared to the device at static condition. The control, static, Transwell culture was prepared with the coculture of the alveolar epithelial cells and capillary endothelial cells on the opposing sides of the membrane. In another study, a human nasal epithelial mucosa-on-chip with a dual-channel structure was developed to model the nasal epithelial barrier against nasal drug delivery [73]. The transport of ibuprofen across the modelled epithelium was evaluated under physiologically relevant flowing fluid conditions in the donor channel. It was observed that the circulating pulsatile fluid flow in the bottom channel resulted in an increase in the drug transport rate compared to the static condition in the chip. This observation was explained by the increase in the convective mass transfer coefficient in the fluid due to the flow condition, which highlights the necessity of conducting nasal drug transport studies under dynamic conditions.

2.3.4. Further Improvements of Drug Delivery Studies by Organ-on-Chips

There remain unexplored areas to improve the relevance of drug-delivery studies by organ-on-chips. For instance, the simulation of realistic drug administration forms, namely the topical delivery of aerosolized solutions or dry powders within the dynamic microenvironments of organ-on-chips, has not been studied. Such simulations enable mimicking of the realistic deposition of fluidized particles to the target tissue and the aerosol characteristics that can influence the efficacy of aerosol drug delivery. These characteristics include particle size and flowrate of aerosolized particles, which can affect the delivered dose [74,75]. Additionally, physiological barriers that adversely affect the efficiency of drug-delivery routes have not been comprehensively studied by organ-on-chips. As a case-in-point, the mucociliary clearance mechanism that limits the residential time of drug particles in human airways is an unexplored area by in vitro models of lower [12,13,76] and upper [73] airways-on-chip. Simulation of this airway defense mechanism by future lung and nasal epithelium-on-chip models can profoundly enhance their physiological relevance and, eventually, their ability to model relevant clinical drug test results in vitro.

3. Advantages of Organ-on-Chips for Drug Testing

The benefits of conducting drug tests by organ-on-chip models can be summarized broadly into three main categories. Firstly, organ-on-chips are engineered to better mimic the in vivo physiology and pathophysiology of human organs. In doing so, organ-on-chips support the coculture of cells in one platform, simulation of the mechanical cues, and mimicking human diseased tissues. Secondly, organ-on-chips can facilitate and accelerate the drug-testing procedure by offering in situ sensing tools in their designs. Thirdly, organ-on-chips cultured with patient-derived cells can enhance personalized medicine.

3.1. Organ-on-Chips Offer Engineered In Vivo-Inspired Microenvironments

The main advantage of organ-on-chips is their engineered structure, which enables mimicking the microenvironments and functionalities of human organs with the in vivo-inspired architecture, multicellular nature, and biomechanical stimuli that may interfere with drug delivery, absorption, metabolism, and cytotoxic responses.

3.1.1. Cellular Coculture and Organ–Organ Crosstalk

Coculture of multiple cell types in organ-on-chips enables the understanding of the inter-organ crosstalk contribution in drug testing [30]. The interface of epithelial tissue and the neighbouring microvascular endothelium is the most commonly reproduced tissue–tissue interface structure in various organ-on-chips to evaluate drug efficacy [12,76–78]. The importance of recapitulating such interfaces lies in their role in regulating the transport of drugs, immune cells, and nutrients that could influence therapeutic outcomes [12]. Importantly, multiple tissue analogs interconnected on-chip with dynamic fluidic channels based on a physiologically correct scale, order, and cell-to-liquid ratio can reproduce the interactions between multiple organs in a whole body [16]. Hence, these devices are expected to simulate the uptake and circulation of therapeutics, the complex process of drug metabolism, the potential toxic or therapeutic effect of one organ's metabolites on the second organ, and the combined responses of several tissues to drugs and toxicants [30,47–49,55]. Such multi-organs-on-chip platforms—designed and scaled based on mathematical pharmacokinetic–pharmacodynamic (PK–PD) models—can represent time-dependent changes in drug concentration (PK) and its physiological effects (PD) in the human body [30]. Therefore, these devices are advanced platforms for more precise prediction of drug efficacy, as well as their potential side effects in a monitored whole-body system compared to the single-organ models [16].

In particular, these multi-organ platforms can mimic metabolism-dependent drug actions, such as testing the efficacy of prodrugs undergoing chemical or enzymatic transformation to form active drug moieties with pharmacological effects [79,80]. For instance, human hepatic carcinoma cells and human mammary epithelial cells were cocultured on a microfluidic chip [62] to investigate amifostine prodrug metabolism and its resultant radioprotection effect on the living-tissue analog. The dual-tissue device was compared with a single-cell model, and a two-fold improvement in the effectiveness of the drug was observed in the system with the two cell types cocultured. In another study, the coculture of hepatic and lung cells on-chip enabled mimicking irinotecan prodrug hydrolysis in the liver and the consequent anticancer effect on the lung cells [29]. This was consistent with the bioactivation of this prodrug in the liver when administered orally in vivo [81].

By involving the interactions between multiple tissues, multi-organs-on-chips can enhance the physiological relevance and predictive capabilities of the simulated PK in vitro compared to mono-cultured organ models [50]. For instance, the first-pass metabolism of orally taken apigenin, a natural flavonoid, was modelled in a liver–gut-on-chip [28]. It was found that this system could improve the physiological relevance of the modelled metabolism process compared to the monoculture gut cells. The type of formed metabolites detected in the gut–liver device was similar to the first-pass metabolism of apigenin in a rat model. This was due to the contribution of both tissues in forming the different metabolites of apigenin.

Evaluating the responses from multiple organs to a novel therapy by multi-organs-on-chip platforms has shown to be promising at predicting the side effects and the off-target cytotoxicity responses to drugs. The undetected side effects of an approved drug put potential patients at risk of acute or chronic poisoning [82]. Therefore, evaluation of drug side effects at the early stages of drug discovery is of growing concern to protect public health and for the pharmaceutical industries, where the application of multi-organs-on-chips for drug testing in a physiologically representative microenvironment could be a potential solution. Promising results have been reported in the literature regarding the evaluation of drug side effects by multi-organs-on-chips [29,47,49]. Sung et al. reported the development of a microscale cell culture analog based on a mathematical PK–PD model with cocultured liver, tumor (colon cancer), and marrow cells to study the toxicity of an anticancer drug, 5-fluorouracil [52]. This device allowed the assessment of different responses to the test drug from each cell type consistent with clinical findings [83,84]. A multi-organs-on-chip system interconnecting liver spheroids with either intestinal epithelial cells or skin biopsies enabled the simulation of oral or systemic routes for troglitazone administration [48]. The simulation of oral drug exposure using this chip allowed biotransformation of the drug to its metabolites by the liver tissue, similar to in vivo results. Further, the simulation of systemic drug administration by this chip could mimic drug uptake by the fatty tissue underneath skin biopsies similar to in vivo. Another device, hosting four human organ analogs by coculturing the human small intestine, skin biopsy, liver, and kidney proximal tubule cells, was reported by Maschmeyer et al. [85]. These four organs-on-chip models enabled the in vitro reproduction of the oral and dermal drug absorption, first-pass hepatic metabolism, secondary metabolism, and renal metabolite excretion, emulating the in vivo absorption, distribution, metabolism, and excretion profiling of tested drugs and their toxicity.

Multi-organs-on-chip shed more insight into understanding drug action when treating human diseases. To demonstrate this, a study by Lee et al. developed a gut–liver chip to recreate hepatic steatosis, a process of abnormal lipid deposition in the liver cells [86]. The device emulated the anti-steatosis effects of butyrate, which acted on the gut tissue by enhancing its barrier function against the transport of free fatty acids. The mechanism underlying the influence of butyrate on the intestinal barrier function is known to be through promoting the assembly of TJs between the gut epithelial cells [87], and the use of this advanced platform enables understanding this.

3.1.2. Simulation of Biomechanical Cues

Numerous organ-on-chips have been designed to simulate mechanical stimuli present in vivo to recreate the in vivo-like dynamic microenvironment of tissues. In this regard, physiologically relevant physical cues such as motion, deformation, fluid flow, strain, and shear stress are simulated on-chip. Kidney-, liver-, and lung-on-chips have shown the dependence of the tissue's physiological properties and functionalities on the mechanical cues. Figure 2 summarizes some of these effects. Hence, drug testing in organ-on-chips integrated with in vivo-inspired biomechanical factors is expected to deliver results closer to in vivo than conventional static cell cultures. As demonstrated, the flow-induced shear stress on primary kidney epithelial cells enhanced epithelial cell polarization and primary cilia formation in a human kidney proximal-tubule-on-chip [14]. Also, hepatic cells responded to the dynamic fluidic condition in a kidney–liver-on-chip by increasing the expression of the metabolism-associated biomarkers [65]. A lung-on-chip also showed that the stretching motion exerted on the primary human pulmonary alveolar epithelial cells could enhance cellular metabolic activity and biomarker secretion [13].

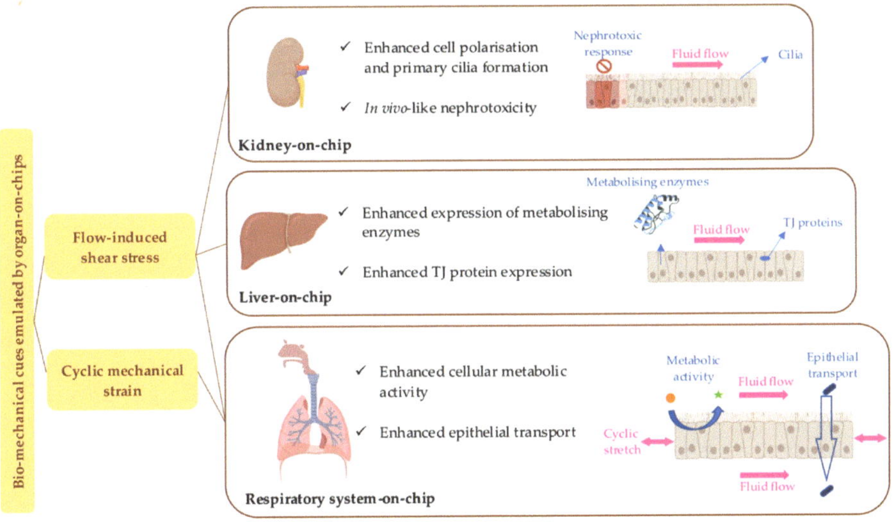

Figure 2. The biomechanical cues emulated by kidney-, liver-, and respiratory-system-on-chips and the summary of the observations in terms of the effects on the cellular morphology, permeability, and drug response. Figure is created with BioRender.com.

One of the important physiological characteristics of tissue-equivalence-on-chip models is the influence of mechanical stimuli on the tissue barrier function. For different tissue types cultured on organ-on-chips, it was reported that exposure to fluid flow could alter the paracellular permeability of tissue analogs on-chip compared to conventional static cultures [13,28,45,88–90]. Paracellular permeability refers to the passage of molecules through the intercellular spaces between adjacent epithelial cells. The tightness of small openings between the cells—the rate-limiting step in this process—is identified by the expression of TJ proteins. The fluid flow induces shear stress to the cells, which is reported to affect the expression of TJ proteins that consequently influence the paracellular permeability [45,88,90]. This is an important consideration when the permeation of drugs across the cell layer models are performed on organ-on-chip models where the obtained results could dictate the efficacy and toxicity of drugs [91,92]. For instance, the fluid-flow-induced shear stress on the liver and gut cells on-chip has been reported to accelerate cell differentiation, enhancing the epithelial barrier function and TJ protein expression, which, as a result, decreases the permeability of the cell layers modelled in these chips [28,31,45]. Another potential reason for the influence of flowing fluid on epithelial permeability is explained by the influence of the flow on the thickness of the unstirred water layer (UWL) [28]. UWL is an aqueous layer adjacent to the biologic solid–liquid interfaces that have a slow laminar flow and acts as an additional diffusion barrier against drug absorption [93–95]. The thickness of the UWL decreases with the fluid flowrate. This in turn increases the drug permeation when the diffusion through the UWL is the rate-limiting step [96,97].

In addition to the fluid flow, biomechanical strains are also found to influence the tissue barrier function in organ-on-chips. Mechanically active tissues—such as the lungs and GI tract tissues, which are exposed to stretches by breathing motions or peristalsis movements, respectively—are experiencing deformations and mechanical strains continuously [98]. A lung-on-chip model showed that the physiological strain on the lung epithelial cells resulted in an increase in the model's permeability to hydrophilic molecules, while the cell layer's integrity, TJs, and morphology remained unaffected [13]. This observation is in agreement with an in vivo investigation, where the increase in the lung volume by applying positive pressure to the airway influenced the permeation of hydrophilic solutes across the respiratory epithelium [99,100]. Further, simultaneous exposure of the gut epithelial cells to the fluid-flow shear stress and cyclic mechanical strain stimulated expression of a 3D villi-like morphology and an enhanced intestinal barrier function in a gut-on-chip [101]. Nevertheless, future studies are still required to elucidate the exact mechanism involved in the increase of permeability by mechanical strain at the molecular level.

Given the influence of biomechanical cues on the biological properties of the kidney, liver, and lung models, drug assays in such mechanically dynamic tissue models can potentially deliver different responses compared to the static models. Jang et al. compared the fluid dynamic cell-culture conditions in a kidney proximal-tubule-on-chip with the static culture in Transwell insert [14]. They found improved epithelial tissue morphology and kidney-specific functionalities under dynamic conditions. The dynamic model eventually showed closer in vivo human-relevant renal toxicity response to cisplatin treatment compared to the conventional static culture. The liver-on-chips with fluid-flow condition have also demonstrated increased expression of metabolizing enzymes and bioactivation of drug compounds through metabolism [26,33,36]. The enhanced metabolic functionality of these dynamic liver models can enhance their sensitivity to drug toxicity compared to the static cultures [33]. In addition, the exposure of the pulmonary epithelium to cyclic mechanical strain by a lung-on-chip was demonstrated to enhance cellular uptake and epithelial transport of the silica nanoparticles into the underlying microvascular channel [12]. This effect eventually accentuated both the toxic and inflammatory responses of the lung analog to silica nanoparticles, which is in agreement with in vivo observations.

3.1.3. Drug Testing Using Modelled Human Diseases On-Chip

Another improvement in organ-on-chip technology that can increase the relevance of preclinical drug testing is its capability to recreate human disease on-chip. These are micro-structured models of human organs, where the organ analogs express key features of human diseases [63]. Diseased tissues are known to respond differently to drugs compared to healthy tissues [63]. In addition, animal models fail to mimic the complications shown in many human disease conditions [102]. Therefore, there is a need to develop disease-on-chips as more reliable models of human diseases to be used for the assessment of effective therapies.

The engineering of disease-on-chips includes novel platform designs to facilitate multiple analyses of human diseases in vitro. This enables understanding of the disease development mechanisms. For example, a study by Zhou et al. developed a liver-injury-on-chip integrated with biosensors for monitoring secreted transforming growth factor-β (TGF-β) triggered by alcohol injury [103]. Another feature of this is the reconfigurable structure that enabled the culture chambers of hepatocytes and stellate cells to be either isolated from each other or connected with the possibility for the two cell types to communicate. This feature enabled studies on the cellular origin of secreted TGF-β and found that alcohol injury causes hepatocytes to commence the secretion of TGF-β molecules, which activate the neighboring stellate cells and trigger additional TGF-β production by the stellate cells during development of alcoholic liver injury. Furthermore, disease-on-chips can pave the way to understand the role of organ–organ communication on the development and progression of diseases [78,104]. This was demonstrated in a study by Lee et al. that simulated hepatic steatosis using a gut–liver chip. The hepatic steatosis was induced on-chip by tumor necrosis factor-α, which increases the permeability of gut epithelium and, as a result, increases the level of lipid permeability to the liver tissue, leading to hepatic steatosis [86]. The knowledge and understanding gained on the mechanism of disease development from the organ-on-chip models will enable effective therapies to be developed.

The capability of organ-on-chips to mimic the dynamic microenvironments of human tissues is an asset to study human diseases, as these mechanically active platforms allow investigating the effect of physical cues on disease development [105] as well as the response of the diseased tissues to the test therapies [76,77]. The above can potentially help predict human response to new therapies with higher accuracy and develop novel therapies to treat human diseases effectively. This was demonstrated in a lung-on-chip model that aims to reproduce the drug-toxicity-induced pulmonary edema observed in human cancer patients treated with interleukin-2. The study showed that cyclic mechanical strain that is associated with the breathing motion in the lung tissue increases the likelihood of pulmonary edema development [105]. Based on this knowledge, a new inhibitor of the transient receptor potential vanilloid 4 ion channels (that are activated by mechanical strain) was tested as a potential preventive treatment. Interestingly, this therapy revealed satisfactory results in inhibition of pulmonary edema.

Taken together, disease-on-chips are reliable models and have shown promise for both understanding human diseases and developing novel effective therapies.

3.2. Integrated Sensing Tools in Organ-on-Chips for In Situ Drug Testing

Another advantage of organ-on-chips is their ability to integrate sensing tools in their designs, which enables in situ monitoring of the chip microenvironment. The sensors integrated into organ-on-chips can collect data from potential changes in the culture environment and cells' biological properties in real-time. This can subsequently enhance the throughput of organ-on-chips used for drug tests [18]. The real-time monitoring of the cellular microenvironments by in situ analytical tools offers several advantages over the conventional analyses conducted off-chip, for example, shorter analysis time, cost-effectiveness due to reduced consumption of solvents, simpler operation, no requirement for sampling, elimination of the sampling errors, and lower risk of contamination, resulting in more reliable and accurate analyses. Several reviews have covered the application of biosensors in organ-on-chips for real-time monitoring of the physiological conditions of modelled tissue, focusing on cellular metabolism, function, and response to stimuli [106,107].

Recently, a liver-on-chip model was incorporated with commercialized amperometric glucose and lactate sensors that enabled the detection of minute shifts from oxidative phosphorylation to anaerobic glycolysis, indicating mitochondrial damage caused by drug concentrations previously considered as safe [19]. Furthermore, the integration of multiple sensors into the structure of organ-on-chips has paved the way for miniaturization and automation of various analyses in one platform without the need for large sample preparations [108]. Commercially available miniature microscopes, although challenging to incorporate in organ-on-chips at the moment, have been found helpful for monitoring the cells or cell-culture environment in real-time [35].

Another potential sensor that could benefit drug testing on organ-on-chips is the TEER measurement electrode. The integration of TEER measurement sensors in organ-on-chips has been reported for BBB-on-chip [109], heart-on-chip [110], pulmonary epithelium, and gut-on-chip models [109–112]. These devices, integrated with gold or platinum (Pt) electrodes, are reported to monitor the differentiation of cells and the formation of TJs in real-time. However, these chips have not been used in the context of drug testing to evaluate how drugs could affect TJ integrity and formation. However, they have the potential to facilitate the monitoring of TJ dynamics while treating the cells with novel drug formulations in terms of detection of toxicity effects of drugs on the cells or modulation of TJ dynamics by permeation-enhancing drug carriers [91].

Currently, there are still a very limited number of studies conducted to incorporate different sensors within organ-on-chips. Hence, there remains enormous potential for further development of versatile platforms that function like a living organ and are integrated with in situ sensing tools for real-time monitoring for pharmaceutical development. An example is the potential of using carbon-based electroanalytical sensors, which have proven to be suitable sensors for the electrochemical detection of numerous pharmaceuticals [113]. Since these sensors can offer reliable detection of drug compounds, integrating them with organ-on-chip designs can be useful in monitoring drug permeability and effects on-chip [73]. A human nasal epithelial mucosa-on-chip was developed—capable of real-time monitoring of both drug transport and TEER by housing the respective sensors in its structure [73]. This dual-channel chip was integrated with a carbonaceous electrode for in situ quantification of ibuprofen that was transported across the nasal epithelium to the acceptor channel. In addition, Pt electrodes were incorporated in the chip for in situ real-time TEER measurements. The presence of both sensors enables this chip to simultaneously measure the drug quantity in the cellular microenvironment and the potential variations in the barrier properties of the cell layer during drug testing. The in situ drug quantification by this chip was validated against high-performance liquid chromatography, which is the current gold standard for the quantification of pharmaceutics. In addition, the TEER measurements for the nasal epithelium model-on-chip were similar to what was reported for excised human nasal mucosa [114]. These features make this chip a potentially reliable alternative to the costly, time-consuming analytical techniques conventionally used for nasal drug assays.

3.3. *Organ-on-Chips Enable Personalized Drug Testing*

While integration of organ-on-chips with biological tissue or fluid specimens from patients, such as primary cells, biofluid samples, and cells derived from induced pluripotent stem cells, will enable emulation of patient-specific physiology, genetics, and biometric parameters on-chip [17,50], it also enables personalized organ-on-chips. These personalized chips can ultimately be used to stratify patients' responses to drug exposure and develop personalized medicine or therapies. While it is challenging to test different drugs on a patient receiving therapies, a personalized organ-on-chip model developed for the specific patient could be of significant value in predicting the most efficient drug to treat their disease. The potential of this personalized model was demonstrated by Xu et al. who developed an in vivo-like tumor microenvironment using a microfluidic chip to assess anticancer drug efficacy on primary cells from fresh lung cancer tissues of eight patients to model individualized clinical treatment [115]. The cells were treated with different anticancer drugs: gefitinib, paclitaxel, and gemcitabine. This chip could mimic different drug sensitivities of the lung cancer tissues from different individuals. It also showed that the drug sensitivity of the patient-derived tissues was different from what was observed in conventionally used cell-line models. The comparison of drug responses from the patient-derived tissues on-chip and clinical trials on the patients should be the focus of future research in this area, helping to validate these devices in predicting clinical outcomes.

4. Disadvantages of Organ-on-Chips in Drug Testing

While there are numerous advantages for the use of organ-on-chips in facilitating drug testing and formulation-development processes, there are also several disadvantages. Polydimethylsiloxane (PDMS) is one of the main materials used for the fabrication of the majority of organ-on-chips. Despite the advantages, such as gas permeability and transparency, that make it a good option for cell culture and microscopic imaging purposes, it has been reported that PDMS can absorb some drug components with hydrophobic behavior [28,107]. This can adversely affect the outcomes of drug testing when hydrophobic compounds are investigated. However, this effect could be prevented by applying coatings on the surface of the PDMS [28].

Although the dynamic environment of organ-on-chips bridge the gap between the static tissue culture and the in vivo counterpart, it could also have potential disadvantages, where adverse effects involving the fluid flow on the cells have been reported, such as the detachment of the cells when exposed to high flow rates [28]. Jang et al. reported a negative influence of the perfusion flow on the matrix-embedded hepatocytes in a chip as they observed the premature death of the cells in the proximity of the perfusion channel. In contrast, the cells that were located distally from the perfusion channel were protected against this effect [33]. Zhao et al. reported that the high flowrate of the supplied medium into their microfluidic device washed the cells away because of the high shear force [116]. On the other hand, a low flowrate adversely affected the viability of the cells as it led to the shortage of nutrients in their chip to culture human non-small-cell lung cancer cells. In light of such observations, optimizing the microchannels' dimensions and the extent of mechanical stimuli should be considered, as they appear to be important requisites to recapitulate physiologically relevant biomechanical cues. Such considerations may mitigate the potential negative effects of the dynamic microenvironments on the cells.

There is also currently no standardization of any of the organ-on-chip designs. Taking the lung tissue as an example, multiple organ-on-chip models with varying dimensions and architectures have been developed [12,13,64,117,118]. The differences in these models include presence [12,13] or absence [64,117,118] of mechanical strain, emulation of 2D [12] or 3D [13] breathing motions, and the type of cultured cells [12,13,117]. These differences can potentially result in different readouts from each of these models for a similar assay. Further, the variables inherent with the different models will not be useful for a highly regulated setting, such as the pharmaceutical industry, where reproducibility and standardization are essential. Consequently, it may not be suitable to compare the findings from different organ-on-chip designs, which highlights the need for the introduction of a standardization system for all organ-on-chip models that recreate the physiology of the same tissue. This highlights the necessity of defining standards for design and operation factors considered in development of organ-on-chip models for a certain organ.

5. Conclusions and Future Outlook

Organ-on-chips are found to be promising alternatives to the conventional cell culture models for enhancing the predictive power of preclinical drug tests. These platforms enable monitoring drug metabolism pathways, toxicity effects on single or multiple connected organs, and drug delivery to a target organ while also improving the time efficacy and reliability of the readouts.

Despite the ongoing progress in the design and development of biomimetic organ-on-chips in a plethora of research studies, this technology has yet to accomplish its ultimate level of advancement: mimicking the complex physiology of the whole human body to be potentially accepted as an alternative to animal models or preclinical trials. Although multi-organs-on-chips are widely studied to recreate human response to numerous drugs, they are far from a full human body-on-chip, with still many organs to be explored. Future research can focus on incorporating several organs in multi-organs-on-chips combined with coculture models that will significantly enhance the accuracy of evaluating drug actions and side effects in the whole human body.

In addition, further work should focus on introducing standardized design factors, such as the ratio of dimensions, fluid volume, flowrate, shear stress, and biomechanical forces, for the development of organ-on-chips which can be used in highly regulated settings (e.g., the pharmaceutical industry). This will help ensure that every novel organ-on-chip design can mimic the physiological features of the tissue according to stated standards. These factors can also be presented in the form of nondimensional numbers in order to reduce the number of parameters needed to categorize the key physical and mechanical attributes of a particular device, as is done in the engineering of drug-delivery devices [119–121]. Hence, researchers could compare the results from different models of one tissue type with the obtained experimental data from multiple studies being used as a valid reference for potential future clinical studies with discrepancies less-likely.

While organ-on-chip models to study the functionality of organs and disease states have progressed significantly over the last decade, they rely on off-chip analysis and imaging techniques, which are labor-intensive and time- and cost-consuming. Integration of continuous, non-invasive, in-site, real-time monitoring of the functionality of the tissues is in significant demand to address this problem. While current efforts are focused on the physical and biochemical cues of the physiological microenvironment, such as cell-secreted molecules and TJ integrity, there remains a significant underexplored area of using these microfluidic platforms equipped with drug sensors for preclinical drug testing in a physiologically relevant environment. This can potentially accelerate the translation of drugs to clinic, where organ-on-chips can play a significant role in the development pipelines in future drug discoveries as more accurate and reliable alternatives to conventional in vitro analyses or animal trials.

Author Contributions: Conceptualization, S.C., H.X.O., A.K., H.G.; writing—original draft preparation, H.G.; writing—assistance with reviewing papers, H.X.; writing—review and editing, S.C., H.X.O., A.K., D.T., P.M.Y.; visualization, H.G.; supervision, S.C., H.X.O. All authors have read and agreed to the published version of the manuscript.

Funding: Daniela Traini is funded by an NHMRC grant APP1173363.

Institutional Review Board Statement: Not applicable.

Informed Consent Statement: Not applicable.

Acknowledgments: H. Gholizadeh acknowledges the Australian Government Research Training Program (RTP) Scholarship to support her PhD study.

Conflicts of Interest: The authors declare no conflict of interest.

References

1. McGonigle, P.; Ruggeri, B. Animal models of human disease: Challenges in enabling translation. *Biochem. Pharmacol.* **2014**, *87*, 162–171. [CrossRef] [PubMed]
2. Khetani, S.R.; Bhatia, S.N. Microscale culture of human liver cells for drug development. *Nat. Biotechnol.* **2008**, *26*, 120–126. [CrossRef] [PubMed]
3. Esch, E.W.; Bahinski, A.; Huh, D. Organs-on-chips at the frontiers of drug discovery. *Nat. Rev. Drug Discoov.* **2015**, *14*, 248–260. [CrossRef] [PubMed]
4. Wu, Q.; Liu, J.; Wang, X.; Feng, L.; Wu, J.; Zhu, X.; Wen, W.; Gong, X. Organ-on-a-chip: Recent breakthroughs and future prospects. *Biomed. Eng. Online* **2020**, *19*, 1–19. [CrossRef]
5. Kimura, H.; Sakai, Y.; Fujii, T. Organ/body-on-a-chip based on microfluidic technology for drug discovery. *Drug Metab. Pharmacok.* **2018**, *33*, 43–48. [CrossRef]
6. Deng, J.; Qu, Y.; Liu, T.; Jing, B.; Zhang, X.; Chen, Z.; Luo, Y.; Zhao, W.; Lu, Y.; Lin, B. Recent organ-on-a-chip advances toward drug toxicity testing. *Development* **2018**, *19*, 20. [CrossRef]
7. Cong, Y.; Han, X.; Wang, Y.; Chen, Z.; Lu, Y.; Liu, T.; Wu, Z.; Jin, Y.; Luo, Y.; Zhang, X. Drug toxicity evaluation based on organ-on-a-chip technology: A review. *Micromachines* **2020**, *11*, 381. [CrossRef]
8. Bhise, N.S.; Ribas, J.; Manoharan, V.; Zhang, Y.S.; Polini, A.; Massa, S.; Dokmeci, M.R.; Khademhosseini, A. Organ-on-a-chip platforms for studying drug delivery systems. *J. Control. Release* **2014**, *190*, 82–93. [CrossRef]
9. Ma, C.; Peng, Y.; Li, H.; Chen, W. Organ-on-a-Chip: A new paradigm for drug development. *Trends Pharmacol. Sci.* **2020**, *42*, 119–133. [CrossRef]
10. Vulto, P.; Joore, J. Adoption of organ-on-chip platforms by the pharmaceutical industry. *Nat. Rev. Drug Discov.* **2021**, *20*, 961–962. [CrossRef]
11. Low, L.A.; Mummery, C.; Berridge, B.R.; Austin, C.P.; Tagle, D.A. Organs-on-chips: Into the next decade. *Nat. Rev. Drug Discov.* **2021**, *20*, 345–361. [CrossRef] [PubMed]
12. Huh, D.; Matthews, B.D.; Mammoto, A.; Montoya-Zavala, M.; Hsin, H.Y.; Ingber, D.E. Reconstituting organ-level lung functions on a chip. *Science* **2010**, *328*, 1662–1668. [CrossRef] [PubMed]
13. Stucki, A.O.; Stucki, J.D.; Hall, S.R.; Felder, M.; Mermoud, Y.; Schmid, R.A.; Geiser, T.; Guenat, O.T. A lung-on-a-chip array with an integrated bio-inspired respiration mechanism. *Lab Chip* **2015**, *15*, 1302–1310. [CrossRef] [PubMed]
14. Jang, K.-J.; Mehr, A.P.; Hamilton, G.A.; McPartlin, L.A.; Chung, S.; Suh, K.Y.; Ingber, D.E. Human kidney proximal tubule-on-a-chip for drug transport and nephrotoxicity assessment. *Integr. Biol.* **2013**, *5*, 1119–1129. [CrossRef]
15. Abaci, H.E.; Shuler, M.L. Human-on-a-chip design strategies and principles for physiologically based pharmacokinetics/pharmacodynamics modeling. *Integr. Biol.* **2015**, *7*, 383–391. [CrossRef]

16. Sung, J.H.; Wang, Y.I.; Narasimhan Sriram, N.; Jackson, M.; Long, C.; Hickman, J.J.; Shuler, M.L. Recent advances in body-on-a-chip systems. *Anal. Chem.* **2018**, *91*, 330–351. [CrossRef]
17. Van Den Berg, A.; Mummery, C.L.; Passier, R.; Van der Meer, A.D. Personalised organs-on-chips: Functional testing for precision medicine. *Lab Chip* **2019**, *19*, 198–205. [CrossRef]
18. Yu, F.; Deng, R.; Tong, W.H.; Huan, L.; Chan Way, N.; IslamBadhan, A.; Iliescu, C.; Yu, H. A perfusion incubator liver chip for 3D cell culture with application on chronic hepatotoxicity testing. *Sci. Rep.* **2017**, *7*, 14528. [CrossRef]
19. Bavli, D.; Prill, S.; Ezra, E.; Levy, G.; Cohen, M.; Vinken, M.; Vanfleteren, J.; Jaeger, M.; Nahmias, Y. Real-time monitoring of metabolic function in liver-on-chip microdevices tracks the dynamics of mitochondrial dysfunction. *Proc. Natl. Acad. Sci. USA* **2016**, *113*, E2231–E2240. [CrossRef]
20. Hornberg, J.J.; Laursen, M.; Brenden, N.; Persson, M.; Thougaard, A.V.; Toft, D.B.; Mow, T. Exploratory toxicology as an integrated part of drug discovery. Part I: Why and how. *Drug Discov. Today* **2014**, *19*, 1131–1136. [CrossRef]
21. Kumar, G.N.; Surapaneni, S. Role of drug metabolism in drug discovery and development. *Med. Res. Rev.* **2001**, *21*, 397–411. [CrossRef] [PubMed]
22. Tscheik, C.; Blasig, I.E.; Winkler, L. Trends in drug delivery through tissue barriers containing tight junctions. *Tissue Barriers* **2013**, *1*, e24565. [CrossRef] [PubMed]
23. Weng, Y.S.; Chang, S.F.; Shih, M.C.; Tseng, S.H.; Lai, C.H. Scaffold-free liver-on-a-chip with multiscale organotypic cultures. *Adv. Mater.* **2017**, *29*, 1701545. [CrossRef] [PubMed]
24. Srinivasan, B.; Kolli, A.R.; Esch, M.B.; Abaci, H.E.; Shuler, M.L.; Hickman, J.J. TEER measurement techniques for in vitro barrier model systems. *J. Lab. Autom.* **2015**, *20*, 107–126. [CrossRef]
25. Chang, R.; Emami, K.; Wu, H.; Sun, W. Biofabrication of a three-dimensional liver micro-organ as an in vitro drug metabolism model. *Biofabrication* **2010**, *2*, 045004. [CrossRef]
26. Liu, X.; Wang, H.; Liang, X.; Roberts, M.S. Hepatic Metabolism in Liver Health and Disease. In *Liver Pathophysiology*; Elsevier: Amsterdam, The Netherlands, 2017; pp. 391–400.
27. Satoh, T.; Sugiura, S.; Shin, K.; Onuki-Nagasaki, R.; Ishida, S.; Kikuchi, K.; Kakiki, M.; Kanamori, T. A multi-throughput multi-organ-on-a-chip system on a plate formatted pneumatic pressure-driven medium circulation platform. *Lab Chip* **2018**, *18*, 115–125. [CrossRef]
28. Bhise, N.S.; Manoharan, V.; Massa, S.; Tamayol, A.; Ghaderi, M.; Miscuglio, M.; Lang, Q.; Zhang, Y.S.; Shin, S.R.; Calzone, G.; et al. A liver-on-a-chip platform with bioprinted hepatic spheroids. *Biofabrication* **2016**, *8*, 014101. [CrossRef]
29. Lee, H.; Kim, D.S.; Ha, S.K.; Choi, I.; Lee, J.M.; Sung, J.H. A pumpless multi-organ-on-a-chip (MOC) combined with a pharmacokinetic–pharmacodynamic (PK–PD) model. *Biotechnol. Bioeng.* **2017**, *114*, 432–443. [CrossRef]
30. Wang, Y.; Wang, H.; Deng, P.; Chen, W.; Guo, Y.; Tao, T.; Qin, J. In situ differentiation and generation of functional liver organoids from human iPSCs in a 3D perfusable chip system. *Lab Chip* **2018**, *18*, 3606–3616. [CrossRef]
31. Theobald, J.; Ghanem, A.; Wallisch, P.; Banaeiyan, A.A.; Andrade-Navarro, M.A.; Taškova, K.; Haltmeier, M.; Kurtz, A.; Becker, H.; Reuter, S.; et al. Liver-kidney-on-chip to study toxicity of drug metabolites. *ACS Biomater. Sci. Eng.* **2018**, *4*, 78–89. [CrossRef]
32. Jalili-Firoozinezhad, S.; Prantil-Baun, R.; Jiang, A.; Potla, R.; Mammoto, T.; Weaver, J.C.; Ferrante, T.C.; Kim, H.J.; Cabral, J.; Levy, O.; et al. Modeling radiation injury-induced cell death and countermeasure drug responses in a human Gut-on-a-Chip. *Cell Death Dis.* **2018**, *9*, 1–14. [CrossRef] [PubMed]
33. Snyder, J.; Hamid, Q.; Wang, C.; Chang, R.; Emami, K.; Wu, H.; Sun, W. Bioprinting cell-laden matrigel for radioprotection study of liver by pro-drug conversion in a dual-tissue microfluidic chip. *Biofabrication* **2011**, *3*, 034112. [CrossRef] [PubMed]
34. Shuler, M.L. Organ-, body-and disease-on-a-chip systems. *Lab Chip* **2017**, *17*, 2345–2346. [CrossRef] [PubMed]
35. Hongmao, S. (Ed.) Chapter 7—In Silico ADMET Profiling: Predictive Models for CYP450, P-gp, PAMPA, and hERG. In *A Practical Guide to Rational Drug Design*; Woodhead Publishing: Cambridge, UK, 2016; pp. 225–268.
36. Larrey, D.; Ursic-Bedoya, J.; Meunier, L. Drug-induced Hepatotoxicity. *Schiff Dis. Liver* **2017**, *32*, 740–773.
37. Theobald, J.; el Maaty, M.A.A.; Kusterer, N.; Wetterauer, B.; Wink, M.; Cheng, X.; Wölfl, S. In vitro metabolic activation of vitamin D3 by using a multi-compartment microfluidic liver-kidney organ on chip platform. *Sci. Rep.* **2019**, *9*, 4616. [CrossRef]
38. Jie, M.; Lin, H.; He, Z.; Liu, H.; Li, H.; Lin, J.M. An on-chip intestine-liver model for multiple drugs absorption and metabolism behavior simulation. *Sci. China Chem.* **2018**, *61*, 236–242. [CrossRef]
39. Kim, S.; LesherPerez, S.C.; Yamanishi, C.; Labuz, J.M.; Leung, B.; Takayama, S. Pharmacokinetic profile that reduces nephrotoxicity of gentamicin in a perfused kidney-on-a-chip. *Biofabrication* **2016**, *8*, 015021. [CrossRef]
40. Mi, S.; Yi, X.; Du, Z.; Xu, Y.; Sun, W. Construction of a liver sinusoid based on the laminar flow on chip and self-assembly of endothelial cells. *Biofabrication* **2018**, *10*, 025010. [CrossRef]
41. Jang, M.; Neuzil, P.; Volk, T.; Manz, A.; Kleber, A. On-chip three-dimensional cell culture in phaseguides improves hepatocyte functions in vitro. *Biomicrofluidics* **2015**, *9*, 034113. [CrossRef]
42. Lee, H.; Chae, S.; Kim, J.Y.; Han, W.; Kim, J.; Choi, Y.; Cho, D.W. Cell-printed 3D liver-on-a-chip possessing a liver microenvironment and biliary system. *Biofabrication* **2019**, *11*, 025001. [CrossRef]
43. Ozkan, A.; Ghousifam, N.; Hoopes, P.J.; Yankeelov, T.E.; Rylander, M.N. In vitro vascularized liver and tumor tissue microenvironments on a chip for dynamic determination of nanoparticle transport and toxicity. *Biotechnol. Bioeng.* **2019**, *116*, 1201–1219. [CrossRef] [PubMed]

44. Musah, S.; Mammoto, A.; Ferrante, T.C.; Jeanty, S.S.; Hirano-Kobayashi, M.; Mammoto, T.; Roberts, K.; Chung, S.; Novak, R.; Ingram, M.; et al. Mature induced-pluripotent-stem-cell-derived human podocytes reconstitute kidney glomerular-capillary-wall function on a chip. *Nat. Biomed.* **2017**, *1*, 69. [CrossRef] [PubMed]
45. Chen, H.J.; Miller, P.; Shuler, M.L. A pumpless body-on-a-chip model using a primary culture of human intestinal cells and a 3D culture of liver cells. *Lab Chip* **2018**, *18*, 2036–2046. [CrossRef] [PubMed]
46. Weber, E.J.; Lidberg, K.A.; Wang, L.; Bammler, T.K.; MacDonald, J.W.; Li, M.J.; Redhair, M.; Atkins, W.M.; Tran, C.; Hines, K.M.; et al. Human kidney on a chip assessment of polymyxin antibiotic nephrotoxicity. *JCI Insight* **2018**, *3*, 123673. [CrossRef]
47. Delalat, B.; Cozzi, C.; Rasi Ghaemi, S.; Polito, G.; Kriel, F.H.; Michl, T.D.; Harding, F.J.; Priest, C.; Barillaro, G.; Voelcker, N.H. Microengineered bioartificial liver chip for drug toxicity screening. *Adv. Funct. Mater.* **2018**, *28*, 1801825. [CrossRef]
48. Ma, C.; Zhao, L.; Zhou, E.M.; Xu, J.; Shen, S.; Wang, J. On-chip construction of liver lobule-like microtissue and its application for adverse drug reaction assay. *Anal. Chem.* **2016**, *88*, 1719–1727. [CrossRef]
49. Choucha-Snouber, L.; Aninat, C.; Grsicom, L.; Madalinski, G.; Brochot, C.; Poleni, P.E.; Razan, F.; Guillouzo, C.G.; Legallais, C.; Corlu, A.; et al. Investigation of ifosfamide nephrotoxicity induced in a liver–kidney co-culture biochip. *Biotechnol. Bioeng.* **2013**, *110*, 597–608. [CrossRef]
50. Choe, A.; Ha, S.K.; Choi, I.; Choi, N.; Sung, J.H. Microfluidic Gut-liver chip for reproducing the first pass metabolism. *Biomed. Microdevices* **2017**, *19*, 4. [CrossRef]
51. Lee, D.W.; Ha, S.K.; Choi, I.; Sung, J.H. 3D gut-liver chip with a PK model for prediction of first-pass metabolism. *Biomed. Microdevices* **2017**, *19*, 100. [CrossRef]
52. Zakhariants, A.; Burmistrova, O.; Shkurnikov, M.Y.; Poloznikov, A.A.; Sakharov, D.A. Development of a Specific Substrate—Inhibitor Panel (Liver-on-a-Chip) for Evaluation of Cytochrome P450 Activity. *Bull. Exp. Biol. Med.* **2016**, *162*, 170–174. [CrossRef]
53. Kimura, H.; Ikeda, T.; Nakayama, H.; Sakai, Y.; Fujii, T. An on-chip small intestine–liver model for pharmacokinetic studies. *J. Lab. Autom.* **2015**, *20*, 265–273. [CrossRef] [PubMed]
54. Mao, S.; Gao, D.; Liu, W.; Wei, H.; Lin, J.M. Imitation of drug metabolism in human liver and cytotoxicity assay using a microfluidic device coupled to mass spectrometric detection. *Lab Chip* **2012**, *12*, 219–226. [CrossRef] [PubMed]
55. Prot, J.-M.; Bunescu, A.; Elena-Herrmann, B.; Aninat, C.; Snouber, L.C.; Griscom, L.; Razan, F.; Bois, F.Y.; Legallais, C.; Brochot, C. Predictive toxicology using systemic biology and liver microfluidic "on chip" approaches: Application to acetaminophen injury. *Toxicol. Appl. Pharmacol.* **2012**, *259*, 270–280. [CrossRef] [PubMed]
56. Foster, A.J.; Chouhan, B.; Regan, S.L.; Rollison, H.; Amberntsson, S.; Andersson, L.C.; Srivastava, A.; Darnell, M.; Cairns, J.; Lazic, S.E.; et al. Integrated in vitro models for hepatic safety and metabolism: Evaluation of a human Liver-Chip and liver spheroid. *Arch. Toxicol.* **2019**, *93*, 1021–1037. [CrossRef]
57. Lee, P.J.; Hung, P.J.; Lee, L.P. An artificial liver sinusoid with a microfluidic endothelial-like barrier for primary hepatocyte culture. *Biotechnol. Bioeng.* **2007**, *97*, 1340–1346. [CrossRef]
58. Esch, M.B.; Mahler, G.J.; Stokol, T.; Shuler, M.L. Body-on-a-chip simulation with gastrointestinal tract and liver tissues suggests that ingested nanoparticles have the potential to cause liver injury. *Lab Chip* **2014**, *14*, 3081–3092. [CrossRef]
59. Maschmeyer, I.; Hasenberg, T.; Jaenicke, A.; Lindner, M.; Lorenz, A.K.; Zech, J.; Garbe, L.A.; Sonntag, F.; Hayden, P.; Ayehunie, S. Chip-based human liver–intestine and liver–skin co-cultures–A first step toward systemic repeated dose substance testing in vitro. *Eur. J. Pharm. Biopharm.* **2015**, *95*, 77–87. [CrossRef]
60. Li, Z.; Jiang, L.; Zhu, Y.; Su, W.; Xu, C.; Tao, T.; Shi, Y.; Qin, J. Assessment of hepatic metabolism-dependent nephrotoxicity on an organs-on-a-chip microdevice. *Toxicol. In Vitro* **2018**, *46*, 1–8. [CrossRef]
61. Sung, J.H.; Kam, C.; Shuler, M.L. A microfluidic device for a pharmacokinetic–pharmacodynamic (PK–PD) model on a chip. *Lab Chip* **2010**, *10*, 446–455. [CrossRef]
62. Shintu, L.; Baudoin, R.; Navratil, V.; Prot, J.M.; Pontoizeau, C.; Defernez, M.; Blaise, B.J.; Domange, C.; Péry, A.R.; Toulhoat, P.; et al. Metabolomics-on-a-chip and predictive systems toxicology in microfluidic bioartificial organs. *Anal. Chem.* **2012**, *84*, 1840–1848. [CrossRef]
63. De Mello, C.P.P.; Carmona-Moran, C.; McAleer, C.W.; Perez, J.; Coln, E.A.; Long, C.J.; Oleaga, C.; Riu, A.; Note, R.; Teissier, S.; et al. Microphysiological heart–liver body-on-a-chip system with a skin mimic for evaluating topical drug delivery. *Lab Chip* **2020**, *20*, 749–759. [CrossRef] [PubMed]
64. Li, Z.; Su, W.; Zhu, Y.; Tao, T.; Li, D.; Peng, X.; Qin, J. Drug absorption related nephrotoxicity assessment on an intestine-kidney chip. *Biomicrofluidics* **2017**, *11*, 034114. [CrossRef] [PubMed]
65. Liebler, D.C.; Guengerich, F.P. Elucidating mechanisms of drug-induced toxicity. *Nat. Rev. Drug Discoov.* **2005**, *4*, 410–420. [CrossRef] [PubMed]
66. Jang, K.-J.; Otieno, M.A.; Ronxhi, J.; Lim, H.K.; Ewart, L.; Kodella, K.R.; Petropolis, D.B.; Kulkarni, G.; Rubins, J.E.; Conegliano, D.; et al. Reproducing human and cross-species drug toxicities using a Liver-Chip. *Sci. Transl. Med.* **2019**, *11*, eaax5516. [CrossRef] [PubMed]
67. Bovard, D.; Sandoz, A.; Luettich, K.; Frentzel, S.; Iskandar, A.; Marescotti, D.; Trivedi, K.; Guedj, E.; Dutertre, Q.; Peitsch, M.C.; et al. A lung/liver-on-a-chip platform for acute and chronic toxicity studies. *Lab Chip* **2018**, *18*, 3814–3829. [CrossRef] [PubMed]
68. Paul, A. Drug Absorption and Bioavailability. In *Introduction to Basics of Pharmacology and Toxicology: Volume 1: General and Molecular Pharmacology: Principles of Drug Action*; Raj, G.M., Raveendran, R., Eds.; Springer: Singapore, 2019; pp. 81–88.

69. Haste, P.; Tann, M.; Persohn, S.; LaRoche, T.; Aaron, V.; Mauxion, T.; Chauhan, N.; Dreher, M.R.; Johnson, M.S. Correlation of technetium-99m macroaggregated albumin and Yttrium-90 glass microsphere biodistribution in hepatocellular carcinoma: A retrospective review of pretreatment single photon emission CT and posttreatment positron emission tomography/CT. *J. Vasc. Interv. Radiol.* **2017**, *28*, 722–730. [CrossRef]
70. NDong, C.; Tate, J.A.; Kett, W.C.; Batra, J.; Demidenko, E.; Lewis, L.D.; Hoopes, P.J.; Gerngross, T.U.; Griswold, K.E. Tumor cell targeting by iron oxide nanoparticles is dominated by different factors in vitro versus in vivo. *PLoS ONE* **2015**, *10*, e0115636. [CrossRef]
71. Petryk, A.A.; Giustini, A.J.; Gottesman, R.E.; Kaufman, P.A.; Hoopes, P.J. Magnetic nanoparticle hyperthermia enhancement of cisplatin chemotherapy cancer treatment. *Int. J. Hyperth.* **2013**, *29*, 845–851. [CrossRef]
72. Terentyuk, G.S.; Maslyakova, G.N.; Suleymanova, L.V.; Khlebtsov, B.N.; Kogan, B.Y.; Akchurin, G.G.; Shantrocha, A.V.; Maksimova, I.L.; Khlebtsov, N.G.; Tuchin, V.V. Circulation and distribution of gold nanoparticles and induced alterations of tissue morphology at intravenous particle delivery. *J. Biophotonics* **2009**, *2*, 292–302. [CrossRef]
73. Gholizadeh, H.; Ong, H.X.; Bradbury, P.; Kourmatzis, A.; Traini, D.; Young, P.; Li, M.; Cheng, S. Real-time quantitative monitoring of in vitro nasal drug delivery by a nasal epithelial mucosa-on-a-chip model. *Expert Opin. Drug Del.* **2021**, *18*, 803–818. [CrossRef]
74. Labiris, N.R.; Dolovich, M.B. Pulmonary drug delivery. Part I: Physiological factors affecting therapeutic effectiveness of aerosolized medications. *Br. J. Clin. Pharmacol.* **2003**, *56*, 588–599. [CrossRef] [PubMed]
75. Cheng, S.; Kourmatzis, A.; Mekonnen, T.; Gholizadeh, H.; Raco, J.; Chen, L.; Tang, P.; Chan, H.K. Does upper airway deformation affect drug deposition? *Int. J. Pharm.* **2019**, *572*, 118773. [CrossRef] [PubMed]
76. Benam, K.H.; Villenave, R.; Lucchesi, C.; Varone, A.; Hubeau, C.; Lee, H.H.; Alves, S.E.; Salmon, M.; Ferrante, T.C.; Weaver, J.C.; et al. Small airway-on-a-chip enables analysis of human lung inflammation and drug responses in vitro. *Nat. Methods* **2016**, *13*, 151–157. [CrossRef] [PubMed]
77. Hassell, B.A.; Goyal, G.; Lee, E.; Sontheimer-Phelps, A.; Levy, O.; Chen, C.S.; Ingber, D.E. Human organ chip models recapitulate orthotopic lung cancer growth, therapeutic responses, and tumor dormancy in vitro. *Cell Rep.* **2017**, *21*, 508–516. [CrossRef] [PubMed]
78. Jain, A.; Barrile, R.; van der Meer, A.D.; Mammoto, A.; Mammoto, T.; De Ceunynck, K.; Aisiku, O.; Otieno, M.A.; Louden, C.S.; Hamilton, G.A.; et al. Primary human lung alveolus-on-a-chip model of intravascular thrombosis for assessment of therapeutics. *Clin. Pharmacol. Ther.* **2018**, *103*, 332–340. [CrossRef] [PubMed]
79. Khosa, A.; Saha, R.N.; Singhvi, G. Chapter 16—Drug delivery to the brain. In *Nanomaterials for Drug Delivery and Therapy*; Grumezescu, A.M., Ed.; William Andrew Publishing: Norwich, NY, USA, 2019; pp. 461–514.
80. Choucha Snouber, L.; Bunescu, A.; Naudot, M.; Legallais, C.; Brochot, C.; Dumas, M.E.; Elena-Herrmann, B.; Leclerc, E. Metabolomics-on-a-chip of hepatotoxicity induced by anticancer drug flutamide and its active metabolite hydroxyflutamide using HepG2/C3a microfluidic biochips. *Toxicol. Sci.* **2013**, *132*, 8–20. [CrossRef]
81. Humerickhouse, R.; Lohrbach, K.; Li, L.; Bosron, W.F.; Dolan, M.E. Characterization of CPT-11 hydrolysis by human liver carboxylesterase isoforms hCE-1 and hCE-2. *Cancer Res.* **2000**, *60*, 1189–1192.
82. Zhao, X.; Chen, L.; Lu, J. A similarity-based method for prediction of drug side effects with heterogeneous information. *Math. Biosci.* **2018**, *306*, 136–144. [CrossRef]
83. Testart-Paillet, D.; Girard, P.; You, B.; Freyer, G.; Pobel, C.; Tranchand, B. Contribution of modelling chemotherapy-induced hematological toxicity for clinical practice. *Crit. Rev. Oncol. Hematol.* **2007**, *63*, 1–11. [CrossRef]
84. Van Kuilenburg, A.B. Dihydropyrimidine dehydrogenase and the efficacy and toxicity of 5-fluorouracil. *Eur. J. Cancer* **2004**, *40*, 939–950. [CrossRef]
85. Maschmeyer, I.; Lorenz, A.K.; Schimek, K.; Hasenberg, T.; Ramme, A.P.; Hübner, J.; Lindner, M.; Drewell, C.; Bauer, S.; Thomas, A.; et al. A four-organ-chip for interconnected long-term co-culture of human intestine, liver, skin and kidney equivalents. *Lab Chip* **2015**, *15*, 2688–2699. [CrossRef] [PubMed]
86. Lee, S.Y.; Sung, J.H. Gut–liver on a chip toward an in vitro model of hepatic steatosis. *Biotechnol. Bioeng.* **2018**, *115*, 2817–2827. [CrossRef] [PubMed]
87. Peng, L.; Li, Z.-R.; Green, R.S.; Holzman, I.R.; Lin, J. Butyrate Enhances the Intestinal Barrier by Facilitating Tight Junction Assembly via Activation of AMP-Activated Protein Kinase in Caco-2 Cell Monolayers. *J. Nutr.* **2009**, *139*, 1619–1625. [CrossRef] [PubMed]
88. Siddharthan, V.; Kim, Y.V.; Liu, S.; Kim, K.S. Human astrocytes/astrocyte-conditioned medium and shear stress enhance the barrier properties of human brain microvascular endothelial cells. *Brain Res.* **2007**, *1147*, 39–50. [CrossRef] [PubMed]
89. Booth, R.; Kim, H. Characterization of a microfluidic in vitro model of the blood-brain barrier (μBBB). *Lab Chip* **2012**, *12*, 1784–1792. [CrossRef] [PubMed]
90. Chi, M.; Yi, B.; Oh, S.; Park, D.J.; Sung, J.H.; Park, S. A microfluidic cell culture device (μFCCD) to culture epithelial cells with physiological and morphological properties that mimic those of the human intestine. *Biomed. Microdevices* **2015**, *17*, 58. [CrossRef]
91. Gholizadeh, H.; Cheng, S.; Pozzoli, M.; Messerotti, E.; Traini, D.; Young, P.; Kourmatzis, A.; Ong, H.X. Smart thermosensitive chitosan hydrogel for nasal delivery of ibuprofen to treat neurological disorders. *Expert Opin. Drug Del.* **2019**, *16*, 453–466. [CrossRef]
92. Ong, H.X.; Jackson, C.L.; Cole, J.L.; Lackie, P.M.; Traini, D.; Young, P.M.; Lucas, J.; Conway, J. Primary air–liquid interface culture of nasal epithelium for nasal drug delivery. *Mol. Pharm.* **2016**, *13*, 2242–2252. [CrossRef]

93. Thomson, A.; Dietschy, J. The role of the unstirred water layer in intestinal permeation. In *Pharmacology of Intestinal Permeation II*; Springer: Berlin/Heidelberg, Germany, 1984; pp. 165–269.
94. Cotton, C.U.; Reuss, L. Measurement of the effective thickness of the mucosal unstirred layer in Necturus gallbladder epithelium. *J. Gen. Physiol.* **1989**, *93*, 631–647. [CrossRef]
95. Pohl, P.; Saparov, S.M.; Antonenko, Y.N. Size of the unstirred layer as a function of the solute diffusion coefficient. *Biophys. J.* **1998**, *75*, 1403–1409. [CrossRef]
96. Naruhashi, K.; Tamai, I.; Li, Q.; Sai, Y.; Tsuji, A. Experimental demonstration of the unstirred water layer effect on drug transport in Caco-2 cells. *J. Pharm. Sci.* **2003**, *92*, 1502–1508. [CrossRef] [PubMed]
97. Lewis, L.D.; Fordtran, J.S. Effect of perfusion rate on absorption, surface area, unstirred water layer thickness, permeability, and intraluminal pressure in the rat ileum in vivo. *Gastroenterology* **1975**, *68*, 1509–1516. [CrossRef]
98. Guenat, O.T.; Berthiaume, F. Incorporating mechanical strain in organs-on-a-chip: Lung and skin. *Biomicrofluidics* **2018**, *12*, 042207. [CrossRef] [PubMed]
99. Mason, G.; Peters, A.; Bagdades, E.; Myers, M.J.; Snook, D.; Hughes, J.M.B. Evaluation of pulmonary alveolar epithelial integrity by the detection of restriction to diffusion of hydrophilic solutes of different molecular sizes. *Clin. Sci.* **2001**, *100*, 231–236. [CrossRef]
100. Marks, J.D.; Luce, J.M.; Lazar, N.M.; Wu, J.N.; Lipavsky, A.; Murray, J.F. Effect of increases in lung volume on clearance of aerosolized solute from human lungs. *J. Appl. Physiol.* **1985**, *59*, 1242–1248. [CrossRef]
101. Kim, H.J.; Huh, D.; Hamilton, G.; Ingber, D.E. Human gut-on-a-chip inhabited by microbial flora that experiences intestinal peristalsis-like motions and flow. *Lab Chip* **2012**, *12*, 2165–2174. [CrossRef]
102. Mak, I.W.; Evaniew, N.; Ghert, M. Lost in translation: Animal models and clinical trials in cancer treatment. *Am. J. Transl. Res.* **2014**, *6*, 114.
103. Zhou, Q.; Patel, D.; Kwa, T.; Haque, A.; Matharu, Z.; Stybayeva, G.; Gao, Y.; Diehl, A.M.; Revzin, A. Liver injury-on-a-chip: Microfluidic co-cultures with integrated biosensors for monitoring liver cell signaling during injury. *Lab Chip* **2015**, *15*, 4467–4478. [CrossRef]
104. Yang, X.; Li, K.; Zhang, X.; Liu, C.; Guo, B.; Wen, W.; Gao, X. Nanofiber membrane supported lung-on-a-chip microdevice for anti-cancer drug testing. *Lab Chip* **2018**, *18*, 486–495. [CrossRef]
105. Huh, D.; Leslie, D.C.; Matthews, B.D.; Fraser, J.P.; Jurek, S.; Hamilton, G.A.; Thorneloe, K.S.; McAlexander, M.A.; Ingber, D.E. A human disease model of drug toxicity–induced pulmonary edema in a lung-on-a-chip microdevice. *Sci. Transl. Med.* **2012**, *4*, 159ra147. [CrossRef]
106. Ferrari, E.; Palma, C.; Vesentini, S.; Occhetta, P.; Rasponi, M. Integrating biosensors in organs-on-chip devices: A perspective on current strategies to monitor microphysiological systems. *Biosensors* **2020**, *10*, 110. [CrossRef] [PubMed]
107. Zhu, Y.; Mandal, K.; Hernandez, A.L.; Kawakita, S.; Huang, W.; Bandaru, P.; Ahadian, S.; Kim, H.J.; Jucaud, V.; Dokmeci, M.R.; et al. State of the art in integrated biosensors for organ-on-a-chip applications. *Curr. Opin. Biomed. Eng.* **2021**, *19*, 100–309. [CrossRef]
108. Zhang, Y.S.; Aleman, J.; Shin, S.R.; Kilic, T.; Kim, D.; Shaegh, S.A.M.; Massa, S.; Riahi, R.; Chae, S.; Hu, N.; et al. Multisensor-integrated organs-on-chips platform for automated and continual in situ monitoring of organoid behaviors. *Proc. Natl. Acad. Sci. USA* **2017**, *114*, E2293–E2302. [CrossRef]
109. Van der Helm, M.W.; Odijk, M.; Frimat, J.-P.; van der Meer, A.D.; Eijkel, J.C.; van den Berg, A.; Segerink, L.I. Direct quantification of transendothelial electrical resistance in organs-on-chips. *Biosens. Bioelectron.* **2016**, *85*, 924–929. [CrossRef]
110. Maoz, B.M.; Herland, A.; Henry, O.Y.; Leineweber, W.D.; Yadid, M.; Doyle, J.; Mannix, R.; Kujala, V.J.; FitzGerald, E.A.; Parker, K.K.; et al. Organs-on-Chips with combined multi-electrode array and transepithelial electrical resistance measurement capabilities. *Lab Chip* **2017**, *17*, 2294–2302. [CrossRef]
111. Henry, O.Y.; Villenave, R.; Cronce, M.J.; Leineweber, W.D.; Benz, M.A.; Ingber, D.E. Organs-on-chips with integrated electrodes for trans-epithelial electrical resistance (TEER) measurements of human epithelial barrier function. *Lab Chip* **2017**, *17*, 2264–2271. [CrossRef]
112. Van der Helm, M.W.; Henry, O.Y.; Bein, A.; Hamkins-Indik, T.; Cronce, M.J.; Leineweber, W.D.; Odijk, M.; van der Meer, A.D.; Eijkel, J.C.; Ingber, D.E.; et al. Non-invasive sensing of transepithelial barrier function and tissue differentiation in organs-on-chips using impedance spectroscopy. *Lab Chip* **2019**, *19*, 452–463. [CrossRef]
113. Uslu, B.; Ozkan, S.A. Electroanalytical Application of Carbon Based Electrodes to the Pharmaceuticals. *Anal. Lett.* **2007**, *40*, 817–853. [CrossRef]
114. Dolberg, A.M.; Reichl, S. Expression of P-glycoprotein in excised human nasal mucosa and optimized models of RPMI 2650 cells. *Int. J. Pharm.* **2016**, *508*, 22–33. [CrossRef]
115. Xu, Z.; Gao, Y.; Hao, Y.; Li, E.; Wang, Y.; Zhang, J.; Wang, W.; Gao, Z.; Wang, Q. Application of a microfluidic chip-based 3D co-culture to test drug sensitivity for individualized treatment of lung cancer. *Biomaterials* **2013**, *34*, 4109–4117. [CrossRef]
116. Zhao, L.; Wang, Z.; Fan, S.; Meng, Q.; Li, B.; Shao, S.; Wang, Q. Chemotherapy resistance research of lung cancer based on micro-fluidic chip system with flow medium. *Biomed. Microdevices* **2010**, *12*, 325–332. [CrossRef] [PubMed]
117. Humayun, M.; Chow, C.-W.; Young, E.W. Microfluidic lung airway-on-a-chip with arrayable suspended gels for studying epithelial and smooth muscle cell interactions. *Lab Chip* **2018**, *18*, 1298–1309. [CrossRef] [PubMed]

118. Punde, T.H.; Wu, W.-H.; Lien, P.C.; Chang, Y.L.; Kuo, P.H.; Chang, M.D.T.; Lee, K.Y.; Huang, C.D.; Kuo, H.P.; Chan, Y.F. A biologically inspired lung-on-a-chip device for the study of protein-induced lung inflammation. *Integr. Biol.* **2015**, *7*, 162–169. [CrossRef]
119. Islam, N.; Cleary, M.J. Developing an efficient and reliable dry powder inhaler for pulmonary drug delivery—A review for multidisciplinary researchers. *Med. Eng. Phys.* **2012**, *34*, 409–427. [CrossRef]
120. Dos Reis, L.G.; Chaugule, V.; Fletcher, D.F.; Young, P.M.; Traini, D.; Soria, J. In-vitro and particle image velocimetry studies of dry powder inhalers. *Int. J. Pharm.* **2021**, *592*, 119966. [CrossRef]
121. Kourmatzis, A.; Cheng, S.; Chan, H.K. Airway geometry, airway flow, and particle measurement methods: Implications on pulmonary drug delivery. *Expert Opin. Drug Del.* **2018**, *15*, 271–282. [CrossRef]

bioengineering

MDPI

Review

A Review of Conventional and Novel Treatments for Osteoporotic Hip Replacements

Fahad Alabdah [1,2], Adel Alshammari [1,2], Araida Hidalgo-Bastida [3] and Glen Cooper [2,*]

1 Engineering College, University of Hail, Hail 55476, Saudi Arabia
2 School of Engineering, University of Manchester, Oxford Road, Manchester M13 9PL, UK
3 Department of Life Sciences, Faculty of Science & Engineering, Manchester Metropolitan University, Manchester M15 6BH, UK
* Correspondence: glen.cooper@manchester.ac.uk

Abstract: Introduction: Osteoporosis is a skeletal disease that severely affects the mechanical properties of bone. It increases the porosity of cancellous bone and reduces the resistance to fractures. It has been reported in 2009 that there are approximately 500 million osteoporotic patients worldwide. Patients who suffer fractures due to fragility cost the National Healthcare Systems in the United Kingdom £4.4 billion in 2018, in Europe €56 billion in 2019, and in the United States $57 billion in 2018. Thus, osteoporosis is problematic for both patients and healthcare systems. Aim: This review is conducted for the purpose of presenting and discussing all articles introducing or investigating treatment solutions for osteoporotic patients undergoing total hip replacement. Methods: Searches were implemented using three databases, namely Scopus, PubMed, and Web of Science to extract all relevant articles. Predetermined eligibility criteria were used to exclude articles out of the scope of the study. Results: 29 articles out of 183 articles were included in this review. These articles were organised into three sections: (i) biomechanical properties and structure of osteoporotic bones, (ii) hip implant optimisations, and (iii) drug, cells, and bio-activators delivery through hydrogels. Discussion: The findings of this review suggest that diagnostic tools and measurements are crucial for understanding the characteristics of osteoporosis in general and for setting patient-specific treatment plans. It was also found that attempts to overcome complications associated with osteoporosis included design optimisation of the hip implant; however, only short-term success was reported, while the long-term stability of implants was compromised by the progressive nature of osteoporosis. Finally, it was also found that targeting implantation sites with cells, drugs, and growth factors has been outworked using hydrogels, where promising results have been reported regarding enhanced osteointegration and inhibited bacterial and osteoclastic activities. Conclusions: These results may encourage investigations that explore the effects of these impregnated hydrogels on osteoporotic bones beyond metallic scaffolds and implants.

Keywords: osteoporosis; hydrogels; total hip replacement; tissue scaffolds

Citation: Alabdah, F.; Alshammari, A.; Hidalgo-Bastida, A.; Cooper, G. A Review of Conventional and Novel Treatments for Osteoporotic Hip Replacements. *Bioengineering* **2023**, *10*, 161. https://doi.org/10.3390/bioengineering10020161

Academic Editors: Ali Zarrabi and Elena A. Jones

Received: 7 November 2022
Revised: 18 January 2023
Accepted: 19 January 2023
Published: 25 January 2023

1. Introduction

Osteoporosis is a skeletal disease that severely affects the mechanical properties of bone. It increases the porosity of cancellous bone and reduces the resistance to fractures [1]. It has been reported that there are approximately 500 million osteoporotic patients worldwide [2]. Patients who suffer fractures due to fragility cost healthcare systems around £4.4 billion in the United Kingdom [3], €56 billion in Europe [4], and $57 billion in the United States [5], annually. Thus, osteoporosis is problematic for both patients and healthcare systems.

The presence of osteoporosis often compromises the success of using orthopaedic devices due to the influence of reduced bone quality on stability, and secondary fractures [6,7]. The lack of mechanical stability results in aseptic loosening which consequently leads to inflammation at the bone–implant interface and, in some cases, leads to periprosthetic fractures. These complications may lead to revision surgeries where the success rate is substantially lower, reported to be 35% at 10 years follow-up [8]. Compared to primary hip replacement surgery, the cost of revision surgery is significantly higher due to the complexity associated with revision technique [9]. Therefore, osteoporosis treatment imposes a significant financial burden on healthcare systems and poses a significant threat to the quality of life and survival of elderly patients.

The complications associated with osteoporosis and total hip arthroplasty have been highlighted in this review in three sections related to biomechanical properties, implant optimization and drug-laden hydrogels, see Figure 1. This review aims to identify and discuss articles introducing and investigating treatment solutions for osteoporotic patients submitted for total hip replacement. Studies obtained from the literature have assessed biomechanical properties and structure of osteoporotic bones, hip implant optimisations, and drug, cells, and bio-activators delivery through hydrogels.

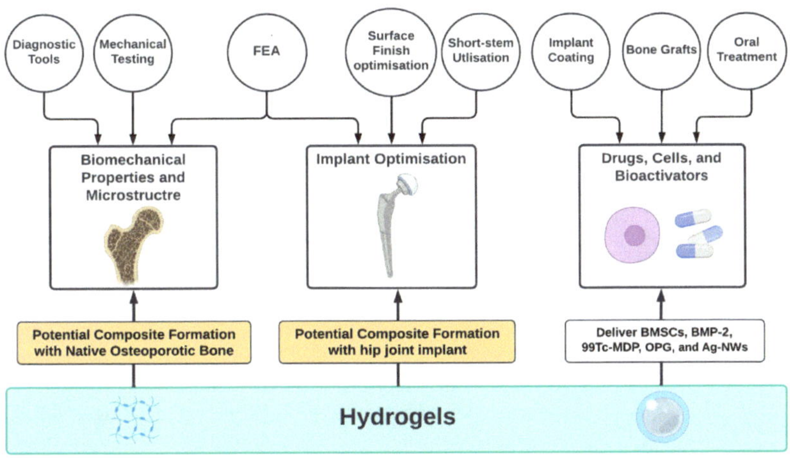

Figure 1. Overview of the review highlighting the importance of osteoporotic bone biomechanics, implant optimisation and drug delivery systems, created with BioRender.com (accessed on 9 January 2023).

2. Materials and Methods

The search was conducted using three search engines: Scopus, PubMed, and Web of Science in April 2022. The method implemented in the search was (Title–Abstract–Keywords) as follows: TITLE-ABS-KEY ("Osteoporosis" AND ("hip implant" OR "hip replacement" OR "joint arthroplasty" OR "joint replacement") AND ("biomechanics" OR "tissue engineering" OR "regenerative medicine" OR "bone implant" OR "tissue scaffold*" OR "hydrogel" OR "modelling" OR "modeling")).

All potential articles generated by the search engines were screened by the title and abstract and were subjected to pre-set inclusion and exclusion criteria. The inclusion criteria for this review were: (a) articles on hip replacements and total hip arthroplasty as a consequence or with the presence of osteoporosis, (b) studies conducted for the purpose of evaluation of existing diagnostic techniques or the establishment if new ones. The exclusion criteria were: (a) studies focused on drug treatments rather than engineering or surgical interventions, (b) studies focused on the immune system, (c) studies on any skeletal parts other than hip, (d) studies in any language other than English, and (e) any document type other than original articles and reviews.

3. Results

The search engines used in this review produced 183 potential sources: 104 by Scopus, 18 by PubMed, and 61 by Web of Science. There were 45 duplicates, 94 excluded by abstract screening, and 15 excluded by reading the full text. Therefore, 29 papers were included in this review. Of those 29 papers there were 25 representing current diagnostic and treatments approaches and 4 articles introduced novel treatment approaches.

3.1. Biomechanical Properties and Microstructure

The asymptomatic nature of osteoporosis encouraged researchers to utilise and assess tools that help characterising this disease to guide treatment approaches. In this review, the results include studies investigating accuracy of diagnostic devices, and the role of mechanical tests and finite element analysis (FEA) in the assessment of osteoporotic bones. These three methods of evaluation are complementary to one another. It is possible to develop FE models with the assistance of diagnostic tools, and FEA can be validated with the assistance of mechanical testing.

3.1.1. Diagnostic Tools

The results obtained from these diagnostic devices were interpreted to understand the characteristics of osteoporosis in general and its behaviour with different variables such as age and gender. The tools varied regarding data acquisition, and accuracy as illustrated in Table 1. Osteoporotic bones can be assessed by analysing the femoral neck on the unaffected side of a simple anterior-posterior X-ray, the severity of osteoporosis can be classified into one of six grades, referred to as Singh Index as illustrated in Figure 2. This index attributable to the rarefaction of trabecular structures [10]. Although this tool was reported to be an inexpensive approach for bone architecture assessment, the assessment acquired by Singh Index would be an estimation rather than accurate [11], whereas Singh Index value was combined with bone mineral density (BMD) evaluation and reported as an acceptable approach to investigate the mechanical competence of bone [12]. In fact, BMD has also been used in combination with another assessment techniques such as velocity ultrasound, and that combination was reported to improve the fracture risk assessment for osteoporotic patients [13]. It was also suggested by Endo et al. [14], that the assessment of osteoporotic bones using magnetic resonance imaging (MRI) could enhance the accuracy of the assessment conducted using BMD only. The value MRI can add to BMD is that it can predict the strength of cancellous bone in addition to the bone quality change [14]. It is worth mentioning that BMD can be assessed using dual energy X-ray absorptiometry (DEXA) which was reported to be the most accurate and reliable technique for the assessment of BMD [15].

Table 1. Diagnostic tools used for bone assessment and their efficiency as reported.

Tool	Use	Results	Reference
Singh Index (SI)	Bone architecture assessment	Inexpensive tool, but not accurate results	[11]
Singh Index (SI) + Bone Mineral Density (BMD)	Mechanical competence and architecture of the bone	Acceptable estimation compared to Singh Index alone	[12]
Velocity Ultrasound + Bone Mineral Density (BMD)	Fracture risk assessment	Improved in comparison with Singh Index alone	[13]
Dual-Energy X-ray Absorptiometry (DEXA)	Evaluate (BMD)	Excellent for the assessment of (BMD)	[15]
Magnetic Resonance Imaging (MRI)	Evaluate (BMD)	Enhance accuracy of (DEXA) results	[14]

Table 1. *Cont.*

Tool	Use	Results	Reference
Low Field Nuclear Magnetic Resonance (LF-NMR), High Resolution Computed Tomography (HR-CT), and micro-CT (μCT)	Evaluate bone porosity and structure	Qualitative and quantitative information that can be used for Finite Elements Analysis	[16]

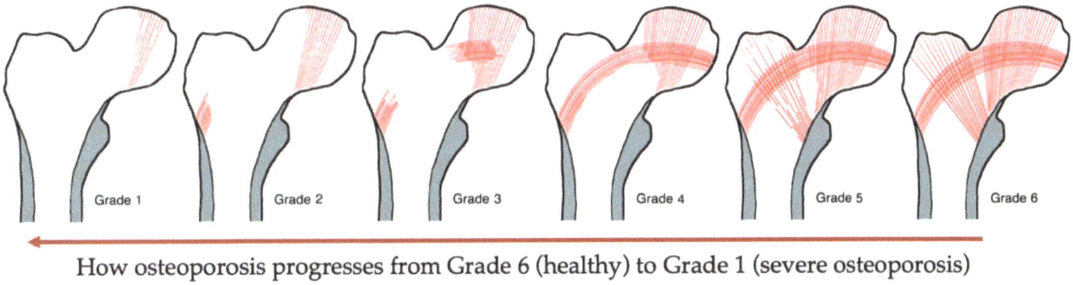

Figure 2. Singh Index grades: Grade 6 The radiograph clearly shows each of the trabecular subgroups. Cancellous bone appears to fill the whole top of the femur. Grade 5: the primary tensile trabecula has been highlighted, and the Ward triangle is clearly visible. Grade 4: the primary tensile trabeculae are significantly diminished, but can still be traced from the lateral cortex to the upper femoral neck. Grade 3: the continuity of the major tensile trabeculae is broken. Grade 2: only the major compressive trabeculae are visible, but other groups have been assimilated. Grade 1: the number and size of the main compressive trabeculae are diminished and no longer prominent. Adapted from reference [17].

Another view to consider is the suggestion, made by Porrelli et al. [16], that morphological information cannot be extracted from DEXA and ultrasonography alone in a qualitative and quantitative manner. It was reported that using a combination of MRI, low field nuclear magnetic resonance (LF-NMR), high resolution computed tomography (HR-CT), and micro-computed tomography (μCT) have enhanced the study of bone porosity and structure. The reason for classifying these techniques as accurate and more informative is due to the ability to build models based on the obtained data for FEA [16].

3.1.2. Mechanical Testing

Mechanical testing in the field of tissue engineering can be conducted for various reasons such as bone stress, strain, stiffness, failure load, and fracture risk assessment as shown in Table 2.

Table 2. Mechanical testing approaches.

Aim	Type of Test	Results	Reference
Determine gender effect on fracture risk	Compression	Males have a bone Young's modulus of 293.68 MPa and an ultimate stress of 8.04 MPa, whereas females have 174.26 MPa and 4.46 MPa for young's modulus and ultimate stress, respectively. Therefore, men have lower fracture risk compared to women.	[18]

Table 2. *Cont.*

Aim	Type of Test	Results	Reference
Evaluate the weightbearing immediately after fixation of posterior wall (PW) fractures	Cyclic loading	With assistance, immediate load bearing is allowable with 50% of PW and 25% of acetabular rim, regardless of PW fixation.	[19]
Investigate effect osteoporosis on bone fracture toughness	Fracture toughness	Fracture toughness decreased with ageing (7.0% each decade, $r = -0.36$, $p = 0.029$), while comparable fracture resistance properties were found in osteoporotic, osteoarthritic and control groups (10% difference for indentation and $p > 0.05$ for fracture properties).	[20]
Introduce synthetic bone that represent osteoporotic cadaveric bones.	Four-point bending, axial compression, and pullout	There was good correlation found between the cadaveric and synthetic bone samples. The *p*-values in all mechanical tests were acceptable, ranging between 0.1–0.9 except in pullout tests ($p = 0.005$).	[21]

In a biomechanical study investigating the bone fragility and mechanical behaviour, compression testing has revealed that men have lower fracture risk compared to women in the presence of osteoporosis in both populations [18]. The findings of this study indicate that gender is one of the variables which must be taken into account while considering a treatment plan for an osteoporotic patient.

Mechanical testing can also be used to assess load bearing with the presence of fractures in addition to the mechanical evaluation of different fixation approaches [19,20]. On a total of six osteoporotic female cadaveric pelvises, Marmor et al. [19] produced posterior wall fractures. After the fracture was created, cyclic loading equal to 1.8 times the body weight was applied. Every specimen was able to withstand the loading with a cup motion of less than 150 µm, which is within the permissible limit. Similarly, Jenkins et al. [20] tested the fracture toughness for three groups: osteoporosis, osteoarthritis, and control group. It appeared that neither osteoporosis nor osteoarthritis have any additional influence on the fracture toughness of the inferomedial femoral neck beyond that which is caused by natural ageing.

In research that aims to create therapeutic options for osteoporosis, it is crucial to possess bone samples that represent this skeletal condition in order to examine and evaluate the approach. Gluek et al. [21] have introduced and evaluated a novel synthetic bone with a mechanical reaction equivalent to that of osteoporotic bone from cadavers.

3.1.3. Finite Element Analysis

Medical engineering has implemented FEA in studies of bone structure, mechanical properties, and assessments of treatment approaches as shown in Table 3. The accuracy of the data obtained by FEA primarily depends on the CT scans from which the models are constructed [22]. Rieger et al. [23] stated that their approach to study and assess bone macrostructure and microstructure has also been used by a number of scholars. They used high-resolution µCT images of fractured femoral heads to produce µFE mesh in order to obtain bone stress and strain. They stated that the mechanical properties of the bone on the macroscopic level can be obtained by the analysis of the microstructure only. Their findings showed that using FEA in addition to numerical calculations based on that FEA as an inversed approach can reveal macroscopic and microscopic mechanical properties of the bone as they reported their results indicating osteoporotic bones have comparable elasticity to healthy ones. However, the only difference identified was the yield stress with a mean of

85.6 ± 16.7 MPa which is lower than yield stress of healthy bones. It is worth mentioning that this approach was suggested to add supportive data to the histomorphometric analysis in the orthopaedic studies.

Table 3. FEA for bone microstructure assessment.

Aim	Bone Model	Software	Results	Reference
Evaluate macroscopic mechanical properties the bone	Virtual trabecular bone biopsy from CT scan	Abaqus 6.9-2	Osteoporotic bones have comparable elasticity to healthy ones, with young's modulus mean (±SD) of 18.92 ± 5.43 GPa. However, the yield stress was found to be lower in osteoporotic bones with a mean (±SD) of (85.6 ± 16.7 MPa).	[23]
Evaluate the influence of plate and rod in osteoporotic and osteoarthritic patients	Virtual subchondral trabecular bone biopsy from CT scan	Scanco Medical Finite Element Software 1.06	Osteoarthritic subchondral bones had higher stiffness with a mean (±SD) of 12,003.56 (±7590.42) kN/mm, while the mean stiffness of osteoporotic bones was 4964.01 (±3778.37) kN/mm. Similarly, the failure load was reported to be higher in osteoarthritic bones compared to osteoporotic ones with 477.7 (±279.56) MPa and 215.89 (±143.73), respectively.	[24]

Similarly, He et al. [24] implemented FEA to compare and understand osteoporosis and osteoarthritis by analysing bone structure and mechanical behaviour. In their study, the bone microstructure was generated through virtual biopsies obtained from μCT scans of the subchondral trabecular bone. The FEA results showed that the plate and rod structures are significantly higher in the osteoarthritis group compared to the osteoporosis group, which consequently the failure load, stiffness, young's modulus, compressive strength, yield strength, and maximum compressive force are reported to be higher in the osteoarthritis group.

3.2. *Implant Optimisation*

The use of hip prothesis has been a huge leap in the treatment of skeletal diseases, as it was stated that primary total hip replacement (THR) conveys more desirable outcomes as a treatment intervention in comparison to other approaches such as open reduction internal fixation regarding the stability of the acetabular component especially for osteoporotic patients [25]. However, the complications associated with this procedure opened an area of research for the purposes of ensuring success in the long-term. Since the major downside of using an artificial hip joint, has been stem instability which consequently leads to further complications. There have been different approaches reported in the literature to enhance implant fixation and long-term stability such as design, and surface finish optimisation. FEA has also been used to investigate factors that lead to complications, the results obtained from these investigations has provided some insights that inspired implant optimisation.

3.2.1. Design Optimisation

Altered implant designs compared to conventional ones were implemented in clinical trials to eliminate aseptic loosening and periprosthetic fractures (Table 4). The implant stem was shortened to be used in THA for osteoporotic patients [26,27]. It was reported that a short, tapered stem can show desirable stability compared to conventional; refer to Figure 3, which shows both short and conventional implants in a scan.

Table 4. Implant design optimisation.

Implant Design	Targeted Complication	Results	Limitations	Reference
Short stem implant	Aseptic loosening	The mean of the Harris Hip Score (HHS) in the two groups increased from 45.0 ± 16 (29–61) and 40.0± 11 (29–51) prior to surgery, to 93 ± 9 (84–100) and 96 ± 7 (89–100), respectively. The survival rate with stem revision for aseptic loosening was 100%.	Some cases with Vancouver B1 and Vancouver B2 fractures were reported in both groups.	[26]
Cementless short metaphyseal fitting stem	implant instability	The mean HHS improved from 48.0 ± 8.0(38.0–61.0) prior to surgery to 91.0 ± 8.0 (85.0–98.0). In addition, there were no postoperative complications such as infection, deep vein thrombosis, loosening, or peri-prosthetic fracture.	Low number of patients, and short follow-up duration.	[27]
Dual-mobility cups in total hip arthroplasty (DM-THA)	Femoral Neck Fractures (FNFs)	The mean HHS increased from 58.62 (+15.79) preopratively to 86.13 (+9.92).	Cases of loosening, revision DM-THA, intra-prosthetic dislocation, migration, tilting, and severe wear were reported in the study.	[28]

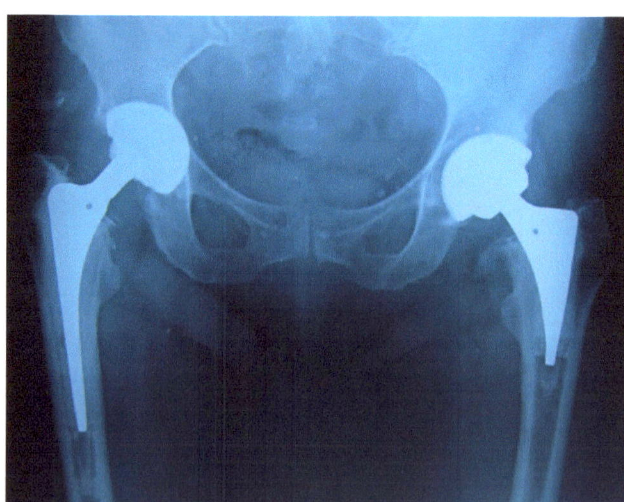

Figure 3. A patient with both conventional implant (Left), and short stem implant (Right) [26].

Although the success rate in Santori et al.'s [26] clinical trial with regards to aseptic loosening was reported to be 100%, there are some cases were periprosthetic fractures occurred; whereas, Zhen et al. [27] reported that the utilisation of short-stem hip joint has eliminated both aseptic loosing and, periprosthetic fractures, and thigh pain, yet their study had some limitations which may influenced their conclusions. The mean duration of the follow-up after operation was (5.5 ± 1.1 year) which was deemed short and insufficient, in addition to the low number of patients.

Implant design alteration was also used to tackle femoral neck fractures (FNFs) phenomena. An approach of using dual-mobility cup in total hip arthroplasty (DM-THA) procedure was conducted and evaluated on osteoporotic Chinese population [28]. The use of DM-THA has shown desirable outcomes regarding dislocation of FNFs, yet in their clinical study there was a need for revision due to loosening. These design manipulations have shown solutions for some of the complications associated with osteoporosis such as acetabular component fixation; however, periprosthetic fractures, and aseptic loosening still existed with those designs. Therefore, scholars have been investigating the effectiveness of implant surface treatment to enhance fixation and stability.

3.2.2. Surface Finish Optimisation

Large area electron beam melting (LAEB) was used to adjust the nanotopography of the titanium alloy surface used for joint implants. The resultant surface roughness with topography Ra of ~40 nm was reported to enhance the osteogenic differentiation in vitro on human skeletal stem cells (SSCs) [29]. However, the mechanical properties, mineralisation, and the bone matrix organisation of an implant treated with LAEB have not been investigated in vivo.

3.2.3. Finite Element Analysis

Conducting FEA for the purpose of anticipating the success rate of hip joint implants was conducted by Rafiq et al. [30] to assess the feasibility of using a cementless implant for osteoporotic patients. The FE algorithm used was simulating stairs-climbing to investigate micromotion at the bone–implant interface. An osteoporotic model was compared to healthy and osteoarthritic models. Poor bone density, stiffness and thin cortical bone in the osteoporotic model allowed an increase in the surface area which compromised bone growth and implant stability observed by micromotion. The analysis findings suggested that cementless implants are predicted to experience loosening on the long-term with osteoporotic host bone.

3.3. *Drugs, Cells, and Bioactivators*

Some studies focused on the effect that osteoporosis has on hip prothesis after the implantation and how that can be reversed by using anti-osteoporosis drugs, stem cells and bio-activators for the purposes of restoring the natural bone remodelling process which is compromised by osteoporosis. These substances have been investigated when delivered orally or as an implant coating. Bone grafts were also investigated for their desirable bioactivity. Yet, the most recent approach found in the literature for drug delivery into bones is hydrogels.

3.3.1. Implant Coating

Several studies have investigated the effect of implant surface treatment on the stability of the bone–implant interface as illustrated in Table 5.

Table 5. Implant coating.

Treatment	Targeted Complication	Results	Limitation	Reference
Zoledronate	Instability and poor bone formation in the bone–implant interface	Bone formation was enhanced by the elimination of osteoclastic activity by Zolendronate. Thus, in comparison to the implant not coated with Zoledronate, coated implants showed significantly higher maximal pullout force ($p < 0.05$) and ($p < 0.01$).	Used with hydroxyapatite coating, which is reported to impair osteoporotic bone ingrowth, consequently long-term survival.	[31,32]
Hydroxyapatite (HA) coated implants	Poor bone–implant ingrowth	The mean osseointegrated implant surface (OIS) in implants coated with HA and uncoated ones were 23.7 and 23.5 in ovariectomised rats, respectively. HA have no effect on osteoporotic bones while it enhances the OIS in healthy bones.	The results of the study indicate that HA-coated implants deteriorate bone ingrowth in the long term.	[33]
Calcium Phosphates coating (CaP) with platelet-rich plasma (PRP)	Implant instability	Enhanced stability, evident by the increase in the maximal push-out force in the group treated with CaP and PRP compared to the control group ($p < 0.05$).	No limitations mentioned in the study	[34]
Surface Large Area Electron Beam melting (LAEB)	Implant surface nanotopography	When compared to the untreated group, the group treated with a cathode voltage of 35 kV and 25 shots showed a significant increase in osteogenic activity (two- to three-fold). This peak was observed to correlate with a surface roughness (R_a) of 44 nm.	The technique has not been investigated in vivo for mechanical interface strength	[29]

It is worth noting that bisphosphonates were reported to increase the fatigue life on bone cement when they are mixed in a powder form [35]. However, different forms of bisphosphonates were compared to Zoledronate, and the results showed that Zoledronate demonstrated desirable enhancement of early bone formation and bone–implant integration, as shown in Figure 4. Gao et al. [32] stated that the benefit of bisphosphonate immersion on the implant surface is that they impact osteoclasts by eradicating their proliferation activity, which is desirable for patients with osteoporosis. Another study suggested that the use of hydroxyapatite-coated implants enhances bone mineralisation, formation, and mechanical stability; however, using such implants may lead to loosening and inflammation in the implant site due to the damage it causes to the bone matrix [33].

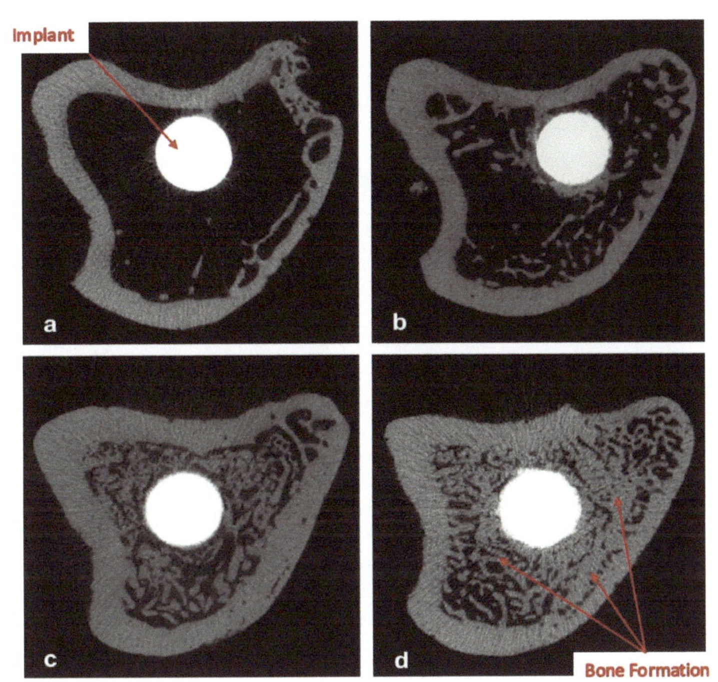

Figure 4. Micro-CT binary images of tibiae with implants 3 months after implantation: (**a**) uncoated implant; (**b**) Pamidronate; (**c**) Ibandronate; (**d**) Zoledronate [32].

Another approach reported to have a potential to eliminate osteoporotic effects and influence THA success which is using calcium phosphates (CaP) to coat the implants [34]. It is worth noting that the CaP-coated implant was investigated along with platelet-rich plasma (PRP) treatment, which may compromise the accuracy of the conclusions made about the effects of CaP coating on its own.

3.3.2. Bone Grafts

The use of bone autografts and allografts in primary and revision arthroplasty have been well established as a surgical solution [36]. It was also stated by Brewster et al. [36] that the purpose of implementing a bone graft is to enhance the cup stability and fixation in the hip, due to their superior ability to withstand complex forces (i.e., normal loads, and shear strain). They noted that osteoporotic specimens such as femoral heads can provide similar properties to healthy ones regarding bone grafts, though with much fewer particles.

3.3.3. Hydrogels within Metallic Scaffolds

Several scholars have studied the feasibility of using hydrogels as drug delivery vehicles for osteoporotic bones owing to their desirable biocompatibility in addition to their ability to provide a medium where impregnated cells can survive during the desired course of release Table 6 [37–40]. Hydrogels have been studied within metallic scaffolds as a treatment approach in fracture site for enhancement of bone remodelling and osteointegration processes.

Table 6. Hydrogels within metallic scaffolds.

Material	Fabrication Process	Impregnated Drugs	Reference
NCECS-PVA and AGPVA	Chemical crosslinking	Autophagy-regulated rapamycin	[37]
Poloxamer 407	Thermosensitive mixture	Bone morphogenetic protein-2 (BMP-2)	[38]
Poloxamer 407	Thermosensitive mixture	Technetium methylenediphosphonate (^{99}Tc-MDP)	[39]
N-carboxyethyl chitosan (N-chitosan)	In situ crosslinking	Bone marrow stem cells (BMSCs) + (BMP-2)	[40]

Hydrogels were preliminarily investigated in vitro and in vivo to evaluate their potential contribution in the development of orthopaedic complications solutions.

The four studies identified in the literature that implemented impregnated hydrogels all used hydrogels in combination with porous 3D printed titanium scaffolds to investigate their effect on osteoporotic bones. They assessed the biocompatibility, cell proliferation, cell differentiation, and mechanical stability of the composite implants.

The hydrogels in those studies were used as drug and cells delivery vehicles, they were impregnated autophagy-regulated rapamycin [37], bone morphogenetic protein-2 (BMP-2) [38,40], and technetium methylenediphosphonate (^{99}Tc-MDP) [39]. In fact, Bai et al. [40] have also impregnated bone marrow stem cells (BMSCs) in the hydrogels with the BMP-2. They were investigated in three groups to distinguish the effect of each compared to the effect of them acting together.

It is crucial to evaluate the biocompatibility of the hydrogels when they are being investigated for biological applications, to understand their effect in promoting cell adhesion and proliferation. In all four studies, it was reported that the hydrogels show good biocompatibility, cell proliferation, and cell differentiation using biological indicators (i.e., Calcein acetoxymethyl ester (AM)/propidium iodide (PI) staining, and Alizarin red staining). Although, Bai et al. [40] reported that there were negligible inflammations after implantation, but the bone–implant site gradually became normal along the course of the study. This inflammatory reaction by the host bone towards the implanted composite scaffolds were not observed in the other studies.

Wang et al. [38] and Cui et al. [39] reported that the gel was formed using a thermosensitive approach. Poloxamer 407 was used as a powder to be added into a solution of sterilised phosphate-buffered saline (PBS)and kept at 4 °C until the solution was transparent. The drug investigated in the studies was added to the solution at the same temperature of 4 °C, the mixtures were subjected to a temperature of 37 °C until gelation was achieved.

However, in Li et al. [37] and Bai et al. [40] chemical and crosslinking approaches were used for the gel formation process. Strong hydrogen bonds between polyvinyl alcohol (PVA), N-carboxyethyl chitosan (NCECS), and agarose (AG) in the form of NCECS-AG, NCECSPVA and AG-PVA solutions, were the main factor of fabricating the hydrogels in Li et al. [37], where gelation was instantly achieved by the intended chemical reaction; whereas, an in-situ crosslinking approach was performed to prepare the solution of N-carboxyethyl chitosan (N-chitosan) and adipic acid dihydraside (ADH) with hyaluronic acid-aldehyde (HA-ALD) in Bai et al. [40]. The hydrogel was formed using a Lab Dancer to achieve homogeneity.

The degradation rate varied in each study which indicated that different materials and preparation methods could alter the degradation process. Li et al. [37] reported that the hydrogels have a slow degradation rate where the process took 36 days to degrade in vitro almost completely. Whereas, in Wang et al. [38] the drug release profiles were observed over the course of 20 days by which 70% of the of the drug was released, and that was a result of both drug diffusion and hydrogel degradation. It was also reported in Wang et al. [38]

that due to protein concentrations, the detection of the drug was difficult, therefore, only 70% of the degradation was detected in that study. In Cui et al. [39] and Bai et al. [40], the hydrogels were reported to completely degrade in 12 and 28 days, respectively. It is noted that thermosensitive hydrogels degrade at a faster rate compared to the ones fabricated via crosslinking.

The four studies have observed the microstructure of the hydrogels after the gelation processes, and that was conducted using scanning electron microscope (SEM). The pore size of the hydrogel was reported to be approximately between 100–200 μm which is favourable to provide space for osseointegration where desired, on the scaffold interface, and inside the pores of metallic scaffold. The advantage of allowing bone formation inside the pores of the scaffold is increasing its stability and attachment with the host bone.

One of the outcomes of the in vivo study conducted in Bai et al. [40] is that impregnating the hydrogels with a combination of BMSCs, and BMP-2 have improved the bone regeneration process compared to the case of implementing the porous scaffold alone as shown in Figure 5. The results obtained by Bai et al. [40] have also revealed that the mechanical properties were significantly increased in the implant impregnated both BMSCs and BMP-2 compared to unfilled porous titanium scaffold ($p < 0.01$), evident by the peak values of the push-out tests. The desirable enhancement in the mechanical properties was achieved by the high level of osteointegration formed between the implanted scaffold and the host bone.

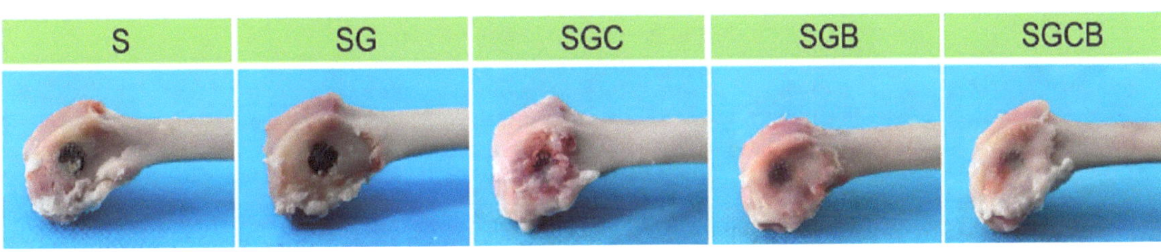

Figure 5. Bone regeneration at the implant interface. (S) scaffold, (SG) scaffold with hydrogels, (SGC) scaffold with bone marrow stem cells (BMSCs) impregnated into hydrogels, (SGB) scaffolds with bone morphogenetic protein-2 (BMP-2) impregnated hydrogels, and (SGCB) scaffolds with BMSCs and BMP-2 impregnated into hydrogels. Figure adapted from reference [40].

Similarly, Li et al. [37], Wang et al. [38], and Cui et al. [39] performed push-out tests and their results indicated higher peak values when porous titanium scaffolds were filled with impregnated hydrogels ($p < 0.01$ or $p < 0.001$). Although each study investigated different substance impregnated in hydrogels, they have comparable osteointegration against titanium scaffolds on their own.

Remarkably, Cui et al. [39] and Wang et al. [38] investigated the osteoclastic activity and found that both osteoprotegerin (OPG) and technetium methylenediphosphonate (^{99}Tc-MDP) inhibit osteoclasts proliferation evident by RANKL expression ($p < 0.001$ or $p < 0.0001$), respectively, compared to the pure titanium scaffolds.

On the other hand, Li et al. [37] investigated the bacterial proliferation by examining the absorbance of *S. aureus* and MRSA. It was reported that the silver nanowires (Ag-NWs) have significant effect that inhibited bacterial proliferation, which means that Ag-NWs can eliminate postoperative inflammations. It was also observed that when porous scaffolds were not filled with hydrogels, bacterial proliferation was greater than in the control group, which indicate grater bacterial reproduction allowed in the voids of the porous structure.

4. Discussion

The asymptomatic nature of osteoporosis increases the chance of bone fracture occurrence due to fragility. Therefore, diagnostic tools are vital for preventive and analytical purposes. BMD has been classified as one of the most important markers of bone quality, and the most effective way to assess BMD is using DEXA scans [16]. It was also found that there have been studies implemented and validated using FEA as an assessment and predictive tool of bone quality and behaviour [23,24]. The understanding of osteoporosis led to attempts to overcome complications regarding fractures, in fact, these attempts were even extended towards the improvement of the device that had been used for osteoporotic patients (i.e., total hip replacement). Although implant design optimisations were reported to improve the postoperative loosening, periprosthetic fractures still existed [26]. These unfavourable results may be attributable to the fact that osteoporosis is a degenerative illness, indicating that optimisation of the implant design may achieve short-term success but will not provide long-term stability.

On the other hand, bisphosphonates, alendronates in particular, have been widely used in daily oral doses as a treatment for osteoporotic patients and shown good results regarding the enhancement of bone remodelling [41–43]. Therefore, the desirable outcomes of oral bisphosphonates treatments encouraged scholars to immerse implants with bisphosphonates as a targeted treatment to increase the stability and fixation of the implant, and that was done using Zoledronate [31,32]. Immersing Zoledronate on the surface of the implant showed higher osteointegration on the bone–implant interface compared to oral dosing of alendronate. It was noted that implants immersed with bisphosphonates were synthesised using titanium alloys and coated with hydroxyapatite before immersion. However, the effect of hydroxyapatite coating was reported to deteriorate bone–implant ingrowth in osteoporotic patients [32,33]. Although they reported that there are undesired effects of hydroxyapatite, their study groups were coated before immersion. Their practice indicates that one cannot be used without the other.

Despite the desirable properties of bone grafts in promoting cup stability and fixation and their biocompatibility, it was reported that the use of a bone grafts has a high infection rate in addition to its effect of blocking revascularisation with the close packing, and rare restoration of muscle attachment [36,44]. Further, in bone remodelling, grafts may degrade within this process in which stability would be compromised [36]. In addition to their limited supply, the use of autografts is not practical for osteoporotic patients due to biomechanical complications associated with osteoporosis.

In studies that investigated hydrogels as treatment approaches with orthopaedic devices, the results are promising and they address the current complications associated with THR [37–40]. As stated earlier, osteoporosis compromises the success rate of THR by the low osseointegration of the bone–implant interface which may lead to aseptic loosening, severe inflammation, secondary fractures, and consequently revision surgeries. Those complications were eliminated by implanting porous metallic scaffolds loaded with hydrogels impregnated with cells, drugs, and growth factors, into osteoporotic bones in vivo. The composite scaffolds with impregnated hydrogels have conveyed significant increases in osteointegration and blocked the undesired osteoclastic activity which causes the excessive bone resorption. The composites have also been impregnated with sliver nanowires which promoted antibacterial activity which inhibited bacterial proliferation and inflammation. Li et al. [37], Wang et al. [38], Cui et al. [39], and Bai et al. [40] investigated hydrogels within the porous metallic scaffolds therefore, mechanical tests were performed to assess osteo-integration on the bone–implant interface.

The promising results reported by Li et al. [37], Wang et al. [38], Cui et al. [39], and Bai et al. [40] of using impregnated hydrogels are a huge leap in the treatment of osteoporotic bones. The significance of this treatment approach is that the impregnated substances work on restoring the biological activities affected by osteoporosis such as the lack of bone ingrowth and osteointegration, and the excessive bone resorption. Yet, these studies neither assessed the mechanical behaviour of the hydrogels nor did they investigate the effect of impregnated hydrogels beyond metallic scaffolds. The influence of those impregnated hydrogels could have a positive effect on osteoporotic bone and form a composite with the native bone similar to the presented composite with metallic scaffolds. The results of such investigations may lead to a preventive treatment approach for osteoporotic patients before a fracture occurs.

5. Conclusions

In this review, it was found that there have been attempts to overcome complications associated with osteoporosis when patients are submitted for total hip replacement due to fragility fractures. The optimisation of the implant design was reported to show short-term stability. However, the degenerative nature of osteoporosis has led to loosening and periprosthetic fractures in the long term. However, stability was improved when bisphosphonates were used on the implant surface to target the implantation site and reverse the osteoporotic effect. Their effect was compromised by the presence of hydroxyapatite which deteriorate bone ingrowth in osteoporotic patients.

Promising results were found in studies that used hydrogels impregnated with cells, drugs, and growth factors within metallic scaffolds. Significant increase in osteointegration was observed in addition to the inhabitation of osteoclastic and bacterial activities. This is an indication that this approach restores biological activities compromised by osteoporosis.

The results from the use of impregnated hydrogels for osteoporosis are promising to improve osteointegration and block excessive osteoclastic activity. This suggests that hydrogels merit further investigation, which could include their use for preventative strategies as well as investigating further novel approaches to improve the outcomes for total hip replacements.

Author Contributions: Conceptualization and supervision, G.C. and A.H.-B.; writing-original draft preparation, F.A., A.A., A.H.-B. and G.C.; writing-review and editing, F.A., A.A., A.H.-B. and G.C. visualization, F.A. All authors have read and agreed to the published version of the manuscript.

Funding: This research was funded through PhD Scholarships from University of Hail, Hail, Saudi Arabia.

Conflicts of Interest: The authors declare no conflict of interest.

References

1. Dickenson, R.P.; Hutton, W.C.; Stott, J.R.R. The mechanical properties of bone in osteoporosis. *J. Bone Jt. Surg. Br.* **1981**, *63*, 233–238. [CrossRef] [PubMed]
2. International Osteoporosis Foundation Epidemiology of Osteoporosis and Fragility Fractures | International Osteoporosis Foundation. Available online: https://www.osteoporosis.foundation/facts-statistics/epidemiology-of-osteoporosis-and-fragility-fractures (accessed on 19 July 2022).
3. National Institute for Health and Care Excellence. *NICE Impact Falls and Fragility Fractures*; National Institute for Health and Care Excellence: London, UK, 2018.
4. Kanis, J.A.; Norton, N.; Harvey, N.C.; Jacobson, T.; Johansson, H.; Lorentzon, M.; McCloskey, E.v.; Willers, C.; Borgström, F. SCOPE 2021: A New Scorecard for Osteoporosis in Europe. *Arch. Osteoporos.* **2021**, *16*, 82. [CrossRef] [PubMed]
5. Lewiecki, E.M.; Ortendahl, J.D.; Vanderpuye-Orgle, J.; Grauer, A.; Arellano, J.; Lemay, J.; Harmon, A.L.; Broder, M.S.; Singer, A.J. Healthcare Policy Changes in Osteoporosis Can Improve Outcomes and Reduce Costs in the United States. *JBMR Plus* **2019**, *3*, e10192. [CrossRef] [PubMed]
6. Bukata, S.v.; Crawford, B.M.; Vallera, C. Orthopedic Aspects of Osteoporosis. In *Marcus and Feldman's Osteoporosis*, 5th ed.; Academic Press: Cambridge, MA, USA, 2021; Volume 2, pp. 1613–1625. ISBN 9780128130735.
7. Kammerlander, C.; Neuerburg, C.; Verlaan, J.J.; Schmoelz, W.; Miclau, T.; Larsson, S. The Use of Augmentation Techniques in Osteoporotic Fracture Fixation. *Injury* **2016**, *47*, S36–S43. [CrossRef]
8. Springer, B.D.; Fehring, T.K.; Griffin, W.L.; Odum, S.M.; Masonis, J.L. Why Revision Total Hip Arthroplasty Fails. *Clin. Orthop. Relat. Res.* **2009**, *467*, 166–173. [CrossRef]
9. Barrack, R.L.; Sawhney, J.; Joe, H.; Cofield, R.H. Cost Analysis of Revision Total Hip Arthroplasty. A 5-Year Followup Study. *Clin. Orthop. Relat. Res.* **1999**, *369*, 175–178. [CrossRef]
10. Singh, M.; Nagrath, A.R.; Maini, P.S. Changes in Trabecular Pattern of the Upper End of the Femur as an Index of Osteoporosis. *J. Bone Jt. Surg. Am.* **1970**, *52*, 457–467. [CrossRef]
11. Wachter, N.J.; Augat, P.; Hoellen, I.P.; Krischak, G.D.; Sarkar, M.R.; Mentzel, M.; Kinzl, L.; Claes, L. Predictive Value of Singh Index and Bone Mineral Density Measured by Quantitative Computed Tomography in Determining the Local Cancellous Bone Quality of the Proximal Femur. *Clin. Biomech.* **2001**, *16*, 257–262. [CrossRef]

12. D'Amelio, P.; Rossi, P.; Isaia, G.; Lollino, N.; Castoldi, F.; Girardo, M.; Dettoni, F.; Sattin, F.; Delise, M.; Bignardi, C. Bone Mineral Density and Singh Index Predict Bone Mechanical Properties of Human Femur. *Connect. Tissue Res.* **2008**, *49*, 99–104. [CrossRef]
13. Njeh, C.F.; Kuo, C.W.; Langton, C.M.; Atrah, H.I.; Boivin, C.M. Prediction of Human Femoral Bone Strength Using Ultrasound Velocity and BMD: An In Vitro Study. *Osteoporos. Int.* **1997**, *7*, 471–477. [CrossRef]
14. Endo, K.; Takahata, M.; Sugimori, H.; Yamada, S.; Tadano, S.; Wang, J.; Todoh, M.; Ito, Y.M.; Takahashi, D.; Kudo, K.; et al. Magnetic Resonance Imaging T1 and T2 Mapping Provide Complementary Information on the Bone Mineral Density Regarding Cancellous Bone Strength in the Femoral Head of Postmenopausal Women with Osteoarthritis. *Clin. Biomech.* **2019**, *65*, 13–18. [CrossRef] [PubMed]
15. Gasbarra, E.; Iundusi, R.; Perrone, F.L.; Saturnino, L.; Tarantino, U. Densitometric Evaluation of Bone Remodelling around Trabecular Metal Primary Stem: A 24-Month Follow-Up. *Aging Clin. Exp. Res.* **2015**, *27*, 69–75. [CrossRef] [PubMed]
16. Porrelli, D.; Abrami, M.; Pelizzo, P.; Formentin, C.; Ratti, C.; Turco, G.; Grassi, M.; Canton, G.; Grassi, G.; Murena, L. Trabecular Bone Porosity and Pore Size Distribution in Osteoporotic Patients—A Low Field Nuclear Magnetic Resonance and Microcomputed Tomography Investigation. *J. Mech. Behav. Biomed. Mater.* **2022**, *125*, 104933. [CrossRef] [PubMed]
17. Kanakaris, N.K.; Lasanianos, N.G. Singh Index for Osteoporosis. In *Trauma and Orthopaedic Classifications: A Comprehensive Overview*; Springer-Verlag London Ltd.: London, UK, 2015; pp. 405–407. ISBN 9781447165729.
18. Vale, A.C.; Aleixo, I.P.; Lúcio, M.; Saraiva, A.; Caetano-Lopes, J.; Rodrigues, A.; Amaral, P.M.; Rosa, L.G.; Monteiro, J.; Fonseca, J.E.; et al. At the Moment of Occurrence of a Fragility Hip Fracture, Men Have Higher Mechanical Properties Values in Comparison with Women. *BMC Musculoskelet. Disord.* **2013**, *14*, 295. [CrossRef] [PubMed]
19. Marmor, M.; Knox, R.; Huang, A.; Herfat, S. Acetabulum Cup Stability in an Early Weight-Bearing Cadaveric Model of Geriatric Posterior Wall Fractures. *J. Orthop. Trauma* **2020**, *34*, 55–61. [CrossRef]
20. Jenkins, T.; Katsamenis, O.L.; Andriotis, O.G.; Coutts, L.v.; Carter, B.; Dunlop, D.G.; Oreffo, R.O.C.; Cooper, C.; Harvey, N.C.; Thurner, P.J.; et al. The Inferomedial Femoral Neck Is Compromised by Age but Not Disease: Fracture Toughness and the Multifactorial Mechanisms Comprising Reference Point Microindentation. *J. Mech. Behav. Biomed. Mater.* **2017**, *75*, 399–412. [CrossRef]
21. Gluek, C.; Zdero, R.; Quenneville, C.E. Evaluating the Mechanical Response of Novel Synthetic Femurs for Representing Osteoporotic Bone. *J. Biomech.* **2020**, *111*, 110018. [CrossRef]
22. Prendergast, P.J. Review Paper Finite Element Models in Tissue Mechanics and Orthopaedic Implant Design. *Clin. Biomech.* **1997**, *12*, 343–366. [CrossRef]
23. Rieger, R.; Auregan, J.C.; Hoc, T. Micro-Finite-Element Method to Assess Elastic Properties of Trabecular Bone at Micro- and Macroscopic Level. *Morphologie* **2018**, *102*, 12–20. [CrossRef]
24. He, Z.; Chu, L.; Liu, X.; Han, X.; Zhang, K.; Yan, M.; Li, X.; Yu, Z. Differences in Subchondral Trabecular Bone Microstructure and Finite Element Analysis-Based Biomechanical Properties between Osteoporosis and Osteoarthritis. *J. Orthop. Translat.* **2020**, *24*, 39–45. [CrossRef]
25. Boelch, S.P.; Jordan, M.C.; Meffert, R.H.; Jansen, H. Comparison of Open Reduction and Internal Fixation and Primary Total Hip Replacement for Osteoporotic Acetabular Fractures: A Retrospective Clinical Study. *Int. Orthop.* **2017**, *41*, 1831–1837. [CrossRef] [PubMed]
26. Santori, N.; Falez, F.; Potestio, D.; Santori, F.S. Fourteen-Year Experience with Short Cemented Stems in Total Hip Replacement. *Int. Orthop.* **2019**, *43*, 55–61. [CrossRef] [PubMed]
27. Zhen, P.; Chang, Y.; Yue, H.; Chen, H.; Zhou, S.; Liu, J.; He, X. Primary Total Hip Arthroplasty Using a Short Bone-Conserving Stem in Young Adult Osteoporotic Patients with Dorr Type C Femoral Bone. *J. Orthop. Surg. Res.* **2021**, *16*, 17. [CrossRef] [PubMed]
28. Zhang, Z.; Xu, G.; Cao, L.; Sun, W.; Zeng, X.; Xiong, N.; Wang, S.; Yu, W.; Liu, Q.; Lin, H. Dual-Mobility Cup Total Hip Arthroplasty for Displaced Femoral Neck Fractures: A Retrospective Study With a Median Follow-Up of 5 Years. *Geriatr. Orthop. Surg. Rehabil.* **2021**, *12*, 1–7. [CrossRef]
29. Goriainov, V.; Cook, R.B.; Murray, J.W.; Walker, J.C.; Dunlop, D.G.; Clare, A.T.; Oreffo, R.O.C. Human Skeletal Stem Cell Response to Multiscale Topography Induced by Large Area Electron Beam Irradiation Surface Treatment. *Front. Bioeng. Biotechnol.* **2018**, *6*, 91. [CrossRef] [PubMed]
30. Rafiq, M.; Kadir, A.; Kamsah, N. The Effect of Bone Properties Due to Skeletal Diseases on Stability of Cementless Hip Stems. *Am. J. Appl. Sci.* **1988**, *6*, 1988–1994.
31. Peter, B.; Gauthier, O.; Laı¨b, S.L.; Bujoli, B.; Me Guicheux, J.; Janvier, P.; Harry Van Lenthe, G.; Mü, R.; Zambelli, P.-Y.; Bouler, J.-M.; et al. Local Delivery of Bisphosphonate from Coated Orthopedic Implants Increases Implants Mechanical Stability in Osteoporotic Rats. *J. Biomed. Mater. Res. A* **2005**, *76*, 133–143. [CrossRef]
32. Gao, Y.; Zou, S.; Liu, X.; Bao, C.; Hu, J. The Effect of Surface Immobilized Bisphosphonates on the Fixation of Hydroxyapatite-Coated Titanium Implants in Ovariectomized Rats. *Biomaterials* **2009**, *30*, 1790–1796. [CrossRef] [PubMed]
33. Eberhardt, C.; Stumpf, U.C.; Kurth, A.H.A. Simulated Osteopenia Impairs Metaphyseal Bone Ingrowth of Metal Implants in an Animal Model. *Eur. J. Trauma* **2005**, *31*, 51–56. [CrossRef]
34. Sun, P.; Wang, Y.; Xu, D.; Gong, K. The Calcium Phosphate Modified Titanium Implant Combined With Platelet-Rich Plasma Treatment Promotes Implant Stabilization in an Osteoporotic Model. *J. Craniofacial Surg.* **2021**, *32*, 603–608. [CrossRef] [PubMed]

35. Lewis, G.; Janna, S. Alendronate in Bone Cement: Fatigue Life Degraded by Liquid, Not by Powder. *Clin. Orthop. Relat. Res.* **2006**, *445*, 233–238. [CrossRef] [PubMed]
36. Brewster, N.T.; Gillespie, W.J.; Howie, C.R.; G Madabhushi, S.P.; Usmani, A.S.; Fairbairn, D.R. Mechanical Considerations in Impaction Bone Grafting. *J. Bone Jt. Surg. Br.* **1999**, *81*, 118–124. [CrossRef]
37. Li, Z.; Zhao, Y.; Wang, Z.; Ren, M.; Wang, X.; Liu, H.; Lin, Q.; Wang, J. Engineering Multifunctional Hydrogel-Integrated 3D Printed Bioactive Prosthetic Interfaces for Osteoporotic Osseointegration. *Adv. Healthc. Mater.* **2022**, *11*, 2102535. [CrossRef] [PubMed]
38. Wang, X.; Li, Z.; Wang, Z.; Liu, H.; Cui, Y.; Liu, Y.; Ren, M.; Zhan, H.; Li, Z.; Wu, M.; et al. Incorporation of Bone Morphogenetic Protein-2 and Osteoprotegerin in 3D-Printed Ti6Al4V Scaffolds Enhances Osseointegration Under Osteoporotic Conditions. *Front. Bioeng. Biotechnol.* **2021**, *9*, 754205. [CrossRef] [PubMed]
39. Cui, Y.; Wang, Z.; Li, Z.; Ji, X.; Yuan, B.; Sun, Y.; Peng, C.; Leng, Y.; Dou, M.; Wang, J.; et al. Functionalized Anti-Osteoporosis Drug Delivery System Enhances Osseointegration of an Inorganic–Organic Bioactive Interface in Osteoporotic Microenvironment. *Mater. Des.* **2021**, *206*, 109753. [CrossRef]
40. Bai, H.; Zhao, Y.; Wang, C.; Wang, Z.; Wang, J.; Liu, H.; Feng, Y.; Lin, Q.; Li, Z.; Liu, H. Enhanced Osseointegration of Three-Dimensional Supramolecular Bioactive Interface through Osteoporotic Microenvironment Regulation. *Theranostics* **2020**, *10*, 4779–4794. [CrossRef]
41. von Knoch, M.; Wedemeyer, C.; Pingsmann, A.; von Knoch, F.; Hilken, G.; Sprecher, C.; Henschke, F.; Barden, B.; Löer, F. The Decrease of Particle-Induced Osteolysis after a Single Dose of Bisphosphonate. *Biomaterials* **2005**, *26*, 1803–1808. [CrossRef]
42. von Knoch, F.; Eckhardt, C.; Alabre, C.I.; Schneider, E.; Rubash, H.E.; Shanbhag, A.S. Anabolic Effects of Bisphosphonates on Peri-Implant Bone Stock. *Biomaterials* **2007**, *28*, 3549–3559. [CrossRef]
43. Migliorati, C.A. Bisphosphanates and Oral Cavity Avascular Bone Necrosis. *J. Clin. Oncol.* **2003**, *21*, 4253–4254. [CrossRef]
44. Hooten, J.P., Jr.; Engh, C.A.; Heekin, R.D.; Vinh, T.N. Structural Bulk Allografts in Acetabular Reconstruction. Analysis of Two Grafts Retrieved at Post-Mortem. *J. Bone Jt. Surg. Br.* **1996**, *78*, 270–275. [CrossRef]

bioengineering

MDPI

Article

Mn-Based Methacrylated Gellan Gum Hydrogels for MRI-Guided Cell Delivery and Imaging

Sílvia Vieira [1,2], Paulina Strymecka [3], Luiza Stanaszek [3], Joana Silva-Correia [1,2], Katarzyna Drela [3], Michał Fiedorowicz [4], Izabela Malysz-Cymborska [5], Miroslaw Janowski [3,6], Rui Luís Reis [1,2], Barbara Łukomska [3], Piotr Walczak [5,6] and Joaquim Miguel Oliveira [1,2,*]

1 3B's Research Group, I3Bs—Research Institute on Biomaterials, Biodegradables and Biomimetics, University of Minho, Headquarters of the European Institute of Excellence on Tissue Engineering and Regenerative Medicine, AvePark—Parque de Ciência e Tecnologia, Zona Industrial da Gandra, Barco, 4805-017 Guimarães, Portugal
2 ICVS/3B's–PT Government Associate Laboratory, 4806-909 Braga/Guimarães, Portugal
3 NeuroRepair Department, Mossakowski Medical Research Centre, Polish Academy of Sciences, 02-106 Warsaw, Poland
4 Small Animal Magnetic Resonance Imaging Laboratory, Mossakowski Medical Research Centre, Polish Academy of Sciences, 02-106 Warsaw, Poland
5 Department of Neurology and Neurosurgery, School of Medicine, Collegium Medicum, University of Warmia and Mazury, 10-082 Olsztyn, Poland
6 Center for Advanced Imaging Research, Department of Diagnostic Radiology and Nuclear Medicine, University of Maryland Marlene and Stewart Greenebaum Comprehensive Cancer Center, University of Maryland, Baltimore, MD 21201, USA
* Correspondence: miguel.oliveira@i3bs.uminho.pt; Tel.: +351-253510931; Fax: +351-253510909

Abstract: This work aims to engineer a new stable injectable Mn-based methacrylated gellan gum (Mn/GG-MA) hydrogel for real-time monitored cell delivery into the central nervous system. To enable the hydrogel visualization under Magnetic Resonance Imaging (MRI), GG-MA solutions were supplemented with paramagnetic Mn^{2+} ions before its ionic crosslink with artificial cerebrospinal fluid (aCSF). The resulting formulations were stable, detectable by T1-weighted MRI scans and also injectable. Cell-laden hydrogels were prepared using the Mn/GG-MA formulations, extruded into aCSF for crosslink, and after 7 days of culture, the encapsulated human adipose-derived stem cells remained viable, as assessed by Live/Dead assay. In vivo tests, using double mutant $MBP^{shi/shi}$/rag2 immunocompromised mice, showed that the injection of Mn/GG-MA solutions resulted in a continuous and traceable hydrogel, visible on MRI scans. Summing up, the developed formulations are suitable for both non-invasive cell delivery techniques and image-guided neurointerventions, paving the way for new therapeutic procedures.

Keywords: gellan gum; cell delivery; manganese; magnetic resonance imaging; injectable hydrogels

Citation: Vieira, S.; Strymecka, P.; Stanaszek, L.; Silva-Correia, J.; Drela, K.; Fiedorowicz, M.; Malysz-Cymborska, I.; Janowski, M.; Reis, R.L.; Łukomska, B.; et al. Mn-Based Methacrylated Gellan Gum Hydrogels for MRI-Guided Cell Delivery and Imaging. *Bioengineering* **2023**, *10*, 427. https://doi.org/10.3390/bioengineering10040427

Academic Editor: Ali Zarrabi

Received: 29 January 2023
Revised: 20 March 2023
Accepted: 23 March 2023
Published: 28 March 2023

1. Introduction

Cell-based therapies hold great potential for the treatment of central nervous system (CNS) diseases. Indeed, the rationale for transplanting stem cells has been already established in small [1] and large [2] animal models.

In this regard, the CNS can be accessed using different approaches, resulting in cell deposition in different areas. Intraparenchymal cell delivery is performed via direct needle injections into the parenchyma, resulting in a spatially precise cell deposition. This route is currently considered the method of choice for a large number of studies aiming for direct cell delivery into the CNS. In this regard, the safety of intraparenchymal cell injection was already confirmed in studies using large animals [3], and also in phase I/II clinical trials [4], for direct cell transplantation into the spinal cord and brain. Delivery of therapeutic agents directly near the vicinities of a stroke poses a great advantage of intraparenchymal cell transplantation [5,6]. However, the hostile environment near the stroke region hampers successful cell transplantation with effective therapeutic effect [7]. To overcome such limitation, cell-laden hydrogels capable of being injected directly into the CNS have been successfully used, showing their potential to promote a pro-survival environment, even near the stroke region [8]. In addition, while significant progress has been made in the CNS field, disseminated or multifocal diseases, where a broad cell distribution is required, pose a particular challenge for cell delivery [9]. Examples of such multifocal disorders include multiple sclerosis or amyotrophic lateral sclerosis (ALS). For the therapy to be effective, cell delivery must be carried to extensive areas of the CNS. Although promising, intraparenchymal delivery is linked with a higher risk when multifocal cell depositions are needed, making room for new, less invasive gateways to the spinal cord [10]. Intrathecal injection is an alternative route that allows a more widespread delivery of cells in the CNS, and it is therefore considered as an attractive access for spinal cord targeting [11]. With this minimally invasive method, it is possible to dispense cell suspensions, or drugs, directly into the cerebrospinal fluid (CSF), leading to a wide distribution within the CNS. Although intrathecal injection is a routine clinical procedure with a low rate of complications, and numerous stem cell delivery studies (both pre-clinical and clinical) strengthened the feasibility of this route [12–14], there are still some bottlenecks that hamper the development of this technique. The cells injected into the CSF as a suspension in a buffer are subject to gravitational sedimentation, with their accumulation in cauda equina or removal within circulating CSF [11,15]. One way to avoid these undesirable effects is to embed the transplanted cells within a biomaterial matrix.

In this regard, injectable hydrogel formulations are particularly promising, mainly due to their (i) soft and pliable nature; (ii) easy transport of nutrients and metabolites; (iii) tissue-like fluidity; (iv) ease of fabrication; and (v) appropriate bio-adhesiveness and integration with biological interfaces [16]. After being injected into the CNS, the hydrogel physically supports the encapsulated cells, protects the biological material from a hostile environment, and it can also assure a widespread distribution along the spinal cord when injected in the intrathecal space [17].

Another challenge of CNS direct cell delivery is the correct hydrogel placement and allocation. Particularly, for intraparenchymal injections, it is hard to predict the hydrogel placement, as well as a possible dilution in the CSF, because the injection is performed into a fluid compartment. Therefore, the possibility to image and monitor the injection procedure in a real-time and non-invasive way adds significant value to the whole therapeutic approach. Such control can allow the establishment of new procedures that avoid excessive injection, or biomaterial misplacement, assuring its correct positioning. Tomographic techniques, and particularly Magnetic Resonance Imaging (MRI), are the most appropriate for image deep structures, such as the intrathecal space or the parenchyma. Using MRI, it is possible to obtain information about the biomaterial/tissue interface as well as biomaterial placement and biodistribution in both pre-clinical and clinical studies. Another advantage is the non-invasive and radiation-free imaging, thus making MRI one of the most appealing imaging modalities [18].

Visualization of the hydrogel using MRI typically requires hydrogel labelling with a contrast agent. Most frequently used contrasts include iron oxide, gadolinium (Gd), or fluorine in a nanoparticulate form. Although not so popular as other contrast agents, Mn^{2+} ions are increasingly studied and used as a positive contrast [19]. The MRI signal is obtained due to the paramagnetic nature of Mn^{2+}, yielding a high contrast on T1-weighted MRI images [20,21]. Manganese presents contrast properties similar to those of the most commonly used Gd. However, its undeniable advantage relies on the fact that it does not accumulate for a long-term period and might be incorporated into the organism or expelled, as it represents one of the necessary microelements [22]. Manganese-based contrasts can not only be used for anatomy analysis but also functional studies [23,24]. Moreover, manganese can be delivered both by systemic and localized routes of delivery [25]. We have previously successfully used Mn^{2+} as a contrast in alginate-based hydrogels. It needs to be stressed though that each hydrogel must be carefully investigated both in terms of its properties and the potential release of manganese ions [26]. Therefore, the concentration of Mn^{2+} must be carefully optimized due to the possible Mn-induced toxicity that might occur when high or repetitive doses of Mn^{2+} are delivered [27,28].

Regardless of the method used to reach the CNS, intraparenchymal or intrathecal, several conditions need to be met to successfully use these methods for delivery of hydrogel-embedded stem cells. The hydrogel needs to withstand the shear forces caused by the injection, but also rapidly cross-link upon reaching the CNS to avoid cell and biomaterial escape to unwanted regions. In this regard, gellan gum (GG) presents great potential as an injectable material [29]. Indeed, this natural polymer has been successfully used for several tissue engineering applications, including intervertebral disc regeneration [30], nanoparticles coating [31], bioprinting of brain-like tissues [32], among others [33]. In its native form, GG is thermo- and ionic-responsive, and therefore fast gelation can occur in situ due to body temperature or the presence of metallic ions in body fluids. Nevertheless, GG needs to be heated up to approximately 50 °C to be water-soluble, making cell encapsulation and delivery a challenging process. Although cells could be mixed with GG near-physiological temperatures, the time required for injection could lead to a decrease in temperature and consequent needle clot, as a result of hydrogel cross-link before injection.

On the other hand, the methacrylation of low acyl GG improves the solubility of this polymer, making it water-soluble at room temperature. Thus, methacrylated GG (GG-MA) can be easily used for cell encapsulation procedures. In addition, GG-MA remains responsive to ionic strength and rapidly cross-links in the presence of metallic ions. Bearing in mind that the CSF is rich in Na^+, Cl^-, HCO_3^-, K^+, Mg^{2+}, and Ca^{2+} [34], it is highly expected that in situ ionic cross-link occurs upon GG-MA injection. More than triggering the desired in situ cross-linking of the hydrogel, the ionic interaction between GG and metallic ions can be used to incorporate Mn^{2+} within the hydrogel matrix for further MRI tracking. Indeed, it has been shown that Mn^{2+} strongly interacts with GG, specifically at the carboxyl group of the D-glucuronate unit [35,36].

Previous work showed that degradable hydrogel blends formed by GG-MA and hyaluronic acid can be combined with Mn^{2+} to support image-guided intrathecal cell delivery [37]. In turn, this work aims to engineer injectable hydrogels with higher stability that can hold the delivered therapeutic agents for longer periods at the injection site. A slower degradation will also prevent a possible deposition of the material at cauda equina. To that end, GG-MA hydrogels were used in combination with Mn^{2+} ions to prepare traceable and stable hydrogels.

2. Materials and Methods

2.1. Preparation of Methacrylated Gellan Gum (GG-MA), $MnCl_2$, and aCSF Solutions

Methacrylated gellan gum (GG-MA) was obtained as previously described by Silva-Correia et al. [30]. Briefly, a solution of low-acyl gellan gum (Gelzan™ CM Gelrite®, Sigma-Aldrich) reacted with glycidyl methacrylate (GMA, 97%, Sigma-Aldrich) overnight at room temperature, with constant control of pH at 8.5 and under vigorous stirring. The reaction products were precipitated by the addition of cold acetone and further purified by dialysis (cellulose membrane, molecular weight cut-off 12 kDa, Sigma-Aldrich) against distilled water, for one week. Then, the obtained GG-MA was frozen at −80 °C, freeze-dried, and the final dry material was stored protected from light, in a dry place, until further use. GG-MA solutions of desired concentrations were prepared by dissolving the dry material in Milli-Q water using gentle agitation. Manganese (II) chloride powder ($MnCl_2$ powder, Sigma-Aldrich) was used to prepare the $MnCl_2$ aqueous solutions used as a supplement for GG-MA hydrogels. Artificial cerebrospinal fluid (aCSF) was prepared following the composition listed in Table 1, and final pH adjustment to 7.3, with NaOH. For in vitro and in vivo assays, dry GG-MA was sterilized by UV light for 30 min in a laminar flow hood. All other materials and solutions were sterilized by filtration (0.22 μm filter).

Table 1. Composition of artificial cerebrospinal fluid (aCSF) solution.

	Concentration (mM)
NaCl	125
KCl	2.5
$MgCl_2{\cdot}6H_2O$	1
NaH_2PO_4	1.25
$CaCl_2{\cdot}2H_2O$	2
$NaHCO_3$	25
Glucose	25

2.2. Preparation of Mn-Based GG-MA Hydrogels

Hydrogel solutions were prepared by mixing 1% (*w*/*v*) GG-MA with $MnCl_2$ solutions to obtain hydrogels with 0.1 or 1 mM of Mn^{2+} and 0.75% (*w*/*v*) GG-MA. A solution of 0.75% (*w*/*v*) of GG-MA was used as control. To better mimic in vivo conditions, further cross-link was obtained by hydrogel interaction with aCSF. For that, Mn-based GG-MA (Mn/GG-MA) hydrogels were poured into cylindrical silicon templates (diameter = 8 mm; height = 2 mm), and cross-linked with an excess of aCSF, added dropwise, for at least 5 min.

2.3. Rheological Studies

The rheologic properties of the developed hydrogels were obtained using a Kinexus Pro+ rheometer (Malvern Instruments, UK), with the acquisition software rSpace. Oscillatory tests were performed using stainless steel (316 grade) parallel plates, with an upper measurement geometry plate with 8 mm of diameter, and a 20 mm lower pedestal with roughened finish (to prevent sample slippage and resulting errors on the experiments). Pre-gel solutions of 0.75% (*w*/*v*) GG-MA, and 0.75% (*w*/*v*) GG-MA supplemented with 0.1 mM or 1 mM $MnCl_2$ were poured in the lower pedestal. The same formulations were also tested after cross-linking with aCSF and for that purpose, hydrogels with 2 mm height and 8 mm of diameter were prepared beforehand with the silicon templates, as mentioned above. For the oscillatory tests, the Linear Viscoelastic Region was previously determined, and then single-frequency oscillation experiments were performed at 0.1 Hz for 30 min. Shear viscosity was determined by rotational experiments, using an upper measurement geometry cone (40 mm diameter and 4° angle). All experiments were performed at 37 °C, and plots are the average of at least 3 experiments.

2.4. Injection Ability Test

The possibility to extrude the Mn/GG-MA pre-gel solutions from a 31 G needle was investigated using in-house injection equipment (Paralab). The measurements were performed with a 10 µL Hamilton syringe (Gastight Syringe Model 1701 RN) coupled with a 31 G needle. The syringe was filled with the Mn/GG-MA solution, as well as aCSF (control). All the materials were extruded using a defined extrusion rate (10 µL/min), and the force used to comply with it was measured and recorded using an appropriate software.

2.5. Permeability Studies

Permeability studies were performed using 70 kDa fluorescein isothiocyanate–dextran molecules (Dextran-FITC, Sigma-Aldrich). Dextran-FITC was dissolved in Milli-Q water and added to the Mn/GG-MA hydrogel solutions at a final concentration of 125 µg/mL. As before, hydrogels were placed inside 8 mm discs and further cross-linked with aCSF to mimic the delivery of the hydrogels into the intrathecal space. The resulting gels were incubated at 37 °C in aCSF, with mechanical shaking. At different timepoints, 350 µL of supernatant was retrieved and replaced by the same amount of fresh aCSF. At the last timepoint, hydrogels were mechanically destroyed, centrifuged, and the resulting supernatant was used to calculate the concentration of FITC-labelled molecules that were retained inside the hydrogels, and also to calculate the total amount of dextran-FITC that was initially encapsulated. The fluorescence emission of FITC-labelled dextran was measured in a microplate reader (Gen 5 2.01, Synergy HT, BioTek) using an excitation wavelength of 485/20 nm and an emission wavelength of 528/20 nm. The concentration of dextran-FITC released from the hydrogels was finally calculated using a calibration curve, obtained by the measurement of the fluorescence emission of dextran-FITC solutions of known concentrations.

2.6. Degradation Profile

The weight loss profile of the Mn-based GG-MA hydrogels upon incubation in aCSF was used to evaluate their degradation along time. For this, the different hydrogel formulations, previously cross-linked with aCSF, were weighed (initial weight, m_i) and incubated in aCSF at 37 °C. All solutions were supplemented with 0.2% (w/v) sodium azide (Sigma-Aldrich) to avoid bacterial contamination. After 1, 3, 24, 72, and 168 h, the hydrogels were retrieved, the liquid excess was gently removed, and the final mass of samples was determined (m_f). Equation (1) was applied to calculate the weight loss ratio at each time point.

$$Degradation\ (\%) = \left(\frac{m_i - m_f}{m_i}\right) \times 100. \quad (1)$$

2.7. Manganese Release Profile—Inductively Coupled Plasma-Optical Emission Spectroscopy (ICP)

The Mn^{2+} release profile from the Mn-based GG-MA hydrogels was quantified using inductively coupled plasma-optical emission spectroscopy (ICP; JY2000 2, Jobin Yvon, Horiba). Hydrogel cylinders, with 8 mm diameter and 2 mm height, were incubated in aCSF at 37 °C with mechanical shaking. At defined timepoints (0.5, 1, 5, 24, and 48 h), the aCSF supernatant was collected, dissolved in nitric acid and injected into the ICP equipment. Manganese (λ_{em} = 259.37 nm) concentrations in the aCSF solutions were obtained by comparison with standard solutions, with a detection limit of 5 ppb.

2.8. Human Derived Adipose Stem Cells (hASCs) Isolation and Culture

Lipoaspirate samples were collected from abdominal regions of healthy human male and female donors with ages between 18 and 57 years, after informed consent, under established cooperative agreements between the Hospital da Senhora da Oliveira (Guimarães, Portugal) and the 3B's Research Group—University of Minho. After collection, samples were digested with collagenase NB 6 GMP Grade (Serva), and centrifuged to remove all liquids (50× g, 5 min). Then, a new collagenase solution was added to the concentrated adipose tissue at a final concentration of 0.2 U/mL, and incubated for 1 h at 37 °C with shaking. Afterwards, the tissue was washed twice with PBS and centrifuged for 10 min at 250× g. Samples were rewashed with PBS, and the cellular pellet was then ready for use. The obtained cell pellet was mixed with MSC growth medium (MSCGM BulletKit, Lonza) and transferred into cell culture dishes. After 48 h, the fibroblast-like cells were selected from the rest of the floating debris, rinsed with PBS, and cultured for the next 7 days or until near confluence. The cells were also immunostained for the MSC typical cell surface markers: CD73, CD90, CD105 (data not shown). ADSC presented positive signal for typical MSC markers as well as negative signal for hematopoietic markers according to ISCT guidelines.

2.9. Cell Encapsulation

Human derived adipose stem cells (hASCs) were grown as monolayers as previously described. At passage 3–4, confluent cells were detached from tissue culture flasks using Trypsin (Gibco®, Life Technologies, New York, NY, USA), and a cell pellet was formed after centrifugation at 1200 rpm for 5 min. The obtained cell pellet was gently resuspended in Mn-based GG-MA pre-gel solution with 0.75% (w/v) GG-MA and 0.1 M $MnCl_2$, to a final cellular density of 1×10^6 cells/mL. Solutions of 0.75% (w/v) GG-MA without $MnCl_2$ were also prepared and used as controls. Then, a 10 μL Hamilton syringe (Gastight Syringe

Model 1701 RN) coupled with a 31 G needle (length = 35 mm) was filled with the cell-leaden solutions, mounted on a stereotaxic syringe pump, and the pre-gel solutions were injected into aCSF using a 10 μL/min extrusion rate. The resulting cell-laden fibres remained in the aCSF for 5 min, after which the aCSF was replaced by 500 μL of cell culture media. The hydrogels were cultured for 1, and 7 days at 37 °C, in a humidified air atmosphere of 5% CO_2.

2.10. Live/Dead Staining

Live/Dead fluorescence assays were performed after 1 and 7 days of culture. LIVE/DEAD Viability/Cytotoxicity Kit for mammalian cells (ThermoFisher Scientific, Rockford, USA) was used to perform the Live/Dead assay, where Calcein-AM-stained live cells and ethidium homodimer-1 (EthD-1) stain dead cells. At each timepoint, the culture medium was removed, 2 μM of Calcein and 4 μM EthD-1 diluted in PBS were added to each well, and samples were then incubated for 20 min at room temperature (RT), protected from light. After the incubation, cells were visualized in the dark using a fluorescence microscope Cell Observer SD (Carl Zeiss, Jena, Germany) in Z-stack mode. Image acquisition was performed at the Laboratory of Advanced Microscopy Techniques, Mossakowski Medical Research Centre, Polish Academy of Sciences. Cell viability was calculated from the ratio between live and dead cells in each fibre, using the ImageJ software (version 2.0.0-rc-69/1.52p).

2.11. Animal Surgeries

All the procedures were performed with the approval of the Ethical Committee (IV Local Committee in Warsaw, 117/2015). Intrathecal injections of Mn/GG-MA hydrogel were performed in double mutant $MBP^{shi/shi}$/rag2 (n = 3) immunocompromised mice. In this regard, animals were anesthetized with 1.5–2% isoflurane in oxygen, and placed in the stereotaxic frame in a concord-like position [38]. A small incision (7–9 mm) was made in the midline at the posterior aspect of the skull, to expose the atlanto-occipital membrane. Then, 10 μL of Mn/GG-MA hydrogel, prepared as described above using 0.1 mM of $MnCl_2$, was injected into the intrathecal or intracerebral space with a speed of 10 μL/min using a Hamilton syringe coupled with a 31G needle. After injection, the needle was left in the same place for an additional 1 min and then slowly withdrawn. Afterwards, the skin was sutured, and the animal was placed in the MRI scanner.

2.12. Phantom Magnetic Resonance Imaging

Tripilot scan was followed by T1 parametric imaging with the 2D Saturation Recovery Spin Echo Sequence with varying repetition times (TR = 200 ms–8000 ms, TE = 9.5 ms, rare factor = 2, NA = 1, FOV = 75 mm × 75 mm, 5 slices 2.0 mm-thick with 2.0 mm gaps, spatial resolution = 586 μm × 586 μm, scan time~13 min) and T2 parametric imaging with the MSME sequence (TR = 5000 ms, TE = 15 ms–480 ms, NA = 1, FOV = 75 mm × 75 mm, 5 slices 2.0 mm-thick with 2.0 mm gaps, spatial resolution = 586 μm × 586 μm, scan time~8 min).

The relative magnetic resonance signal of T1-weighted images of GG-MA phantoms with different concentrations of manganese ions was analysed using the ImageJ software (version 2.0.0-rc-69/1.52p). Identical regions of interest (ROI) were outlined on MR images and the signal intensity was then measured.

2.13. In Vivo Magnetic Resonance Imaging

MR imaging was performed immediately after intrathecal and intracerebral (intraparenchymal) injection of the hydrogel, and 24 h after surgery. For the imaging, animals were anesthetized with isoflurane (1.5–2% in oxygen) and positioned head prone in an MRI-compatible water-heated bed. Body temperature and respiration rate were monitored throughout the study with MRI-compatible probes (SA Instruments, Stony Brook, NY, USA). A 7T MR scanner (BioSpec 70/30 USR, Bruker, Ettlingen, Germany) equipped with a transmit cylindrical radiofrequency coil (8.6 cm inner diameter, Bruker) and a mouse brain dedicated receive-only array surface coil (2 × 2 elements, Bruker) was used, and the structural imaging protocol was performed as previously described [39]. Briefly, a T1-weighted 3D FLASH sequence (TR = 12 ms; TE = 4 ms; flip angle, FA = 18; NA = 10; field of view, FOV = 15 mm × 15 mm × 15 mm, spatial resolution = 117 μm isotropic, scan time~25 min) was implemented. Structural imaging was followed by T1 parametric imaging with the 2D Saturation Recovery Spin Echo Sequence with varying repetition times (TR = 410 ms–8000 ms, TE = 22 ms, rare factor = 4, NA = 3, FOV = 20 mm × 20 mm, 8 slices 0.8 mm thick with no gaps, spatial resolution = 156 μm × 156 μm, scan time ~ 23 min) and T2 parametric imaging with the MSME sequence (TR = 5000 ms, TE = 13 ms–416 ms, NA = 1, FOV = 20 mm × 20 mm, 8 slices 0.8 mm-thick with no gaps, spatial resolution = 156 μm × 156 μm, scan time~8 min).

2.14. *Statistical Analysis*

Results are presented as mean ± standard deviation, when appropriate. When applicable, the experimental data were analysed using a single-factor analysis of variance (one-way ANOVA) to assess the statistical significance of the results, followed by post hoc Tukey tests. Statistical significance was set at a p-value of ≤ 0.05. All statistical analysis was performed using GraphPad Prism version 7.0a.

3. Results and Discussion

3.1. *Mn-Based GG-MA Hydrogels*

Mn-based GG-MA hydrogels (Mn/GG-MA) were prepared using methacrylated gellan gum (GG-MA) as a hydrogel matrix (Figure 1A). Solutions of GG-MA were mixed with $MnCl_2$ solutions to obtain the desired final concentration of $MnCl_2$ (0, 0.1, and 1 mM), in a final concentration of 0.7% (w/w) in GG-MA. All formulations were studied by rheological techniques to assess their mechanical properties. Time-sweep curves, plotted in Figure 1B, were performed along 30 min to study a possible gelation effect over time. Particularly, the evolution of the storage modulus (G′), loss modulus (G″), and phase angle (δ) upon the addition of ionic solutions was considered. While G′ is regarded as the stiffness of the material, G″ represents the liquid-like behaviour of the hydrogel. Therefore, these parameters allow the determination of the elastic (G′) and viscous (G″) character of the tested hydrogels, and the ratio between these parcels (δ) providing a quantitative measure of the material mechanical properties [40].

Without the addition of $MnCl_2$, 0.75% (w/v) GG-MA hydrogels showed a very low G′ value of 0.65 ± 0.18 Pa, which was similar to the obtained G″ value (0.35 ± 0.09 Pa). Considering the higher G′ value over G″, one can assume this formulation as viscoelastic solid, as confirmed by the phase angle value of 28.51 ± 4.58 ° that is between the purely elastic (δ = 0°) and purely viscous (δ = 90°) values [41,42] (Supplementary Information, Figure S1). The addition of 0.1 mM $MnCl_2$ leads to a modest increase in G′ value to 0.80 ± 0.23 Pa, with G″ following the same trend (0.39 ± 0.10 Pa). In turn, GG-MA hydrogels with 1 mM $MnCl_2$ showed a considerably high G′ and G″ values after the 30 min sweep due to the electrostatic interaction of the Mn^{2+} ions with the carboxylic groups of GG-MA. Interestingly, both parameters increased over time, showing a time-dependent interaction between GG-MA polymeric networks and the Mn^{2+} ions. After the time sweep, a G′ and G″ value of 34.54 ± 18.78 Pa and 5.34 ± 1.71 Pa was registered, respectively. As expected, the measured phase angle was lower for hydrogels with 1 mM, 9.69 ± 1.81° (Supplementary Information, Figure S1A), showing that the lag time between strain and shear stress decreases upon addition of $MnCl_2$. Hence, one can assume that this formulation has already a "gel-like" character, although with insufficient mechanical stability.

Although the presence of 1 mM $MnCl_2$ seems to have changed the rheological properties of the GG-MA hydrogel, with a slight increase in its stiffness, none of the formulations acquired a defined, self-supporting shape. It is known that GG-MA cross-links in the presence of ions, particularly divalent ions such as Mn^{2+}. However, the obtained results showed that the ionic concentration used to prepare the hydrogels was not sufficient to overcome the intramolecular electrostatic repulsions of the carboxylic groups in the GG-MA network, thus giving origin to a "weak gel" [43].

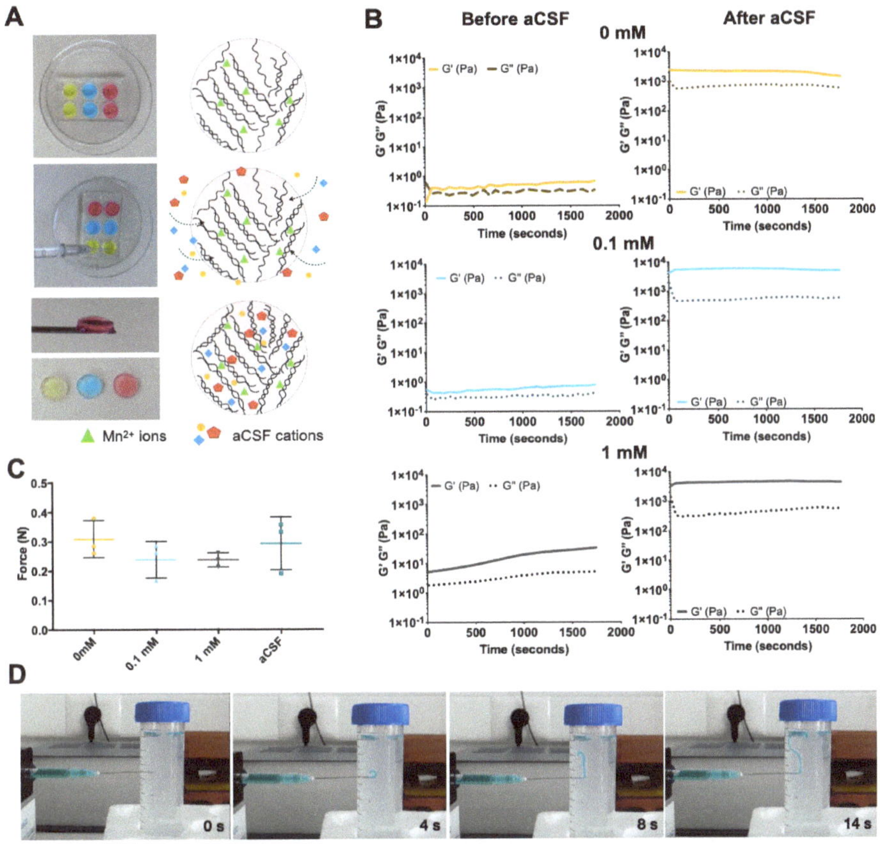

Figure 1. Preparation and rheological characterization of the Mn-based methacrylated gellan gum (GG-MA) hydrogels. (**A**)—Schematic representation of Mn-based GG-MA hydrogel preparation. Pre-gel solutions are poured into silicon templates, followed by the addition of aCSF, resulting in cylindrical hydrogels; (**B**)—oscillatory time sweeps (30 min) of the different hydrogel formulations, before and after addition of aCSF; (**C**) injection force registered after 30 s of hydrogel injection into tube filled with aCSF. Average ± SD, n = 3. (**D**) Time-lapse of hydrogel injection into aCSF. Injection was performed with an 18G needle and with colored hydrogel for better visualization.

The addition of artificial cerebrospinal fluid (aCSF), a salt-rich solution, to the GG-MA led to a more efficient crosslink of the hydrogel network. The ions present in the aCSF can interact with the GG-MA polymer, resulting in a chemical bonding between the divalent ions (e.g., Ca^{2+} or Mg^{2+}) and the polymeric network. Additionally, the presence of monovalent ions, such as K^+ or Na^+, also contributed to hydrogel cross-link by screening the electrostatic repulsion between the GG-MA ionized carboxylate groups [44]. As a result, the Mn/GG-MA hydrogels are ionically crosslinked by the addition of aCSF. Indeed, after incubation in aCSF for 24 h, the Mn/GG-MA hydrogel acquired a definite shape, possible to be handled. Alongside, G' values significantly increased to 1463.02 ± 666.19 Pa, 5157.52 ± 2330.58 Pa and 4490.19 ± 1488.04 Pa, for solutions with 0 mM, 0.1 mM and 1 mM, respectively, confirming the increase in the material stiffness upon aCSF-driven crosslinking. An increase in G'' values was also noticed together with the changes in G'.

However, significant differences on the phase angle, thus in the ratio G''/G', were noticed only for GG-MA hydrogels (0 mM $MnCl_2$) and hydrogels supplemented with 0.1 mM $MnCl_2$, when compared to values without aCSF (Supplementary Information, Figure S1A). Although all formulations showed a viscoelastic nature, the phase angle value was smaller for the GG-MA hydrogels after aCSF incubation, confirming a more elastic behaviour of these hydrogels.

The noticed changes in the rheological properties confirm the feasibility of using Mn/GG-MA hydrogels as in situ cross-linking materials for injectable applications into the CSF. The addition of $MnCl_2$ in the concentrations tested was not sufficient to fully crosslink the GG-MA hydrogel, making the formulations suitable for injection. However, when in contact with aCSF, a solution similar to the CSF found in the intraparenchymal space, a self-supporting, defined, viscoelastic hydrogel is formed. Therefore, the CSF can be used to ionically crosslink the Mn/GG-MA hydrogels in situ, without the use of further stimuli.

3.2. Injection Ability

From the obtained rheology data, it is possible to infer that the ions present in the aCSF physically cross-link all Mn/GG-MA hydrogel formulations. Yet, it is still necessary to study the hydrogel response to high shear rates, similar to what happens during injection, as well as to measure the force needed to inject the material.

Using steady-state shear measurements, it was possible to observe a decrease in viscosity as shear rate increases, typical of shear-thinning solutions (Supplementary Information, Figure S1B). The non-Newtonian behaviour, noticed for all formulations, strongly suggests the feasibility of hydrogel injection. This capacity to decrease viscosity as a response to increasing shear is highly advantageous for injectable formulations [45].

Regardless, a biomaterial is only clinically relevant for injection if the injection force needed is not damaging to embedded cells and it is feasible in a clinical scenario. Therefore, the force applied to extrude the Mn/GG-MA formulations into aCSF was compared with the force needed to extrude a control solution (aCSF). The assay was performed using 31G needle, as this type of needle size is frequently used for injection in neurological applications, in particular when small animals are used. As plotted in Figure 1C, the pre-gel solutions were easily extruded at low forces: 0.309 ± 0.063 N for GG-MA only; 0.238 ± 0.063 N for 0.1 mM; and 0.238 ± 0.024 N for 1 mM. These values are similar to ones obtained when extruding aCSF, 0.294 ± 0.090 N, thus showing that the tested Mn/GG-MA hydrogels are suitable for non-invasive procedures using injection, even when needles with small diameters are used. No needle clotting was observed during the experiment, and the maximal force used to inject the hydrogel in each experiment was similar for all formulations, with no significant differences observed. Interestingly, after injection into a tube filled with aCSF, the pre-gel solution rapidly cross-links, forming a well-defined fibre structure (Figure 1D). This is particularly important for intrathecal injections, as it shows that Mn-based hydrogels will not disperse within the aCSF due to delayed cross-link leading to undesired stem cell sedimentation.

3.3. Hydrogel Permeability

Hydrogel permeability is a key feature for the success of cell-based therapies. Mostly, it is of utmost importance that nutrients and signalling molecules, such as the neurotrophic factors released by transplanted cells, do not face additional diffusional barriers that may hinder their release from the hydrogel to the nervous tissue.

To study the hydrogel permeability, 70 kDa dextran molecules coupled with FITC were mixed with pre-gel solutions followed by cross-linking and incubation in aCSF for 7 days. As depicted in Figure 2A, for all formulations, most of the dextran molecules were released within the first 2 days of incubation. The release profile of dextran is not affected by the concentration of Mn^{2+} used, as no significant differences were detected between the tested formulations. Thenceforth, it is predictable that molecules within this size range, as most of the key neurotrophic factors released by MSCs, will freely diffuse between the encapsulated cells and the surrounding environment. These include the glial cell-line derived neurotrophic factor (GDNF, 30 kDa), insulin growth factor type-1 (IGF-1, 7.65 kDa), brain-derived neurotrophic factor (BDNF, 14 kDa), and neural growth factor (NGF, 27 kDa), among others, which are of particular importance to the therapeutic effect of MSCs in ALS symptoms [46,47]. Consequently, one can assume that the hydrogels herein proposed will not interfere with the delivery of the aforementioned neurotrophic factors.

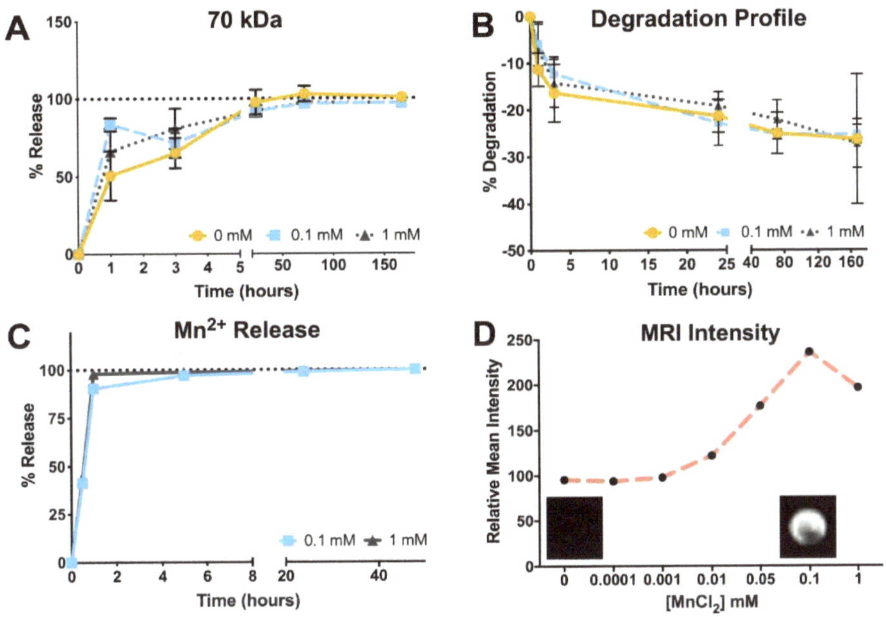

Figure 2. Release and degradation profiles of Mn-based GG-MA hydrogels. (**A**) Release profile of 70 kDa Dextran-FITC, from the hydrogel to surrounding aCSF; (**B**) degradation profile of Mn-based hydrogels along time; (**C**) time-dependent Mn^{2+} release from the prepared hydrogels. Results presented as average $\pm$ SD, n = 5. (**D**) relative mean intensity of the magnetic resonance signal of T1-weighted imaging obtained with phantoms of GG-MA hydrogels supplemented with different concentrations of manganese ions (ex vivo).

3.4. Degradation Profile

The degradation profiles of the different hydrogels were also monitored for 7 days while incubated in aCSF, as plotted in Figure 2B. After 1 day of incubation, there was a mass loss of 21.36 $\pm$ 3.53% for GG-MA only and 22.86 $\pm$ 4.77% and 19.22 $\pm$ 3.04% for hydrogels with 0.1 and 1 mM $MnCl_2$, respectively. During the following 2 days, mass loss was less evident, and the final hydrogel weight variations were 25.31 $\pm$ 1.98% on hydrogels supplemented with 0.1 mM $MnCl_2$ and 27.30 $\pm$ 5.17% when 1 mM $MnCl_2$ was used. Although herein the hydrogels were incubated in aCSF, the obtained degradation profile is aligned with previous studies on GG hydrogel stability in PBS. Different works already pointed out a high stability of GG-based hydrogel, both in the low acyl form or after methacrylation, with a weight loss of only 20% after incubation in PBS for up to 90 days [30,48,49]. When compared with GG-MA only hydrogels, the difference in degradation is not significant, showing again that the introduction of $MnCl_2$ ions does not change the GG-MA properties in a substantial way. The obtained formulations are also more stable when compared to the hydrogel blends previously developed, using GG-MA and hyaluronic acid [37]. Therefore, while the blends with hyaluronic acid may be more suitable for strategies requiring the in situ delivery and migration of cells, the Mn/GG-MA hydrogels can be used as cell depots, assuring a correct position of the therapeutic agents for long periods of time. More importantly, the observed stability suggests that Mn/GG-MA hydrogels can hold the delivered cells for long periods, avoiding the undesired sedimentation at cauda equina.

3.5. Manganese Release Profile and Magnetic Resonance Imaging

The possibility to perform image-guided interventions is highly appealing, as it can facilitate the precise deposition of the cell-laden hydrogel into the intrathecal space or parenchyma. With that being said, it is expected that the Mn^{2+} ions present in the GG-MA matrix remain in the hydrogel long enough to be detected during the procedure of gel injection. Thus, the Mn^{2+} release profile was obtained via ICP detection. As plotted in Figure 2C, Mn^{2+} ions are quickly released from the hydrogel. Specifically, after 30 min of incubation in aCSF, roughly 50% of the Mn^{2+} ions were released to the surrounding media ($45 \pm 4\%$ and $46 \pm 1\%$ for 0.1 and 1 mM, respectively). The total release was noticed after 1 h of incubation, for both formulations. It was already shown that Mn^{2+} ions interact with the carboxyl group of the D-glucuronate unit of GG [35,36]. However, when in contact with the aCSF, other divalent ions start to diffuse into the hydrogel and to interact with the polymeric network leading to hydrogel crosslink [43,50]. The Mn^{2+} ions are then substituted by other divalent ions, such as Ca^{2+} and Mg^{2+}, and consequently released from the hydrogel to the aCSF.

Although the total release of Mn^{2+} was noticed after 1 h of incubation, for both formulations, it should be taken into consideration that Mn^{2+} starts to diffuse only when in contact with the CSF solution. Hence, besides the burst release of the paramagnetic ions, the timeframe until a complete release is achieved is enough for tracking the delivery, placement, and initial biodistribution of the hydrogel within the CSF space. Rapid clearance of the Mn^{2+} contrast signal may be considered advantageous in situations when T1 MRI is further utilized for diagnostic purposes to follow treatment outcomes [51,52].

In order to verify appropriate Mn^{2+} concentration for Mn/GG-MA hydrogel visualization in MRI, phantoms of GG-MA with different concentrations of $MnCl_2$ were prepared followed by T1 and T2 weighted MR imaging acquisition (Figure 2D). The MR signal was detected with an $MnCl_2$ concentration as low as 0.01 mM. The signal intensity increased with $MnCl_2$ concentration, reaching an intensity peak of 0.1 mM, with a 2.4-fold increase in MR signal compared to aCSF. Interestingly, the signal intensity decreased for higher concentrations (1 mM $MnCl_2$), which can be attributed to a T2 effect that results in signal loss [20].

The possibility to retrieve an MRI signal with low $MnCl_2$ concentrations ensures that, besides the rapid Mn^{2+} release from the hydrogels, the biomaterial is still identifiable by MRI during the transplant procedure. In addition, the use of low Mn^{2+} concentrations prevents possible deleterious effects caused by manganese-induced toxicity [53]. From the obtained results, the formulation using 0.1 mM $MnCl_2$ was selected to be used in further studies, as it provides the highest MR signal.

3.6. Cell Viability: hASCs Encapsulation

In vitro biocompatibility was studied by encapsulating human-derived adipose stem cells (hASCs) into the developed hydrogels. To better mimic a possible in vivo scenario, hydrogels were extruded into aCSF using a 31G needle, which is typically used for small animal studies (Figure 3A). Considering the aforementioned MRI results, in vitro biocompatibility was assessed using hydrogels prepared with 0.1 mM $MnCl_2$, while GG-MA only hydrogels were used as controls. As plotted in Figure 3B, the presence of $MnCl_2$ did not significantly impair cell viability along the 7 days of culture. As expected, hydrogels were successfully extruded as fibres, and their shape was maintained throughout the culture time, as depicted in Figure 3C. As a proof-of-concept, 18G needles were also used to extrude hydrogels with 0.1 mM $MnCl_2$, envisioning future large-animal studies (Supplementary Information, Figure S2). Similarly, cells remained viable after being encapsulated within the hydrogel matrix, with $92 \pm 3\%$ of cells viable 7 days after extrusion.

Considering the potential of mesenchymal stem cells [54] and in particular adipose stem cells [55,56] on the treatment of multifocal CNS diseases, the obtained results pave the way to the use of the developed Mn/GG-MA hydrogels as physical support in non-invasive cell-based therapies.

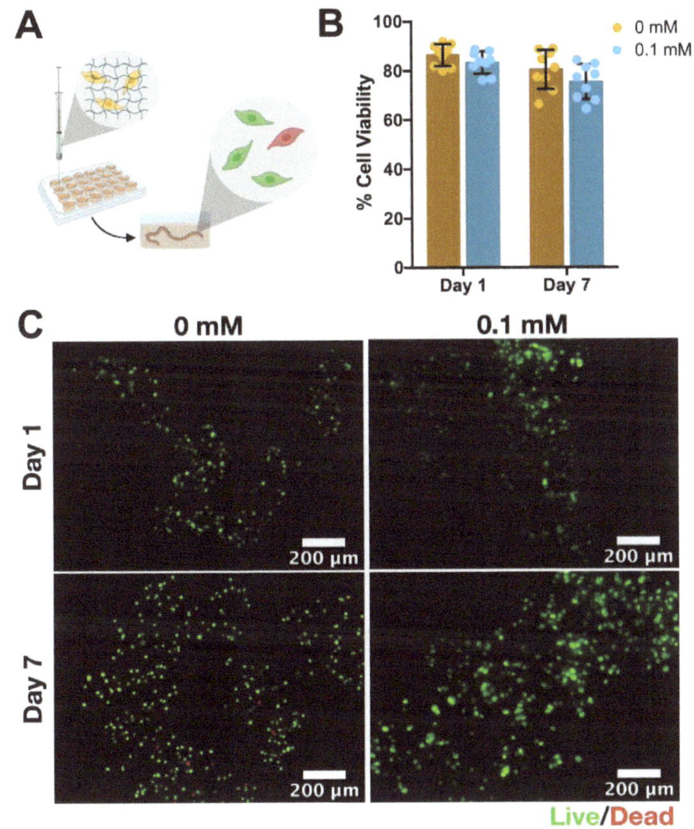

Figure 3. Cell encapsulation within Mn-based GG-MA hydrogel fibers. (**A**) Schematic representation of the experimental outline. Cells were injected directly into aCSF using a Hamilton syringe coupled with a 31G needle, and then cultured for different periods of time; (**B**) cell viability after 1 and 7 days of encapsulation in hydrogels with 0.1 mM $MnCl_2$ and GG-MA only (used as control). Results presented as average ± SD, n = 3; (**C**) fluorescence microscopy images of Live/Dead staining after 1 and 7 days of incubation. Live cells showed as green and dead cells as red. Scale bar: 200 µm. Schematic representation created with BioRender.com.

3.7. *In Vivo Magnetic Resonance Imaging*

The feasibility of using the developed Mn-based GG-MA hydrogels for image-guided cell delivery was further investigated in vivo. For that, hydrogels were injected into the intrathecal space and intraparenchymally in double mutant $MBP^{shi/shi}$/rag2 mice, and followed by MR imaging, as schematically represented in Figure 4A,C.

T1 MRI revealed that hydrogels are visible as a hyperintense signal in comparison to the surrounding tissue (Figure 4(B-ii), arrows) directly after transplantation, confirming the in vitro MR results. However, none of the injected GG-MA hydrogels could be visualized 24 h post-transplantation (Figure 4(B-iii)), which is in accordance with the Mn^{2+} release profile previously discussed. Indeed, the CSF circulation in the intrathecal space is likely responsible for the elution of the Mn^{2+} ions, thus decreasing the hyperintense signal with time. Hence, the engineered hydrogel can be a potential tool for the correct placement of cells along the spinal cord, using a considerably lower amount of $MnCl_2$ as compared to other Mn-based hydrogels [20]. As showed before, the $MnCl_2$ concentration used to prepare the hydrogels is not deleterious for cells and the rapid release to the CSF prevents a local accumulation of Mn^{2+} ions, which might be harmful to the local neuronal cells, and induce Mn-related toxicity [28]. Mn/GG-MA hydrogels were also visible when hydrogels were injected directly into the parenchyma (Figure 4C,D). MRI-guided intraparenchymal cell delivery is also feasible using the designed hydrogel system, which might be useful for stem cell delivery after a stroke event [8]. As shown in Figure 4D, the MR signal was more intense when the hydrogel was injected into the parenchyma. This shows the importance of the CSF on the clearance of the Mn^{2+} from the hydrogel, since in the parenchyma, the ionic diffusion between the hydrogel and the fluid is reduced, thence increasing the signal. Therefore, one should be aware that Mn^{2+} concentration must be optimized for each of the intended cell delivery routes for better imaging outcomes.

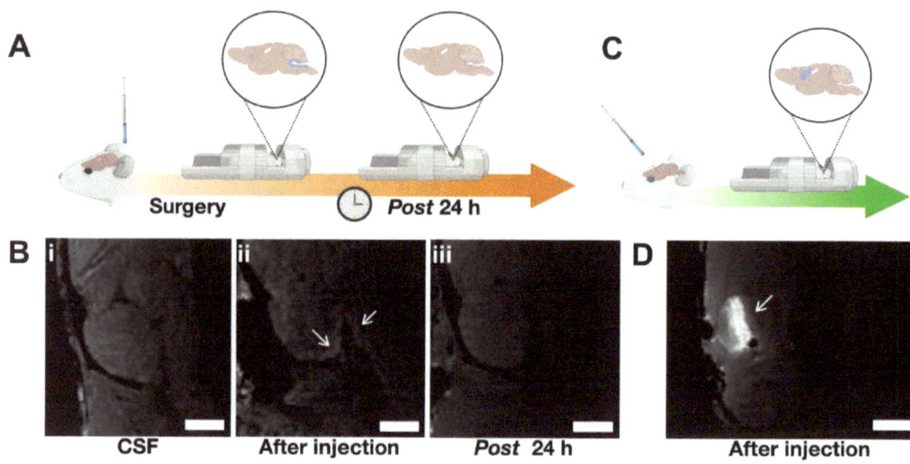

Figure 4. Magnetic resonance T1 weighted imaging of Mn-based GG-MA hydrogels. (**A**) Schematic representation of the in vivo assessment of Magnetic Resonance Imaging (MRI) signal; (**B**) T1 weighted image of (**i**) control $MBP^{shi/shi}$/rag2 mouse; (**ii**) hyperintense signal (arrows) on MR image represent Mn-based GG-MA hydrogel directly after intrathecal transplantation in $MBP^{shi/shi}$/rag2 mice (approximately 20 min after injection); (**iii**) MR scan 24 h after transplantation, where hydrogels are not visible in intrathecal space; (**C**) schematic drawing of intraparenchymal injection and MR scanning; (**D**) MR scan after transplantation of Mn-based GG-MA hydrogel in the parenchyma of $MBP^{shi/shi}$/rag2 mice with prominent hyperintensity (arrow) showing biodistribution of Mn-based GG-MA hydrogel. Scale bar: 2 mm. Schematic representations created with BioRender.com.

4. Conclusions

The present work aimed to design a new stable and traceable hydrogel for image-guided intrathecal cell delivery. By taking advantage of the inherent affinity of methacrylated gellan gum to divalent ions, it was possible to incorporate paramagnetic Mn^{2+} ions within the hydrogel network. Such Mn^{2+} labelled hydrogels can then be used for MRI-guided injection to visualize their biodistribution. The resulting hydrogel was easily injectable, forming distinct, stable, and biocompatible fibres upon injection in a simulated cerebrospinal fluid environment. While the hydrogel formulation would prevent cell removal due to sedimentation or CSF circulation, Mn^{2+} allows real-time visualization of hydrogel position, avoiding an incorrect placement, as evaluated by in vitro and in vivo MR imaging. Thus, the Mn-based methacrylated gellan gum hydrogels developed herein hold great potential on minimally invasive intrathecal cell delivery, tackling two of the main drawbacks reported to date, and possibly open up other applications as a cell/drug delivery matrix for neurological applications when imaging guidance is required.

Supplementary Materials: The following supporting information can be downloaded at: https://www.mdpi.com/article/10.3390/bioengineering10040427/s1, Figure S1: Rheologic characterization of Mn-based hydrogels; Figure S2: Cell encapsulation in GG-MA hydrogel supplemented with 0.1 mM.

Author Contributions: The manuscript was written through contributions of all authors. S.V., L.S., I.M.-C., M.J., P.W. and J.M.O. conceived and planned the experiments. S.V. and P.S. optimized and developed the hydrogels. S.V., P.S., L.S. and K.D. performed the in vitro tests. L.S., I.M.-C. and M.F. optimized and performed the MRI studies. J.M.O., R.L.R., M.J., J.S.-C., B.Ł. and P.W. supervised the experiments and provided financial support. All authors have read and agreed to the published version of the manuscript.

Funding: Sílvia Vieira acknowledges the FCT Ph.D. scholarship (SFRH/BD/102710/2014). J. Miguel Oliveira and J. Silva-Correia acknowledge the FCT grants under the Investigator FCT program (IF/01285/2015 and IF/00115/2015, respectively). The authors also acknowledge the funds provided under the project NanoTech4ALS, funded under the EU FP7 M-ERA.NET program, and ESF (POWR.03.02.00-00-I028/17-00).

Institutional Review Board Statement: The animal study protocol was approved by the Ethical Committee of Mossakowski Medical Research Centre, Polish Academy of Sciences (IV Local Committee in Warsaw, 117/2015).

Informed Consent Statement: Informed consent was obtained from all subjects involved in the study.

Data Availability Statement: The data presented in this study are available on request from the corresponding author.

Conflicts of Interest: The authors declare no conflict of interest.

References

1. Lyczek, A.; Arnold, A.; Zhang, J.; Campanelli, J.T.; Janowski, M.; Bulte, J.W.; Walczak, P. Transplanted human glial-restricted progenitors can rescue the survival of dysmyelinated mice independent of the production of mature, compact myelin. *Exp. Neurol.* **2017**, *291*, 74–86. [CrossRef]
2. Golubczyk, D.; Malysz-Cymborska, I.; Kalkowski, L.; Janowski, M.; Coates, J.R.; Wojtkiewicz, J.; Maksymowicz, W.; Walczak, P. The Role of Glia in Canine Degenerative Myelopathy: Relevance to Human Amyotrophic Lateral Sclerosis. *Mol. Neurobiol.* **2019**, *56*, 5740–5748. [CrossRef] [PubMed]
3. Gutierrez, J.; Lamanna, J.J.; Grin, N.; Hurtig, C.V.; Miller, J.H.; Riley, J.; Urquia, L.; Avalos, P.; Svendsen, C.N.; Federici, T.; et al. Preclinical Validation of Multilevel Intraparenchymal Stem Cell Therapy in the Porcine Spinal Cord. *Neurosurgery* **2015**, *77*, 604–612. [CrossRef]
4. Glass, J.D.; Hertzberg, V.S.; Boulis, N.M.; Riley, J.; Federici, T.; Polak, M.; Bordeau, J.; Fournier, C.; Johe, K.; Hazel, T.; et al. Transplantation of spinal cord-derived neural stem cells for ALS: Analysis of phase 1 and 2 trials. *Neurology* **2016**, *87*, 392–400. [CrossRef] [PubMed]
5. Huang, H.; Qian, K.; Han, X.; Li, X.; Zheng, Y.; Chen, Z.; Huang, X.; Chen, H. Intraparenchymal Neural Stem/Progenitor Cell Transplantation for Ischemic Stroke Animals: A Meta-Analysis and Systematic Review. *Stem. Cells Int.* **2018**, *2018*, 4826407. [CrossRef] [PubMed]
6. Chen, L.; Xi, H.; Huang, H.; Zhang, F.; Liu, Y.; Chen, D.; Xiao, J. Multiple cell transplantation based on an intraparenchymal approach for patients with chronic phase stroke. *Cell Transpl.* **2013**, *22* (Suppl. S1), S83–S91. [CrossRef] [PubMed]
7. Kelly, S.; Bliss, T.M.; Shah, A.K.; Sun, G.H.; Ma, M.; Foo, W.C.; Masel, J.; Yenari, M.A.; Weissman, I.L.; Uchida, N.; et al. Transplanted human fetal neural stem cells survive, migrate, and differentiate in ischemic rat cerebral cortex. *Proc. Natl. Acad. Sci. USA* **2004**, *101*, 11839–11844. [CrossRef]
8. Zhong, J.; Chan, A.; Morad, L.; Kornblum, H.I.; Fan, G.; Carmichael, S.T. Hydrogel matrix to support stem cell survival after brain transplantation in stroke. *Neurorehabil. Neural Repair* **2010**, *24*, 636–644. [CrossRef]
9. Goldman, S.A. Stem and Progenitor Cell-Based Therapy of the Central Nervous System: Hopes, Hype, and Wishful Thinking. *Cell Stem Cell* **2016**, *18*, 174–188. [CrossRef]
10. Boltze, J.; Arnold, A.; Walczak, P.; Jolkkonen, J.; Cui, L.; Wagner, D.C. The Dark Side of the Force—Constraints and Complications of Cell Therapies for Stroke. *Front. Neurol.* **2015**, *6*, 155. [CrossRef]

11. Oliveira, J.M.; Carvalho, L.; Silva-Correia, J.; Vieira, S.; Majchrzak, M.; Lukomska, B.; Stanaszek, L.; Strymecka, P.; Malysz-Cymborska, I.; Golubczyk, D.; et al. Hydrogel-based scaffolds to support intrathecal stem cell transplantation as a gateway to the spinal cord: Clinical needs, biomaterials, and imaging technologies. *NPJ Regen. Med.* **2018**, *3*, 8. [CrossRef] [PubMed]
12. Nabavi, S.M.; Arab, L.; Jarooghi, N.; Bolurieh, T.; Abbasi, F.; Mardpour, S.; Azimyian, V.; Moeininia, F.; Maroufizadeh, S.; Sanjari, L.; et al. Safety, Feasibility of Intravenous and Intrathecal Injection of Autologous Bone Marrow Derived Mesenchymal Stromal Cells in Patients with Amyotrophic Lateral Sclerosis: An Open Label Phase I Clinical Trial. *Cell J.* **2019**, *20*, 592–598. [PubMed]
13. Barczewska, M.; Grudniak, M.; Maksymowicz, S.; Siwek, T.; Ołdak, T.; Jezierska-Woźniak, K.; Gładysz, D.; Maksymowicz, W. Safety of intrathecal injection of Wharton's jelly-derived mesenchymal stem cells in amyotrophic lateral sclerosis therapy, *Neural Regen. Res.* **2019**, *14*, 313–318.
14. Habisch, H.J.; Janowski, M.; Binder, D.; Kuzma-Kozakiewicz, M.; Widmann, A.; Habich, A.; Schwalenstocker, B.; Hermann, A.; Brenner, R.; Lukomska, B.; et al. Intrathecal application of neuroectodermally converted stem cells into a mouse model of ALS: Limited intraparenchymal migration and survival narrows therapeutic effects. *J. Neural. Transm.* **2007**, *114*, 1395–1406. [CrossRef]
15. Chotivichit, A.; Ruangchainikom, M.; Chiewvit, P.; Wongkajornsilp, A.; Sujirattanawimol, K. Chronic spinal cord injury treated with transplanted autologous bone marrow-derived mesenchymal stem cells tracked by magnetic resonance imaging: A case report. *J. Med. Case Rep.* **2015**, *9*, 79. [CrossRef]
16. Gasperini, L.; Mano, J.F.; Reis, R.L. Natural polymers for the microencapsulation of cells. *J. R. Soc. Interface* **2014**, *11*, 20140817. [CrossRef] [PubMed]
17. Malysz-Cymborska, I.; Golubczyk, D.; Kalkowski, L.; Burczyk, A.; Janowski, M.; Holak, P.; Olbrych, K.; Sanford, J.; Stachowiak, K.; Milewska, K.; et al. MRI-guided intrathecal transplantation of hydrogel-embedded glial progenitors in large animals. *Sci. Rep.* **2018**, *8*, 16490. [CrossRef]
18. Oliveira, E.P.; Malysz-Cymborska, I.; Golubczyk, D.; Kalkowski, L.; Kwiatkowska, J.; Reis, R.L.; Oliveira, J.M.; Walczak, P. Advances in bioinks and in vivo imaging of biomaterials for CNS applications. *Acta Biomater.* **2019**, *95*, 60–72. [CrossRef]
19. Gale, E.M.; Wey, H.Y.; Ramsay, I.; Yen, Y.F.; Sosnovik, D.E.; Caravan, P. A Manganese-based Alternative to Gadolinium: Contrast-enhanced MR Angiography, Excretion, Pharmacokinetics, and Metabolism. *Radiology* **2018**, *286*, 865–872. [CrossRef]
20. Morch, Y.A.; Sandvig, I.; Olsen, O.; Donati, I.; Thuen, M.; Skjak-Braek, G.; Haraldseth, O.; Brekken, C. Mn-alginate gels as a novel system for controlled release of Mn2+ in manganese-enhanced MRI. *Contrast Media Mol. Imaging* **2012**, *7*, 265–275. [CrossRef]
21. Addisu, K.D.; Hailemeskel, B.Z.; Mekuria, S.L.; Andrgie, A.T.; Lin, Y.C.; Tsai, H.C. Bioinspired, Manganese-Chelated Alginate-Polydopamine Nanomaterials for Efficient in Vivo T1-Weighted Magnetic Resonance Imaging. *ACS Appl. Mater. Interfaces* **2018**, *10*, 5147–5160. [CrossRef] [PubMed]
22. Wang, J.; Wang, H.; Ramsay, I.A.; Erstad, D.J.; Fuchs, B.C.; Tanabe, K.K.; Caravan, P.; Gale, E.M. Manganese-Based Contrast Agents for Magnetic Resonance Imaging of Liver Tumors: Structure-Activity Relationships and Lead Candidate Evaluation. *J. Med. Chem.* **2018**, *61*, 8811–8824. [CrossRef] [PubMed]
23. Panich, A.M.; Shames, A.I.; Aleksenskii, A.E.; Yudina, E.B.; Vul, A.Y. Manganese-grafted detonation nanodiamond, a novel potential MRI contrast agent. *Diam. Relat. Mater.* **2021**, *119*, 108590. [CrossRef]
24. Panich, A.M.; Salti, M.; Aleksenskii, A.E.; Kulvelis, Y.V.; Chizhikova, A.; Vul, A.Y.; Shames, A.I. Suspensions of manganese-grafted nanodiamonds: Preparation, NMR, and MRI study. *Diam. Relat. Mater.* **2023**, *131*, 109591. [CrossRef]
25. Cloyd, R.A.; Koren, S.A.; Abisambra, J.F. Manganese-Enhanced Magnetic Resonance Imaging: Overview and Central Nervous System Applications with a Focus on Neurodegeneration. *Front. Aging Neurosci.* **2018**, *10*, 403.
26. Araszkiewicz, A.M.; Oliveira, E.P.; Svendsen, T.; Drela, K.; Rogujski, P.; Malysz-Cymborska, I.; Fiedorowicz, M.; Reis, R.L.; Oliveira, J.M.; Walczak, P.; et al. Manganese-Labeled Alginate Hydrogels for Image-Guided Cell Transplantation. *Int. J. Mol. Sci.* **2022**, *23*, 2465. [CrossRef]
27. Roth, J.A.; Ganapathy, B.; Ghio, A.J. Manganese-induced toxicity in normal and human B lymphocyte cell lines containing a homozygous mutation in parkin. *Toxicol. Vitr.* **2012**, *26*, 1143–1149. [CrossRef]
28. Ding, D.; Roth, J.; Salvi, R. Manganese is toxic to spiral ganglion neurons and hair cells in vitro. *Neurotoxicology* **2011**, *32*, 233–241. [CrossRef]
29. Oliveira, J.T.; Santos, T.C.; Martins, L.; Picciochi, R.; Marques, A.P.; Castro, A.G.; Neves, N.M.; Mano, J.F.; Reis, R.L. Gellan gum injectable hydrogels for cartilage tissue engineering applications: In vitro studies and preliminary in vivo evaluation. *Tissue Eng. Part A* **2010**, *16*, 343–353. [CrossRef]
30. Silva-Correia, J.; Oliveira, J.M.; Caridade, S.G.; Oliveira, J.T.; Sousa, R.A.; Mano, J.F.; Reis, R.L. Gellan gum-based hydrogels for intervertebral disc tissue-engineering applications. *J. Tissue Eng. Regen. Med.* **2011**, *5*, e97–e107. [CrossRef]
31. Vieira, S.; Vial, S.; Maia, F.R.; Carvalho, M.; Reis, R.L.; Granja, P.L.; Oliveira, J.M. Gellan gum-coated gold nanorods: An intracellular nanosystem for bone tissue engineering. *RSC Adv.* **2015**, *5*, 77996–78005. [CrossRef]
32. Lozano, R.; Stevens, L.; Thompson, B.C.; Gilmore, K.J.; Gorkin, R.; Stewart, E.M.; Panhuis, M.I.H.; Romero-Ortega, M.; Wallace, G.G. 3D printing of layered brain-like structures using peptide modified gellan gum substrates. *Biomaterials* **2015**, *67*, 264–273. [CrossRef]
33. Stevens, L.R.; Gilmore, K.J.; Wallace, G.G.; Panhuis, M.I.H. Tissue engineering with gellan gum. *Biomater. Sci.* **2016**, *4*, 1276–1290. [CrossRef] [PubMed]

34. Spector, R.; Snodgrass, S.R.; Johanson, C.E. A balanced view of the cerebrospinal fluid composition and functions: Focus on adult humans. *Exp. Neurol.* **2015**, *273*, 57–68. [CrossRef]
35. Kawahara, S.; Yoshikawa, A.; Hiraoki, T.; Tsutsumi, A. Interactions of paramagnetic metal ions with gellan gum studied by ESR and NMR methods. *Carbohydr. Polym.* **1996**, *30*, 29–133. [CrossRef]
36. Tsutsumi, A.; Ya, D.; Hiraoki, T.; Mochiku, H.; Yamaguchi, R.; Takahashi, N. ESR studies of Mn(II) binding to gellan and carrageenan gels. *Food Hydrocoll.* **1993**, *7*, 427–434. [CrossRef]
37. Vieira, S.; Strymecka, P.; Stanaszek, L.; Silva-Correia, J.; Drela, K.; Fiedorowicz, M.; Malysz-Cymborska, I.; Rogujski, P.; Janowski, M.; Reis, R.L.; et al. Methacrylated gellan gum and hyaluronic acid hydrogel blends for image-guided neurointerventions. *J. Mater. Chem. B* **2020**, *8*, 5928–5937. [CrossRef]
38. Janowski, M.; Kuzma-Kozakiewicz, M.; Binder, D.; Habisch, H.J.; Habich, A.; Lukomska, B.; Domanska-Janik, K.; Ludolph, A.C.; Storch, A. Neurotransplantation in mice: The concorde-like position ensures minimal cell leakage and widespread distribution of cells transplanted into the cisterna magna. *Neurosci. Lett.* **2008**, *430*, 169–174. [PubMed]
39. Fiedorowicz, M.; Orzel, J.; Kossowski, B.; Welniak-Kaminska, M.; Choragiewicz, T.; Swiatkiewicz, M.; Rejdak, R.; Bogorodzki, P.; Grieb, P. Anterograde Transport in Axons of the Retinal Ganglion Cells and its Relationship to the Intraocular Pressure during Aging in Mice with Hereditary Pigmentary Glaucoma. *Curr. Eye Res.* **2018**, *43*, 539–546. [CrossRef]
40. Fratini, M.; Abdollahzadeh, A.; DiNuzzo, M.; Salo, R.A.; Maugeri, L.; Cedola, A.; Giove, F.; Grohn, O.; Tohka, J.; Sierra, A. Multiscale Imaging Approach for Studying the Central Nervous System: Methodology and Perspective. *Front. Neurosci.* **2020**, *14*, 72. [CrossRef]
41. Yan, C.; Pochan, D.J. Rheological properties of peptide-based hydrogels for biomedical and other applications. *Chem. Soc. Rev.* **2010**, *39*, 3528–3540. [CrossRef]
42. Malkin, A.Y.; Isayev, A.I. *Rheology: Concepts, Methods, and Applications*; Chemtec Publishing: Scarborough, ON, Canada, 2017.
43. Morris, E.R.; Nishinari, K.; Rinaudo, M. Gelation of gellan—A review. *Food Hydrocoll.* **2012**, *28*, 373–411. [CrossRef]
44. Bacelar, A.H.; Silva-Correia, J.; Oliveira, J.M.; Reis, R.L. Recent progress in gellan gum hydrogels provided by functionalization strategies. *J. Mater. Chem. B* **2016**, *4*, 6164–6174. [CrossRef] [PubMed]
45. Chen, M.H.; Wang, L.L.; Chung, J.J.; Kim, Y.H.; Atluri, P.; Burdick, J.A. Methods to Assess Shear-Thinning Hydrogels for Application As Injectable Biomaterials. *ACS Biomater. Sci. Eng.* **2017**, *3*, 3146–3160. [CrossRef]
46. Forostyak, S.; Sykova, E. Neuroprotective Potential of Cell-Based Therapies in ALS: From Bench to Bedside. *Front. Neurosci.* **2017**, *11*, 591. [CrossRef] [PubMed]
47. Chen, K.S.; Sakowski, S.A.; Feldman, E.L. Intraspinal stem cell transplantation for amyotrophic lateral sclerosis. *Ann. Neurol.* **2016**, *79*, 342–353. [CrossRef] [PubMed]
48. Xu, Z.H.; Li, Z.Q.; Jiang, S.; Bratlie, K.M. Chemically Modified Gellan Gum Hydrogels with Tunable Properties for Use as Tissue Engineering Scaffolds. *ACS Omega* **2018**, *3*, 6998–7007. [CrossRef]
49. Silva-Correia, J.; Zavan, B.; Vindigni, V.; Silva, T.H.; Oliveira, J.M.; Abatangelo, G.; Reis, R.L. Biocompatibility evaluation of ionic- and photo-crosslinked methacrylated gellan gum hydrogels: In vitro and in vivo study. *Adv. Healthc. Mater.* **2013**, *2*, 568–575. [CrossRef]
50. Grasdalen, H.; Smidsrød, O. Gelation of gellan gum. *Carbohydr. Polym.* **1987**, *7*, 371–393. [CrossRef]
51. Turner, M.R.; Modo, M. Advances in the application of MRI to amyotrophic lateral sclerosis. *Expert Opin. Med. Diagn.* **2010**, *4*, 483–496. [CrossRef]
52. Grolez, G.; Moreau, C.; Danel-Brunaud, V.; Delmaire, C.; Lopes, R.; Pradat, P.F.; El Mendili, M.M.; Defebvre, L.; Devos, D. The value of magnetic resonance imaging as a biomarker for amyotrophic lateral sclerosis: A systematic review. *BMC Neurol.* **2016**, *16*, 155. [CrossRef] [PubMed]
53. Malheiros, J.M.; Paiva, F.F.; Longo, B.M.; Hamani, C.; Covolan, L. Manganese-Enhanced MRI: Biological Applications in Neuroscience. *Front. Neurol.* **2015**, *6*, 161. [CrossRef] [PubMed]
54. Gugliandolo, A.; Bramanti, P.; Mazzon, E. Mesenchymal Stem Cells: A Potential Therapeutic Approach for Amyotrophic Lateral Sclerosis? *Stem Cells Int.* **2019**, *2019*, 3675627. [CrossRef] [PubMed]
55. Kim, K.S.; Lee, H.J.; An, J.; Kim, Y.B.; Ra, J.C.; Lim, I.; Kim, S.U. Transplantation of human adipose tissue-derived stem cells delays clinical onset and prolongs life span in ALS mouse model. *Cell Transplant.* **2014**, *23*, 1585–1597. [CrossRef]
56. Walker, C.L. Adipose-derived stem cell conditioned medium for the treatment of amyotrophic lateral sclerosis: Pre-clinical evidence and potential for clinical application. *Neural Regen. Res.* **2019**, *14*, 1522–1524. [CrossRef]

bioengineering

MDPI

Article

De Novo Design of Imidazopyridine-Tethered Pyrazolines That Target Phosphorylation of STAT3 in Human Breast Cancer Cells

Akshay Ravish [1,†], Rashmi Shivakumar [1,†], Zhang Xi [2,†], Min Hee Yang [3], Ji-Rui Yang [4], Ananda Swamynayaka [5], Omantheswara Nagaraja [5], Mahendra Madegowda [5], Arunachalam Chinnathambi [6], Sulaiman Ali Alharbi [6], Vijay Pandey [4,7], Gautam Sethi [8,*], Kwang Seok Ahn [3,*], Peter E. Lobie [2,4,7,*] and Basappa Basappa [1,*]

1 Laboratory of Chemical Biology, Department of Studies in Organic Chemistry, University of Mysore, Manasagangotri, Mysore 570006, India
2 Shenzhen Bay Laboratory, Shenzhen 518055, China
3 Department of Science in Korean Medicine, Kyung Hee University, Seoul 02447, Republic of Korea
4 Tsinghua Berkeley Shenzhen Institute, Tsinghua Shenzhen International Graduate School, Tsinghua University, Shenzhen 518055, China
5 Department of Studies in Physics, University of Mysore, Manasagangotri, Mysore 570006, India
6 Department of Botany and Microbiology, College of Science, King Saud University, P.O. Box 2455, Riyadh 11451, Saudi Arabia
7 Institute of Biopharmaceutical and Health Engineering, Tsinghua Shenzhen International Graduate School, Tsinghua University, Shenzhen 518055, China
8 Department of Pharmacology, Yong Loo Lin School of Medicine, National University of Singapore, Singapore 117600, Singapore
* Correspondence: phcgs@nus.edu.sg (G.S.); ksahn@khu.ac.kr (K.S.A.); pelobie@sz.tsinghua.edu.cn (P.E.L.); salundibasappa@gmail.com (B.B.)
† These authors contributed equally to this work.

Citation: Ravish, A.; Shivakumar, R.; Xi, Z.; Yang, M.H.; Yang, J.-R.; Swamynayaka, A.; Nagaraja, O.; Madegowda, M.; Chinnathambi, A.; Alharbi, S.A.; et al. De Novo Design of Imidazopyridine-Tethered Pyrazolines That Target Phosphorylation of STAT3 in Human Breast Cancer Cells. *Bioengineering* **2023**, *10*, 159. https://doi.org/10.3390/bioengineering10020159

Academic Editors: Ali Zarrabi and Liang Luo

Received: 14 December 2022
Revised: 2 January 2023
Accepted: 8 January 2023
Published: 24 January 2023

Abstract: In breast cancer (BC), STAT3 is hyperactivated. This study explored the design of imidazopyridine-tethered pyrazolines as a de novo drug strategy for inhibiting STAT3 phosphorylation in human BC cells. This involved the synthesis and characterization of two series of compounds namely, 1-(3-(2,6-dimethylimidazo [1,2-a]pyridin-3-yl)-5-(3-nitrophenyl)-4,5-dihydro-1H-pyrazol-1-yl)-2-(4-(substituted)piperazin-1-yl)ethanone and N-substituted-3-(2,6-dimethylimidazo[1,2-a]pyridin-3-yl)-5-(3-nitrophenyl)-4,5-dihydro-1H-pyrazoline-1-carbothioamides. Compound **3f** with 2,3-dichlorophenyl substitution was recognized among the tested series as a lead structure that inhibited the viability of MCF-7 cells with an IC_{50} value of 9.2 μM. A dose- and time-dependent inhibition of STAT3 phosphorylation at Tyr705 and Ser727 was observed in MCF-7 and T47D cells when compound **3f** was added in vitro. Calculations using density functional theory showed that the title compounds HOMOs and LUMOs are situated on imidazopyridine-pyrazoline and nitrophenyl rings, respectively. Hence, compound **3f** effectively inhibited STAT3 phosphorylation in MCF-7 and T47D cells, indicating that these structures may be an alternative synthon to target STAT3 signaling in BC.

Keywords: human breast cancer; STAT3; imidazopyridine; pyrazolines; DFT; de novo design

1. Introduction

Breast cancer (BC) is the most common cancer in women [1–4]. As shown in numerous studies, various drugs such as tamoxifen, vinflunine, anastrazole, 5-fluorouracil, Doxorubicin, paclitaxel/docetaxel, ribociclib, olaparib, and other drugs have been shown to possess therapeutic potential for women with BC depending on the stage and the specific molecular subtype [5–12]. Genes that encode transcription factors are directly involved in breast cancer progression, proliferation, apoptosis, metastasis, and chemotherapy resistance [13,14]. STAT3 is one such transcription factor that harbors six functional domains, including the terminal-NH_2 domain, the coiled-coil domain, the DNA-binding domain, the SRC homology 2 domain, and the transactivation domain [15,16]. STAT3 is activated by both tyrosine and serine phosphorylation and translocates into the nucleus to regulate transcription [17–19].

De novo drug design retains the potential for the discovery of new and potent lead molecules for oncology [20]. Pharmacology has heavily utilized imidazopyridine scaffolds in drug development [21]. Ten approved drugs contain imidazopyridine, and another 12 are in active clinical development [22,23]. A synthetic imidazopyridine compound 16, was synthesized, tested, and determined to reduce the level of phospho-STAT3 and downstream signaling cascades in hepatocellular carcinoma cells, which was attributed to an increase in SHP-1A [24]. Furthermore, the design and synthesis of an imidazopyridine scaffold-bearing compound (P3971) has been identified as a potent STAT3 inhibitor with an IC50 value of 350 nM and demonstrated significant antiproliferative activity against a variety of cancerous cell lines including HCT116 and H460 [25]. Additionally, pyrazoles may provide better pharmacological effects, and have also generated a number of drugs such as pyrazofurin, celecoxib, ramifenazone, lonazolac, and rimonabant [26–30]. We reported the synthesis of pyridine-fused pyrazoles as a STAT3 inhibitor and an inhibitor of cancer cell growth [31]. Additionally, the pyrazole-based compound MNS1-Leu inhibited IL-6-induced STAT3 phosphorylation in patient-derived HGG cells without adversely affecting Akt, STAT1, JAK2, or ERK1/2 phosphorylation [32]. Compound C6 was also discovered to be a STAT3-specific inhibitor that had the strongest anti-proliferation activities against breast cancer cells with an IC_{50} value of 160 nM [33]. In light of this, we developed a conventional de novo design and identified the essential output structures comprising imidazopyridine, pyrazole, pyrrole, and proposed a series of imidazopyridine-tethered pyrazolines (ITP) that could target STAT3 in breast cancer cells based on synthetic accessibility and chemical stability (Figure 1).

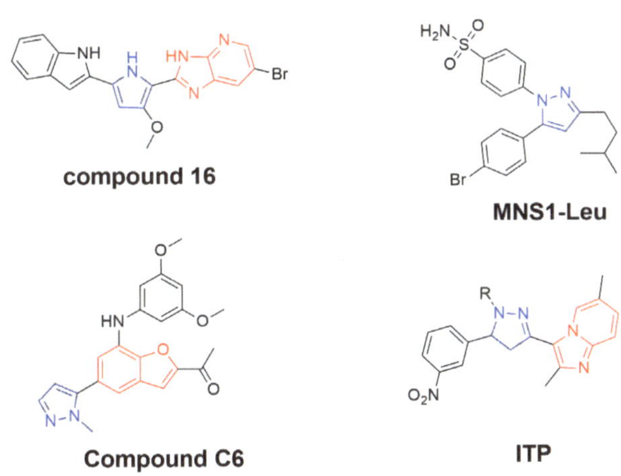

Figure 1. Conventional de novo design of novel ITPs as STAT3 inhibitors in cancer cells.

2. Results and Discussion

Using the literature method, we initially synthesized an intermediate (1-[2-methyl-imidazo[1,2-α] pyridine-3-yl]ethanone) by reacting 2-aminopridine with 3-bromopentane-2,4-dione [34]. Further, the chalcone [(E)-1-(2,6-dimethylimidazo[1,2-α]pyridin-3-yl)-3-(3-nitrophenyl)prop-2-en-1-one] was synthesized by the condensation reaction between the intermediateate and 3-nitro-benzaldehyde in presence of alcoholic NaOH, using the literature protocol [35]. The 2-pyrazoline compound **1** [(2,6-dimethyl-3-(5-(3-nitrophenyl)-4,5-dihydro-1H-pyrazol-3-yl)imidazo[1,2-a]pyridine)] was synthesized by refluxing the chalcone and hydrazine hydrate in ethanol solvent. The synthesis of compound **2** [(1-(3-(2,6-dimethylimidazo[1,2-a]pyridin-3-yl)-5-(3-nitrophenyl)-4,5-dihydro-1H-pyrazol-1-yl)] was synthesized by reacting compound **1** with ethanol via acidic dehydration reaction. Furthermore, the synthesis of 2-pyrazoline compounds (**3a–i**) [1-(3-(2,6-dimethylimidazo[1,2-a]pyridin-3-yl)-5-(3-nitrophenyl)-4,5-dihydro-1H-pyrazol-1-yl)-2-(4-substitutted-piperazin-1-yl)ethanones] was prepared by reacting the compound **1** with chloroactetyl chloride, and substituted-piperazines under basic condition. The compounds **4(a–h)** [3-(2,6-dimethylimidazo pyridin-3-yl)-N-substituted-5-(3-nitrophenyl)-4,5-dihydro-1H-pyrazole-1-carbothioamide] were synthesized by reacting compound **1** with substituted-isothiocyanates under basic condition (Scheme 1, Table 1). All synthesized compounds were characterized by ^{1}H NMR, ^{13}C NMR, and LCMS techniques. The NMR spectral peaks were assigned to all the compounds, which were consistent with the theoretical calculations.

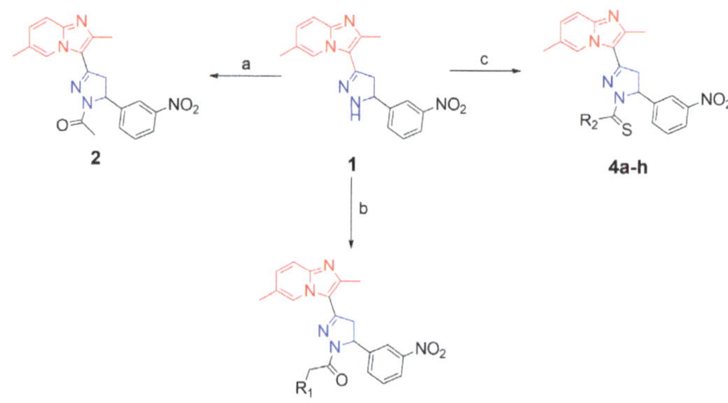

Scheme 1. Synthesis of 2-pyrazolines: reagents and conditions: (**a**) AcOH, EtOH, reflux; (**b**) (i) chloroactetyl chloride, toluene, 0–5 °C, (ii) substituted piperazines, acetone, NEt_3, reflux; (**c**) substituted isothiocyanates, toluene, NEt_3, reflux.

Table 1. Physical data of 2-pyrazolines.

Compound Code	R_1 or R_2	Melting Point in °C	Yield in %
3a	Piperidone	170–172	97
3b	1-Boc piperazine	166–168	96
3c	Piperazine	176–178	35
3d	1-acetyl piperazine	208–210	98
3e	1-(4-chlorophenyl)piperazine	180–182	96
3f	1-(2,3-dichlorophenyl)piperizine	122–124	95
3g	1-(4-chloro-2-fluorophenyl)piperazine	196–198	95
3h	1-(3,4-difluorophenyl)piperazine	192–194	96
3i	1-(2-fluorophenyl)piperazine	160–162	98
4a	Aniline	210–212	98
4b	4-nitrophenylaniline	230–232	95
4c	Cyclohexylamine	204–206	98
4d	4-chlorophenylaniline	240–242	98
4e	3-chlorophenylaniline	242–244	98
4f	4-methylphenylaniline	256–258	97
4g	3,5-bis(triflourophenyl)aniline	230–232	98%
4h	n-butylamine	246–248	96%

The goal herein was to assess the ability of imidazopyridine-tethered pyrazolines to inhibit cell viability in mammary carcinoma cells (using the Alamar Blue assay), since pyridine-fused pyrazoles were previously shown to inhibit cell viability in various breast cancer cells [36,37]. Results of the study indicated that compounds such as (**3f**), (**3e**), (**4g**), and **3g** inhibited MCF-7 cell viability with IC_{50} values of 9.27, 13.24, and 10.90 μM, respectively (Figure 2). These compounds, which contained substitutions of 2,3-dichlorophenyl, 4-chlorophenyl, 2,5-disubstituted trifluorophenyl, 2-fluoro,4-chloro-phenyl groups as the side chain, exhibited relative inhibition of MCF-7 cell viability. The viability of T47D, BT-474, and SK-BR-3 cells was inhibited in a dose-dependent manner by the lead compound **3f** indicating that the compound was efficacious to decrease the viability of human breast cancer cells (Table 2; Supplementary Figures). The low chemical stability of thio-urea-based compounds may account for the inactivity of most thiourea-conjugated pyrazolines (excluding 4g) against TNBC cells. Further, compounds **3f**, **3h**, and **4b** showed significant effect against TNBC cells, but were not toxic to the normal MCF-10A cells, indicating their selectivity towards cancer cell cytotoxicity.

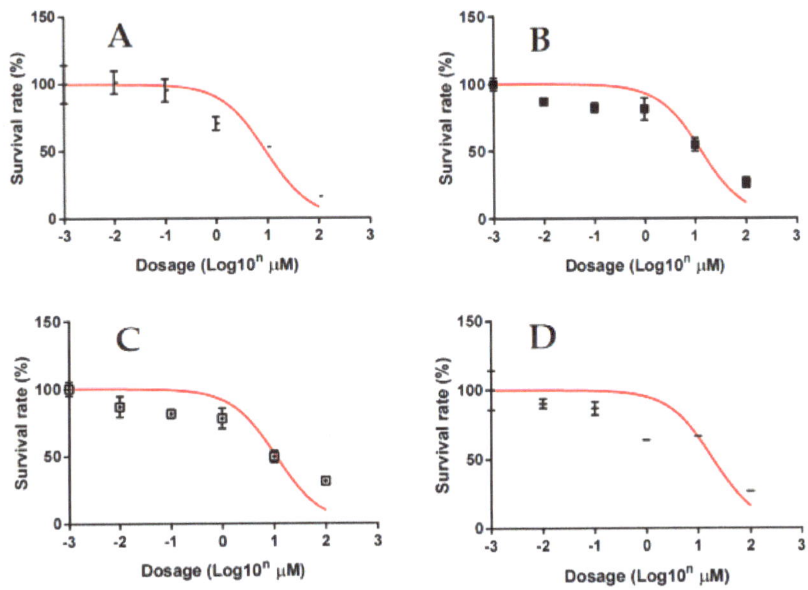

Figure 2. Inhibitory effect of compounds **3f** (**A**), **3e** (**B**), **4g** (**C**), and **3g** (**D**) on the viability of MCF-7 cells. MCF-7 cells were treated with various concentrations of 2-pyrazolines for 72 h and the viability of cells was analyzed by Alamar Blue assays. The results are presented as mean ± S.E.M. of triplicate determinations.

Further, the in silico mode-of-action analysis was performed for lead compound **3f** using CHEMBL's latest version as described by kalakoti et al., [38]. For this purpose, the smile format of compound **3f** was added into the similarity searching engine of CHEMBL, which yielded 14,856 bio-activity profiles and which gave the choice of organism, cell type, and also the predicted human targets in the similarity ranking order [39]. The analysis of the results sheet identified STAT3 as a target for compound **3f** with a ranking of 4783 indicating the predicted score would be reasonable to test in in vitro functional studies.

Table 2. Effect of newly synthesized imidazopyridine-tethered pyrazolines on various breast cancer/carcinoma cell viability.

Entry	IC50 (μM)				
	MCF-7	**T47D**	**BT-474**	**SK-BR-3**	**MCF-10A**
2	>100	>100	97.82±1.76	>100	>100
3a	22.81 ± 1.36	23.31 ± 1.28	4.109 ± 0.41	>100	>100
3b	22.29 ± 1.35	15.15 ± 1.13	21.29 ± 1.32	28.28 ± 1.34	>100
3c	29.27 ± 1.47	39.95 ± 1.49	26.75 ± 1.37	95.17 ± 1.79	>100
3d	21.03 ± 1.32	>100	8.04 ± 0.47	32.14 ± 1.46	>100
3e	13.24 ± 1.12	29.9 ± 1.05	17.6 ± 0.87	39.13 ± 1.46	38.34 ± 1.44
3f	9.27 ± 0.97	23.51 ± 1.25	13.91 ± 1.32	>100	>100
3g	19.40 ± 1.29	86.55 ± 1.95	63.16 ± 1.75	99.96 ± 1.98	8.69 ± 0.69
3h	19.56 ± 1.29	12.5 ± 1.01	3.061 ± 0.32	38.81 ± 1.42	6.27 ± 0.26
3i	63.69 ± 1.80	>100	>100	>100	>100
4a	81.37 ± 1.91	>100	>100	>100	>100
4b	65.36 ± 1.82	45.61 ± 1.43	7.259 ± 0.54	69.02 ± 1.41	>100
4c	46.08 ± 1.66	>100	52.83 ± 1.54	>100	>100
4d	>100	>100	>100	>100	>100
4e	>100	>100	>100	>100	>100
4f	56.92 ± 1.39	>100	>100	>100	>100
4g	10.90 ± 1.01	>100	>100	>100	>100
4h	>100	54.42 ± 1.68	>100	>100	90.59 ± 1.98
Olaparib	3.28 ± 0.68				

Therefore, in vitro experiments were performed to evaluate the effect of compound **3f** on the phosphorylation of STAT3 in MCF-7 and T47D cells, utilizing our laboratory protocol [40]. For this purpose, Western blot analysis was performed after the treatment of MCF-7 and T47D cells with compound **3f** (0, 1, 3, 5 μM) and using p-STAT3 (Tyr705), p-STAT3 (Ser727), and STAT3 antibodies for blotting. Analysis of the results indicated that compound **3f** reduced the phosphorylation level of STAT3 at Tyr705 and Ser727 in a dose-dependent manner without affecting the total STAT3 protein expression (Figure 3A). Furthermore, compound **3f** also reduced the phosphorylation of STAT3 at Tyr705 and Ser727 in a time-dependent manner in both MCF-7 and T47D cells (Figure 3B). These results clearly suggest that compound **3f** decreased the constitutive phosphorylation of STAT3 in ER+ breast cancer cell lines.

In order to understand the lead structure specificity in the bioactivity of the ITP compounds, DFT calculations were performed. From the frontier molecular orbital (FMOs) theory, HOMO and LUMO are the most influential factors in bioactivity. HOMO has the priority to provide electrons, while LUMO can accept electrons. Moreover, the difference in energy between these two FMOs can be used to predict the strength and stability of molecular complexes [41]. Figure 4 shows the molecular orbital of compounds, while Table 2 lists the calculated global chemical reactivity descriptor parameters of compounds.

HOMO-LUMO levels indicate the interactions between the compound and the protein target. Usually, the HOMO of the compound interacts with the LUMO of the target for binding, and vice versa. A higher HOMO and lower LUMO energies of the molecule imply greater target stability and binding. The lower HOMO-LUMO gap indicates that the lead has lower kinetic stability or higher chemical reactivity and polarizability. Compounds **4g** and **3f** have the most considerable ionization potential among the compounds, 6.905 eV and 6.764 eV, respectively. All of the synthesized molecules are stable within the permitted limits. Molecules with high polarizability are chemically more soft or reactive, related to chemical hardness. Figure 4 shows contour plots of the HOMO and LUMO for compounds **3f** and **4g**. The green and red contours surrounding the atoms represent the negative and positive lobes of wave functions, respectively. The green and red contours encircling the atoms depict the wave function's negative and positive lobes. It is clear from the plots that the HOMO is localized on the dimethylimidazole-pyridin, pyrazoline sites, and O and S atoms of all the molecules. In contrast, LUMO is localized on the nitrophenyl ring in all the molecules.

The distribution of electrostatic potential (EP) over atomic sites is represented by the molecular electrostatic potential (MEP) profile, which can be connected to the partial charge distribution, the electronegativity of atoms in lead molecules, and their interactions. The MEP plots of compounds **3f** and **4g** are shown in Figure 5, where the EP varies from the negative (red) to the positive value (blue) in the sequence given by the color spectrum: red (negative) < orange < yellow < green < blue (positive). The negative EPs are located on the O atomic sites of C=O and N-O2, which indicate that these sites are electron-rich. The positive EPs are seen on H atoms, particularly for H attached to the N, indicating this is an electron-deficient site. Depending on the nature of EPs, these sites would prefer to bind to sites having the opposite potential in the binding pocket or hydrogen bonding interaction (Table 3). For example, the electron-rich C=O should combine with positively charged protons of amino acid residues present in the binding pocket.

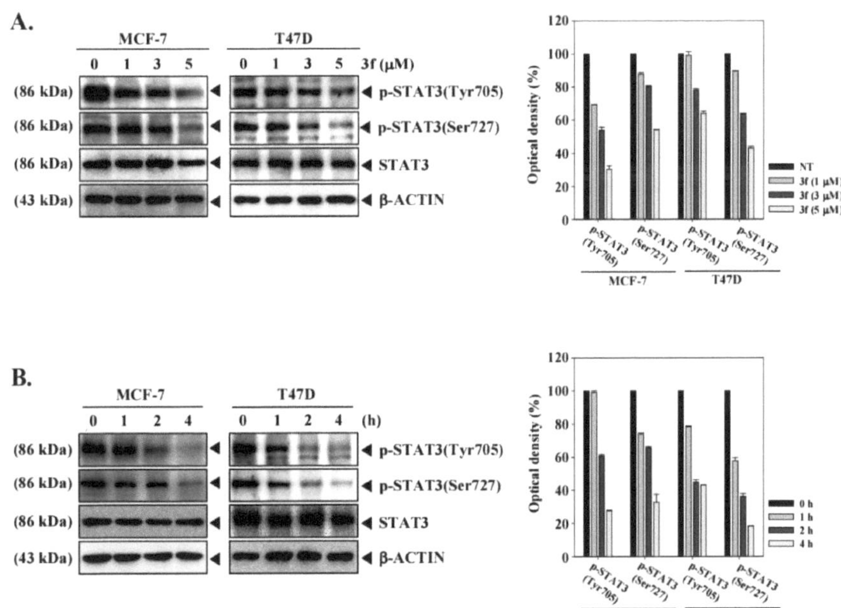

Figure 3. The lead compound **3f** decreases the phosphorylation of STAT3 in MCF-7 and T47D cells in a concentration (**A**) and time-dependent (**B**) manner. MCF-7 and T47D cells were treated with the indicated concentration of **3f** for 4 h. MCF-7 and T47D cells were treated with 5 μM of compound for the indicated time intervals. Thereafter, equal amounts of lysates were analyzed by Western blot analysis using antibodies against p-STAT3(Tyr705), p-STAT3(Ser727), or STAT3. The expression of each protein was quantified. β-ACTIN was used as input control.

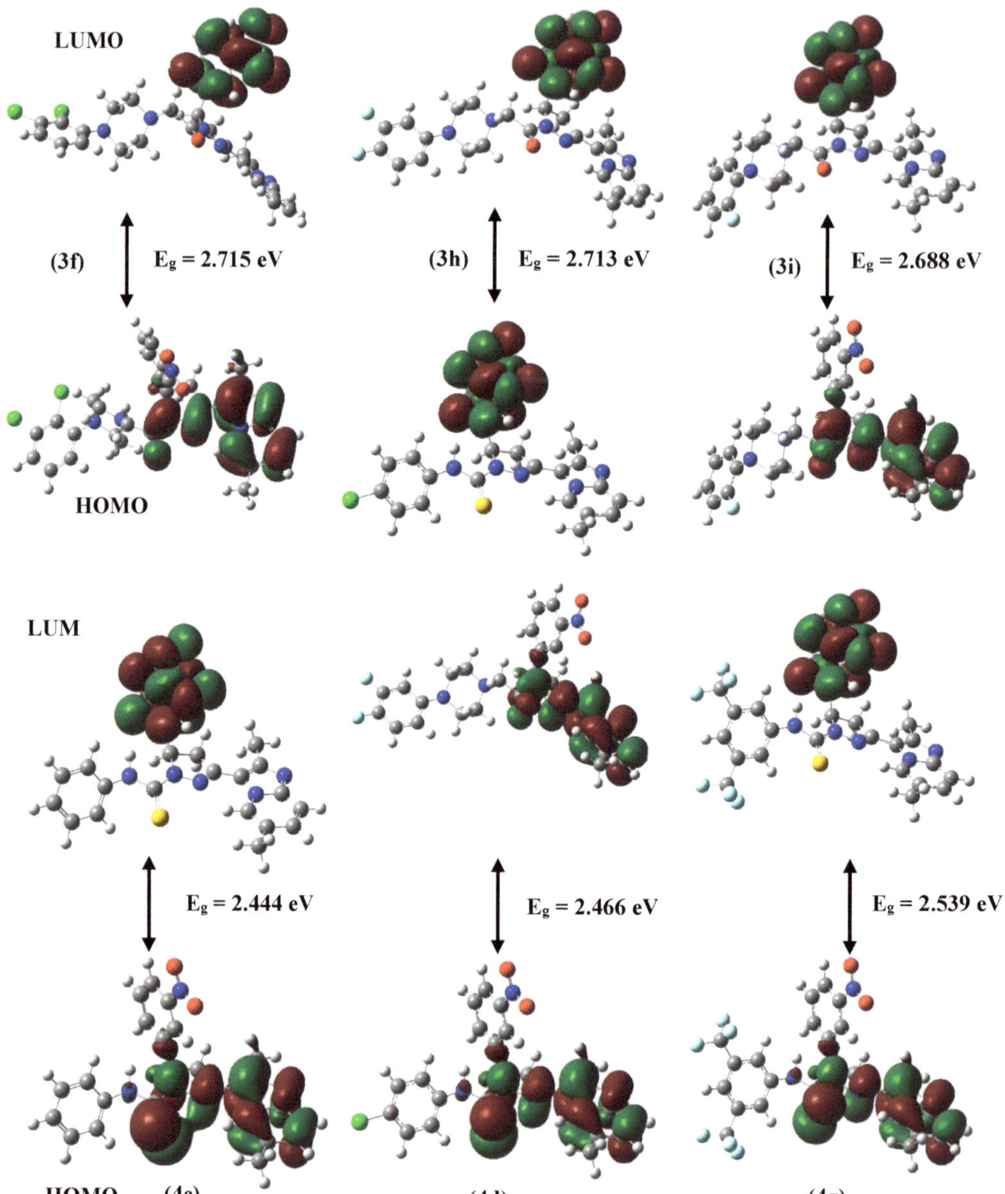

Figure 4. Computed molecular orbital plots for synthesized molecules.

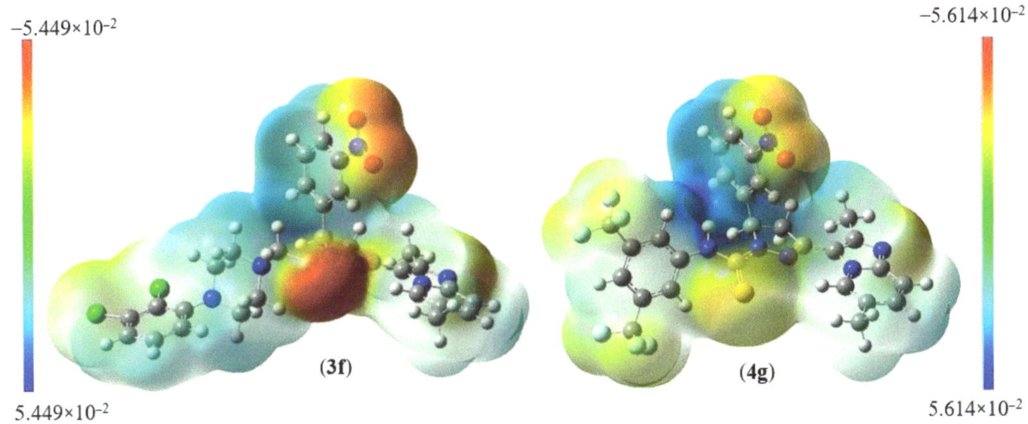

Figure 5. Molecular electrostatic charge distribution plots for **3f** and **4g**.

Table 3. FMOs energy with global chemical parameters values of synthesized compounds.

Global Parameters	3f	3h	3i	4c	4d	4g
E_{HOMO} (eV)	−5.764	−5.799	−5.704	−5.645	−5.733	−5.905
E_{LUMO} (eV)	−3.049	−3.086	−3.016	−3.201	−3.267	−3.366
$\Delta E_{LUMO\text{-}HOMO}$ (eV)	2.715	2.713	2.688	2.444	2.466	2.539
Ionization potential (I) (eV)	5.764	5.799	5.704	5.645	5.733	5.905
Electron affinity (A) (eV)	3.049	3.086	3.016	3.201	3.267	3.366
Hardness (η) (eV)	1.357	1.356	1.344	1.222	1.233	1.269
Softness (S) $(eV)^{-1}$	0.368	0.368	0.372	0.409	0.405	0.393
Chemical potential (μ) (eV)	−4.406	−4.442	−4.360	−4.423	−4.500	−4.635
Electronegativity (χ) (eV)	4.406	4.442	4.360	4.423	4.500	4.635
Electrophilicity (ψ) (eV)	7.151	7.274	7.072	8.004	8.211	8.463

The rigid docking method analyzed the synthesized compounds **3f** and **4g** [42]. AutoDock4.2 was used to determine the orientation of inhibitors bound to STAT3 (PDB ID: 1BG1) and the conformation with the highest binding energy value for each molecule. The binding modes of STAT3 inhibitors were analyzed using the PyMOL software to identify new STAT3 inhibitors. The binding site at the SH2 domain of STAT3 was described by Becker et al. [43]. It was used to elucidate the interactions that contributed to the compounds' binding affinity to STAT3.

The promising binding modes of **3f** and **4g** at the SH2 domain of the STAT3 protein were analyzed. Figures 6 and 7 show the ligand and receptor complex poses with the highest binding energy. The binding energies of **3f** and **4g** to the SH2 domain of STAT3 were observed to be −9.27 kcal/mol and −6.95 kcal/mol, respectively, indicating that the molecule has a high affinity for the target. The binding patterns of the lead molecules **3f** and **4g** were studied. Both molecules bound to the same site on the receptor molecule (STAT3) and exhibited similar interactions with the vital amino acids of the SH2 domain of STAT3. The docking results showed that the ketone group of **3f** forms a hydrogen bond with Leu706 of the SH2 domain, and the nitro group of the same molecule interacts with ARG688 via a salt bridge. One of the oxygens in the nitro group of **4g** forms a hydrogen bond and the other oxygen participated in the interaction through a salt bridge with the same ARG688 (Figure 8). To summarize, the presence of nitro groups (Figures 8 and 9) in the molecules and structural flexibility facilitates its interaction with the SH2 domain of STAT3.

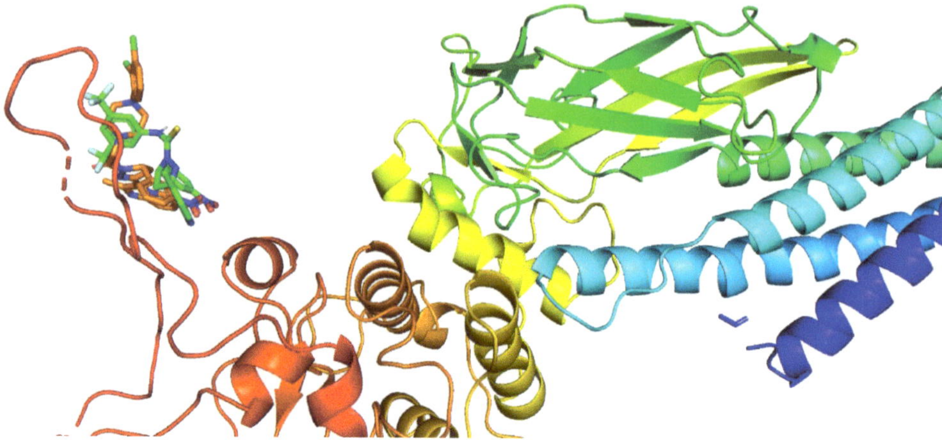

Figure 6. Cartoon representation of the SH2 domain of STAT3 with compounds **3f** and **4g** at the binding pocket.

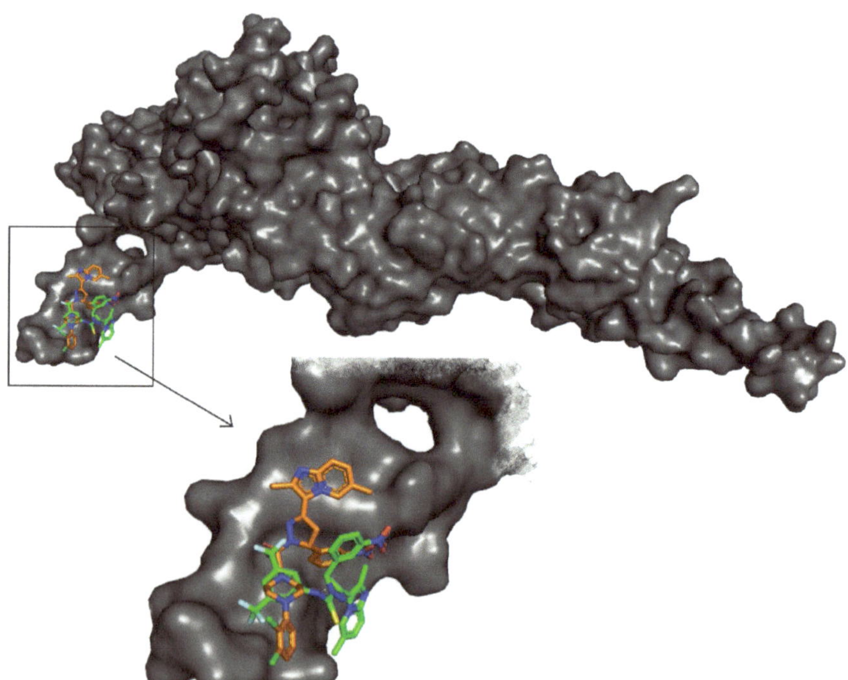

Figure 7. Surface view of STAT3 and the molecules (**3f** and **4g**) and a zoomed view of the molecules in the SH2 domain of STAT3.

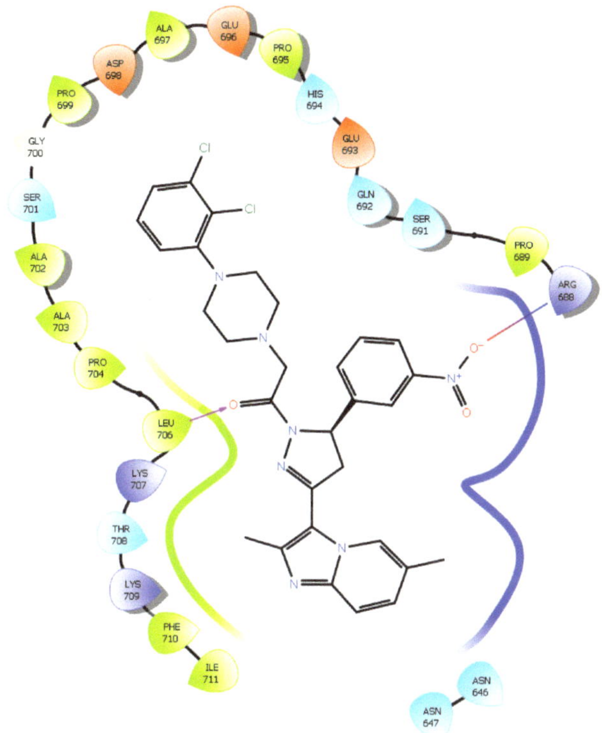

Figure 8. Two-dimensional (2D) interaction diagram of compound **3f** in the SH2 domain of STAT3.

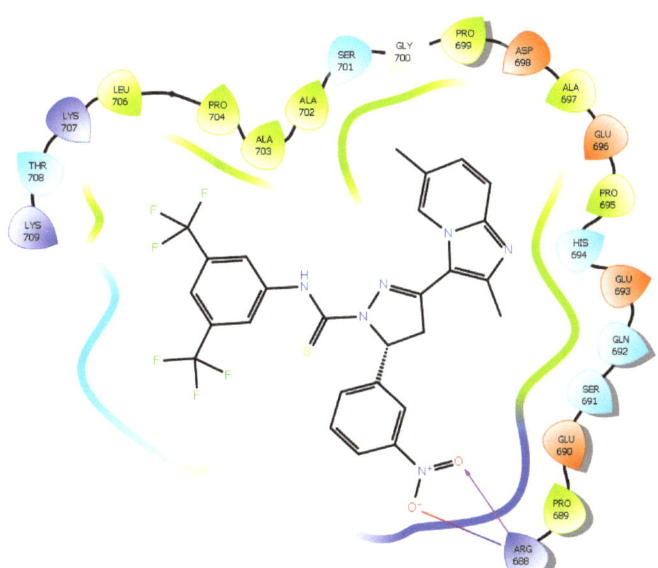

Figure 9. Two-dimensional (2D) interaction diagram of compound **4g** in the SH2 domain of STAT3.

Finally, we attempted to understand the effect of compound **3f** on the blocking of pSTAT3 into nuclear functionally since pSTAT3 dimers could enter the nucleus to give transcription. For this purpose, we conducted an immunocytochemistry assay using MCF-7 and T47D cells. We observed that compound **3f** inhibited the nuclear translocation of STAT3 in MCF-7 and T47D cells. We also analyzed the distribution of phospho-STAT3 in the nucleus and cytoplasm using fluorescent-labeled antibodies. Compound **3f** could block the nuclear translocation of pSTAT3 in MCF-7 and T47D cells, as shown in Figure 10A,B, respectively.

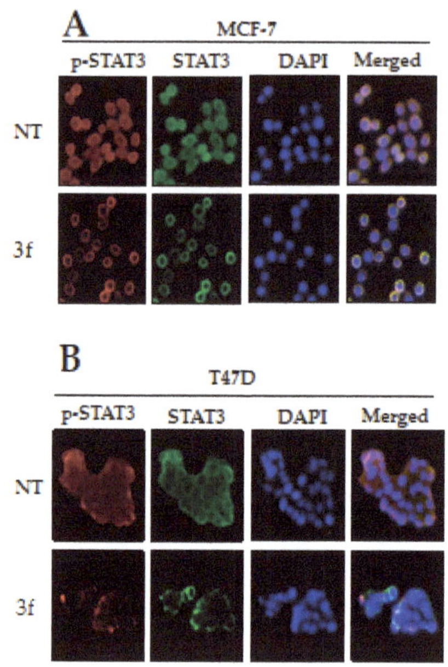

Figure 10. The nuclear localization of STAT3 in MCF-7 (**A**) and T47D (**B**) cells was inhibited by compound **3f**. MCF-7 and T47D cells were treated with the compound **3f** (5 μM), and then analyzed for intracellular distribution by immunocytochemistry. The results shown are representative of three independent experiments. The fluorescence intensity of p-STAT3 and STAT3 was performed. The merged image indicates the overlapping of p-STAT3/STAT3/DAPI images. The results shown are representative of three independent experiments.

3. Materials and Methods

All chemicals and solvents for chemistry were purchased from Sigma-Aldrich and TCI chemicals, INDIA. The completion of the reaction was monitored by pre-coated silica gel TLC plates. ^{1}H and ^{13}C NMR were recorded on an Agilent NMR spectrophotometer (400 MHz); TMS was used as an internal standard and $CDCl_3$ was used as a solvent, chemical shifts are expressed as ppm.

3.1. *General Procedure for the Synthesis of Imidazole-Pyridine Substituted Pyrazoline Derivatives*

1-(2,6-dimethylimidazo[1,2-a]pyridin-3-yl)ethenone reacting with 3-nitrobenzaldehyde in 30% KOH gives chalcone which on treating with hydrazine hydrate gives 2-pyrazolines. The synthesis of the compound was reported in earlier reports and complies with the reported molecule [34,35].

3.2. *Synthesis of 2-Pyrazoline Derivatives* (**3a–i**) *from 2,6-Dimethyl-3-(5-(3-nitrophenyl)-4,5-dihydro-1H-pyrazol-3-yl)imidazo[1,2-a]pyridine*

One mmol of (**1**) was dissolved in toluene and kept in an ice bath. After 15 min triethylamine (1.5 mmol) was added in portions followed chloroacetyl chloride (1.5 mmol) and the reaction was monitored by TLC. After the completion of the reaction, toluene was distilled off under reduced pressure and the residue was extracted with chloroform. The crude solid was used directly in the next reaction without further purification. The crude solid (1 mmol) and substituted piperazines (1 mmol) were dissolved in acetone and refluxed overnight with triethylamine (1 mmol). TLC was monitored and acetone was distilled off, the crude obtained was extracted with chloroform and recrystallized using DCM-Hexane to afford substituted pyrazoline derivatives(**3a–i**).

3.3. *Synthesis of 2-Pyrazoline Derivatives* (**4a–h**) *from 2,6-Dimethyl-3-(5-(3-nitrophenyl)-4,5-dihydro-1H-pyrazol-3-yl)imidazo[1,2-a]pyridine*

Pyrazoline derivative (**1**) (1 mmol) and substituted isothiocyanates (1.2 mmol) were dissolved in toluene and refluxed overnight with triethylamine (1 mmol). The reaction was monitored by TLC and reaction mass was filtered, washed with hexane to remove unreacted isothiocyanate, and dried in hot air oven to yield pure substituted thiourea derivatives (**4a–h**).

3.4. *1-(3-(2,6-Dimethylimidazo[1,2-a]pyridin-3-yl)-5-(3-nitrophenyl)-4,5-dihydro-1H-pyrazol-1-yl)ethanone* (**2**)

Yellow solid; MP: 166–168 °C; yield: 92%;^{1}H NMR (400 MHz, $CDCl_3$): δ 9.28 (s, 1H), 8.14–8.11 (m, 2H), 7.62 (d, J = 7.6 Hz, 1H), 7.56–7.52 (m, 2H), 7.30–7.23 (m, 1H), 5.65 (dd, J = 11.7, 4.7 Hz, 1H, CH), 4.06 (dd, J = 11.6, 5.6 Hz, 1H, CH_2), 3.35 (dd, J = 17.2, 4.8 Hz, 1H, CH_2), 2.59 (s, 3H, CH_3), 2.50 (s, 3H, CH_3), 2.44 (s, 3H, CH_3); ^{13}C NMR (100 MHz, $CDCl_3$): δ 173.40, 153.78, 153.32, 151.49, 150.77, 148.72, 137.10, 135.07, 134.93, 131.57, 128.63, 127.91, 125.82, 121.01, 118.30, 62.92(CH), 49.65(CH_2), 27.07(CH_3), 23.77(CH_3), 21.48(CH_3); MS = 377.1488, *m/z* = 378.1891 $[M + 1]^+$

3.5. *1-(2-(3-(2,6-Dimethylimidazo[1,2-a]pyridin-3-yl)-5-(3-nitrophenyl)-4,5-dihydro-1H-pyrazol-1-yl)-2-oxoethyl)piperidin-4-one* (**3a**)

Yellow solid; ^{1}H NMR (400 MHz, $CDCl_3$): δ 9.21 (s, 1H), 8.21–8.06 (m, 2H), 7.64 (d, J = 7.4 Hz, 1H), 7.54–7.53 (m, 2H), 7.27 (d, J = 9.8 Hz, 1H), 5.66 (dd, J = 11.5, 4.6 Hz, 1H, CH), 4.05 (dd, J = 17.1, 11.9 Hz, 1H, CH), 3.96 (d, J = 16.2 Hz, 1H, $COCH_2$), 3.83 (d, J = 16.2 Hz, 1H, $COCH_2$), 3.36 (dd, J = 17.2, 4.6 Hz, 1H, CH), 3.09–2.86 (m, 4H, $(CH_2)_2$), 2.59 (s, 3H, CH_3), 2.51 (t, J = 5.6 Hz, 4H, $(CH_2)_2$), 2.44 (s, 3H, CH_3); ^{13}C NMR (100 MHz, $CDCl_3$): δ 213.42, 172.00, 153.82, 150.90, 148.42, 137.20, 135.18, 135.07, 131.44, 128.70, 128.05, 125.73, 121.17, 118.15, 82.39, 82.07, 81.76, 63.71, 63.19, 58.41, 49.26, 46.30(($CH_2)_2$), 23.84(CH_3), 21.58(CH_3); MS = 474.2016, *m/z* = 475.2095 $[M + 1]^+$

3.6. *Tert-butyl-4-(2-(3-(2,6-dimethylimidazo[1,2-a]pyridin-3-yl)-5-(3-nitrophenyl)-4,5-dihydro-1H-pyrazol-1-yl)-2-oxoethyl)piperazine-1-carboxylate* (**3b**)

Yellow solid; ^{1}H NMR (400 MHz, cdcl3): δ 9.21 (s, 1H), 8.13 (d, J = 8.2 Hz, 1H), 8.09 (s, 1H), 7.62 (d, J = 7.8 Hz, 1H), 7.59–7.47 (m, 2H), 7.29–7.25 (m, 1H), 5.64 (dd, J = 11.8, 4.9 Hz, 1H, CH), 4.04 (dd, J = 17.3, 11.9 Hz, 1H, CH_2), 3.82 (d, J = 16.0 Hz, 1H, $COCH_2$), 3.69 (d, J = 16.0 Hz, 1H, $COCH_2$), 3.48 (s, 4H, $(CH_2)_2$), 3.35 (dd, J = 17.3, 5.0 Hz, 1H, CH_2), 2.65–2.59 (m, 4H, $(CH_2)_2$), 2.59 (s, 3H, CH_3), 2.44 (s, 3H, CH_3), 1.44 (s, 9H, $(CH_3)_3$); ^{13}C NMR (100 MHz, $CDCl_3$): δ 171.94, 159.69, 153.80, 153.68, 151.91, 150.87, 148.48, 137.18, 135.12, 135.03, 131.48, 128.67, 127.99, 125.79, 121.12, 118.20, 84.70(C(CH_3)), 64.68(CH_2), 63.15(CH), 58.33$(CH_2)_2$, 50.48$(CH_2)_2$, 49.23(CH_2), 33.45$(CH_3)_3$, 23.80(CH_3), 21.53(CH_3).

3.7. 1-(3-(2,6-Dimethylimidazo[1,2-a]pyridin-3-yl)-5-(3-nitrophenyl)-4,5-dihydro-1H-pyrazol-1-yl)-2-(piperazin-1-yl)ethanone (**3c**)

Yellow; ^{1}H NMR (400 MHz, $CDCl_3$): δ 9.20 (s, 1H), 8.14 (d, *J* = 7.5 Hz, 1H), 8.10 (s, 1H), 7.70–7.48 (m, 3H), 7.28 (s, 1H), 5.65 (d, *J* = 7.2 Hz, 1H, CH), 4.04 (dd, *J* = 12.0, 16.4 Hz, 1H, CH_2), 3.78 (dd, *J* = 16.0, 16.0 Hz, 2H, $COCH_3$), 3.36 (d, *J* = 16.8 Hz, 1H, CH_2), 3.08 (s, 2H, $(CH_2)_2$), 2.81 (d, *J* = 18.0 Hz, 2H, $(CH_2)_2$), 2.60 (s, 3H, CH_3), 2.47 (s, 3H, CH_3), 1.26 (s, 1H, NH).

3.8. 2-(4-Acetylpiperazin-1-yl)-1-(3-(2,6-dimethylimidazo[1,2-a]pyridin-3-yl)-5-(3-nitrophenyl)-4,5-dihydro-1H-pyrazol-1-yl)ethanone (**3d**)

Brown solid; ^{1}H NMR (400 MHz, $CDCl_3$): δ 9.20 (s, 1H), 8.20–8.05 (m, 2H), 7.62–7.53 (m, 3H), 7.28–7.26 (m, 1H), 5.64 (d, J = 7.3 Hz, 1H, CH), 4.09–3.99 (m, 1H, CH_2), 3.85 (d, J = 15.9 Hz, 1H, $COCH_2$), 3.67 (t, J = 26.2 Hz, 4H, $(CH_2)_2$), 3.51 (s, 1H, $COCH_2$), 3.39–3.32 (m, 1H, CH_2), 2.66 (d, J = 14.7 Hz, 4H, $(CH_2)_2$), 2.59 (s, 3H, $COCH_3$), 2.44 (s, 3H, CH_3), 2.07 (s, 3H, CH_3); ^{13}C NMR (100 MHz, $CDCl_3$): δ 168.90, 166.74, 148.78, 148.66, 146.99, 145.81, 143.35, 132.13, 130.08, 130.03, 126.38, 123.67, 122.96, 120.66, 116.07, 113.10, 59.34(CH), 58.11(CH_2), 53.31$(CH_2)_2$, 46.18$(CH_2)_2$, 41.35(CH_2), 21.17(CH_3), 18.74(CH_3), 16.42(CH_3); MS = 503.2281, *m/z* = 504.2384 $[M + 1]^+$

3.9. 2-(4-(4-Chlorophenyl)piperazin-1-yl)-1-(3-(2,6-dimethylimidazo[1,2-a]pyridin-3-yl)-5-(3-nitrophenyl)-4,5-dihydro-1H-pyrazol-1-yl)ethanone (**3e**)

Brown solid; ^{1}H NMR (400 MHz, $CDCl_3$): δ 9.25 (s, 1H), 8.14 (d, *J* = 8.0 Hz, 1H), 8.11 (s, 1H), 7.64 (d, *J* = 7.2 Hz, 1H), 7.63–7.46 (m, 2H), 7.26 (s, 1H), 7.17 (d, *J* = 8.0 Hz, 2H), 6.82 (d, *J* = 8.4 Hz, 2H), 5.66 (d, *J* = 7.2 Hz, 1H, CH), 4.05 (dd, *J* = 12.4, 16.0 Hz, 1H, CH_2), 3.90 (d, *J* = 15.6 Hz, 1H, $COCH_2$), 3.74 (d, *J* = 16.0 Hz, 1H, $COCH_2$), 3.37 (dd, *J* = 2.8, 16.8 Hz, 1H, CH), 3.23 (s, 4H, $(CH_2)_2$), 2.86 (d, *J* = 16..0 Hz, 4H, $(CH_2)_2$), 2.60 (s, 3H, CH_3), 2.46 (s, 3H, CH_3); ^{13}C NMR (100 MHz, $CDCl_3$): δ 166.80, 154.41, 148.67, 148.40, 146.71, 145.68, 143.38, 132.11, 129.99, 126.42, 124.35, 123.63, 122.86, 122.44, 122.36, 120.72, 118.90, 116.06, 115.95, 115.86, 113.11, 59.60(CH), 58.06(CH_2), 53.65$(CH_2)_2$, 50.23$(CH_2)_2$, 44.12(CH_2), 18.69(CH_3), 16.34(CH_3); MS = 571.21, *m/z* = 572.26 $[M + 1]^+$

3.10. 2-(4-(2,3-Dichlorophenyl)piperazin-1-yl)-1-(3-(2,6-dimethylimidazo[1,2-a]pyridin-3-yl)-5-(3-nitrophenyl)-4,5-dihydro-1H-pyrazol-1-yl)ethanone (**3f**)

Yellow solid; ^{1}H NMR (400 MHz, cdcl3): δ 9.26 (s, 1H), 8.15–8.13 (m, 2H), 7.65 (d, J = 7.4 Hz, 1H), 7.55–7.51 (m, 2H), 7.26–7.9 (m, 1H), 7.13 (d, J = 6.2 Hz, 2H), 6.95 (d, J = 6.3 Hz, 1H), 5.67 (dd, J = 11.6, 4.6 Hz, 1H, CH), 4.12–4.01 (m, 1H, CH_2), 3.90 (d, J = 15.9 Hz, 1H, $COCH_2$), 3.75 (d, J = 15.8 Hz, 1H, $COCH_2$), 3.37 (dd, J = 17.2, 4.6 Hz, 1H, CH_2), 3.12 (s, 4H, $(CH_2)_2$, 2.88 (d, J = 18.6 Hz, 4H, $(CH_2)_2$), 2.60 (s, 3H, CH_3), 2.47 (s, 3H, CH_3); ^{13}C NMR (100 MHz, $CDCl_3$): δ 167.01, 151.21, 148.77, 148.58, 146.78, 145.82, 143.50, 133.96, 132.17, 130.05, 129.96, 127.51, 127.43, 126.48, 124.60, 123.63, 122.92, 120.76, 118.68, 116.06, 113.19, 59.67(CH_2), 58.12(CH_2), 53.74$(CH_2)_2$, 51.18$(CH_2)_2$, 44.19(CH_2), 18.78(CH_3), 16.48(CH_3); MS = 605.17, *m/z* = 606.23 $[M + 1]^+$

3.11. 2-(4-(4-Chloro-2-fluorophenyl)piperazin-1-yl)-1-(3-(2,6-dimethylimidazo[1,2-a]pyridin-3-yl)-5-(3-nitrophenyl)-4,5-dihydro-1H-pyrazol-1-yl)ethanone (**3g**)

White solid; ^{1}H NMR (400 MHz, $CDCl_3$): δ 9.25 (s, 1H), 8.18–8.08 (m, 2H), 7.64 (d, J = 7.6 Hz, 1H), 7.59–7.47 (m, 2H), 7.29–7.26 (m, 1H), 7.03–7.00 (m, 2H), 6.84 (t, J = 8.7 Hz, 1H), 5.66 (dd, J = 11.7, 4.8 Hz, 1H, CH_2), 4.05 (dd, J = 17.2, 11.9 Hz, 1H, CH_2), 3.88 (d, J = 15.9 Hz, 1H, $COCH_3$), 3.73 (d, J = 15.9 Hz, 1H, $COCH_3$), 3.36 (dd, J = 17.3, 4.9 Hz, 1H, CH_2), 3.13 (s, 4H, $(CH_2)_2$), 2.86 (td, J = 10.7, 5.9 Hz, 4H, $(CH_2)_2$), 2.60 (s, 3H, CH_3), 2.45 (s, 3H, CH_3); ^{13}C NMR (100 MHz, $CDCl_3$): δ 166.92, 148.74, 148.61, 146.81, 145.82, 143.46, 138.96, 138.88, 132.19, 130.06, 129.95, 126.47, 124.45, 123.60, 122.93, 120.75, 119.59,

116.87, 116.63, 116.07, 113.17,59.68(CH), 58.12(CH_2), 53.60($(CH_2)_2$), 50.32($(CH_2)_2$), 44.19(CH_2), 18.76(CH_3), 16.50(CH_3); MS = 589.2004, *m/z* = 590.2119 $[M + 1]^+$

3.12. 2-(4-(3,4-Difluorophenyl)piperazin-1-yl)-1-(3-(2,6-dimethylimidazo[1,2-a]pyridin-3-yl)-5-(3-nitrophenyl)-4,5-dihydro-1H-pyrazol-1-yl)ethanone (**3h**)

Brown solid; ^1H NMR (400 MHz, $CDCl_3$): δ 9.24 (s, 1H), 8.11 (s, 2H), 7.63–7.53 (m, 3H), 7.25 (s, 1H), 7.30–7.27 (m, 1H), 6.72–6.64 (m, 1H), 6.56 (d, J = 8.2 Hz, 1H), 5.66 (dd, J = 11.8, 4.9 Hz, 1H, CH), 4.05 (dd, J = 17.3, 11.9 Hz, 1H, CH_2), 3.90 (d, J = 15.9 Hz, 1H, $COCH_3$), 3.75 (d, J = 15.9 Hz, 1H, $COCH_3$), 3.39–3.33 (m, 1H, CH_2), 3.19 (s, 4H, $(CH_2)_2$), 2.86 (d, J = 15.5 Hz, 4H, $(CH_2)_2$), 2.60 (s, 3H, CH_3), 2.46 (s, 3H, CH_3);^{13}C NMR (100 MHz, $CDCl_3$): δ 166.60, 148.71, 148.51, 146.83, 145.71, 143.30, 132.10, 130.01, 129.98, 126.38, 123.60, 122.90, 122.89, 120.63, 117.15, 116.98, 116.00, 113.07, 111.44, 105.40, 105.20, 77.21, 76.90, 76.58, 59.34, 58.05, 53.20, 49.28, 44.13, 18.70, 16.37.

3.13. 1-(3-(2,6-Dimethylimidazo[1,2-a]pyridin-3-yl)-5-(3-nitrophenyl)-4,5-dihydro-1H-pyrazol-1-yl)-2-(4-(2-fluorophenyl)piperazin-1-yl)ethanone (**3i**)

Brown solid; ^1H NMR (400 MHz, $CDCl_3$): δ 9.25 (s, 1H), 8.14–8.12 (m, 2H), 7.65–7.49 (m, 3H), 7.28–7.24 (m, 1H), 7.04–6.90 (m, 4H), 5.66 (dd, J = 11.5, 4.7 Hz, 1H, CH), 4.04 (dd, J = 17.1, 11.9 Hz, 1H, CH_2), 3.89 (d, J = 15.9 Hz, 1H, $COCH_3$), 3.75 (d, J = 15.9 Hz, 1H, $COCH_2$), 3.36 (dd, J = 17.1, 4.4 Hz, 1H, CH_2), 3.17 (s, 4H, $(CH_2)_2$), 2.88 (d, J = 17.8 Hz, 4H, $(CH_2)_2$), 2.60 (s, 3H, CH_3), 2.45 (s, 3H, CH_3); ^{13}C NMR (100 MHz, $CDCl_3$): δ 166.80, 154.41, 148.67, 148.40, 146.71, 145.68, 143.38, 132.11, 129.99, 126.42, 124.35, 123.63, 122.86, 122.44, 122.36, 120.72, 118.90, 116.06, 115.95, 115.86, 113.11, 59.60(CH), 58.06(CH_2), 53.65($(CH_2)_2$), 50.23($(CH_2)_2$), 44.12(CH_2), 18.69(CH_3), 16.34(CH_3).

3.14. 3-(2,6-Dimethylimidazo[1,2-a]pyridin-3-yl)-5-(3-nitrophenyl)-N-phenyl-4,5-dihydro-1H-pyrazoline-1-carbothioamide (**4a**)

White solid; ^1H NMR (400 MHz, cdcl3): δ 9.13 (s, 1H), 9.04 (s, 1H), 8.18–8.09 (m, 2H), 7.66 (d, J = 7.7 Hz, 3H), 7.56 (dd, J = 17.5, 8.5 Hz, 2H), 7.37 (t, J = 7.8 Hz, 2H), 7.29 (d, J = 8.8 Hz, 1H), 7.19 (s, 1H), 6.25 (dd, J = 11.6, 3.6 Hz, 1H, CH), 4.15 (dd, J = 17.2, 11.7 Hz, 1H, CH_2), 3.39 (dd, J = 17.2, 3.7 Hz, 1H, CH_2), 2.60 (s, 3H, CH_3), 2.46 (s, 3H, CH_3); ^{13}C NMR (100 MHz, $CDCl_3$): δ 173.18, 149.96, 148.64, 147.58, 146.05, 143.87, 138.53, 132.10, 130.31, 129.86, 128.73, 125.92, 125.55, 124.00, 123.73, 122.69, 120.75, 116.29, 112.82, 61.22(CH), 44.57(CH_2), 18.81(CH_3), 16.72(CH_3); MS = 470.1525, *m/z* = 471.1626 $[M + 1]^+$

3.15. 3-(2,6-Dimethylimidazo[1,2-a]pyridin-3-yl)-5-(3-nitrophenyl)-N-(4-nitrophenyl)-4,5-dihydro-1H-pyrazole-1-carbothioamide (**4b**)

Orange solid; ^1H NMR (400 MHz, $CDCl_3$): δ 10.13 (s, 1H), 9.37 (s, 1H), 9.13 (s, 1H), 8.14 (s, 1H), 8.01-7.94 (m, 3H), 7.54–7.60 (t, *J* = 10.0 Hz, 2H), 7.34 (d, 2H), 6.23 (dd, *J* = 2.0, 2.0 Hz, 1H, CH), 4.19 (dd, *J* = 1.0, 1.0 Hz, 1H, CH_2), 3.84–3.76 (m, 1H, CH_2), 2.63 (s, 3H, CH_3), 2.51 (s, 3H, CH_3); ^{13}C NMR (100 MHz, $CDCl_3$): δ 173.18, 149.96, 148.64, 147.58, 146.05, 143.87, 138.53, 132.10, 130.31, 129.86, 128.73, 125.92, 125.55, 124.00, 123.73, 122.69, 120.75, 116.29, 112.82, 61.22(CH), 44.57(CH_2), 18.81(CH_3), 16.72(CH_3); MS = 515.1376, *m/z* = 516.2025 $[M + 1]^+$

3.16. N-Cyclohexyl-3-(2,6-dimethylimidazo[1,2-a]pyridin-3-yl)-5-(3-nitrophenyl)-4,5-dihydro-1H-pyrazoline-1-carbothioamide (**4c**)

Yellow solid; ^1H NMR (400 MHz, $CDCl_3$): δ 9.10 (s, 1H), 8.08 (s, 2H), 7.64–7.48 (m, 3H), 7.27–7.29 (m, 1H), 7.21 (d, J = 8.1 Hz, 1H), 6.15 (dd, J = 11.7, 3.8 Hz, 1H, CH), 4.07 (dd, J = 17.1, 11.7 Hz, 1H, CH_2), 3.34 (dd, J = 17.1, 3.9 Hz, 1H, CH_2), 2.58 (s, 3H, CH_3), 2.46 (s, 3H, CH_3), 2.09 (s, 1H, CH), 1.73 (s, 4H, $(CH_2)_2$), 1.54–1.45 (m, 2H, CH_2), 1.45–1.29 (m, 4H, $(CH_2)_2$); ^{13}C NMR (100 MHz, $CDCl_3$): δ 168.90, 166.74, 148.78, 148.66, 146.99, 145.81, 143.35, 132.13, 130.09, 126.38, 123.67, 122.96, 120.66, 116.07, 113.10, 59.34(CH), 58.11(CH),

53.31$(CH_2)_2$, 46.18$(CH_2)_2$, 44.18(CH_2), 41.35(CH_2), 18.74(CH_3), 16.42(CH_3); MS = 476.20, *m/z* = 477.25 [M + 1]$^+$

3.17. *N-(4-Chlorophenyl)-3-(2,6-dimethylimidazo[1,2-a]pyridin-3-yl)-5-(3-nitrophenyl)-4,5-dihydro-1H-pyrazoline-1-carbothioamide* (**4d**)

Yellow solid; ^{1}H NMR (400 MHz, $CDCl_3$): δ 9.11 (s, 1H), 8.97 (s, 1H), 8.15 (d, J = 6.1 Hz, 2H), 7.67 (d, J = 7.6 Hz, 1H), 7.63–7.52 (m, 4H), 7.34 (d, J = 8.7 Hz, 2H), 7.25 (s, 1H), 6.23 (dd, J = 11.5, 3.5 Hz, 1H, CH), 4.16 (dd, J = 17.2, 11.7 Hz, 1H, CH_2), 3.41 (dd, J = 17.2, 3.6 Hz, 1H, CH_2), 2.61 (s, 3H, CH_3), 2.47 (s, 3H, CH_3); ^{13}C NMR (100 MHz, CDCl3): δ 173.18, 149.96, 148.64, 147.58, 146.05, 143.87, 138.53, 132.10, 130.31, 129.86, 128.73, 125.92, 125.55, 124.00, 123.73, 122.69, 120.75, 116.29, 112.82, 61.22(CH), 44.57(CH_2), 18.81(CH_3), 16.72(CH_3); MS = 476.1681, *m/z* = 485.2469 [M + 1]$^+$

3.18. *N-(3-Chlorophenyl)-3-(2,6-dimethylimidazo[1,2-a]pyridin-3-yl)-5-(3-nitrophenyl)-4,5-dihydro-1H-pyrazoline-1-carbothioamide* (**4e**)

Yellow solid; ^{1}H NMR (400 MHz, $CDCl_3$): δ 9.11 (s, 1H), 9.03 (s, 1H), 8.17–8.11 (m, 2H), 7.77 (t, J = 2.0 Hz, 1H), 7.66 (d, J = 7.8 Hz, 1H), 7.60 (d, J = 9.0 Hz, 1H), 7.55 (dd, J = 7.0, 1.7 Hz, 2H), 7.31 (dd, J = 13.8, 4.8 Hz, 2H), 7.18–7.12 (m, 1H), 6.22 (dd, J = 11.6, 3.8 Hz, 1H, CH), 4.16 (dd, J = 17.3, 11.7 Hz, 1H, CH_2), 3.40 (dd, J = 17.3, 3.9 Hz, 1H, CH_2), 2.61 (s, 3H, CH_3), 2.48 (s, 3H, CH_3).; ^{13}C NMR (100 MHz, $CDCl_3$): δ 167.01, 151.21, 148.77, 148.58, 146.78, 145.82, 143.50, 133.96, 132.17, 130.05, 129.96, 127.51, 127.43, 126.48, 124.60, 123.63, 122.92, 120.76, 118.68, 116.06, 113.19, 53.74(CH), 51.18(CH_2), 18.78(CH_3), 16.48(CH_3).

3.19. *3-(2,6-Dimethylimidazo[1,2-a]pyridin-3-yl)-5-(3-nitrophenyl)-N-(p-tolyl)-4,5-dihydro-1H-pyrazoline-1-carbothioamide* (**4f**)

White solid; ^{1}H NMR (400 MHz, $CDCl_3$): δ 9.13 (s, 1H), 8.93 (s, 1H), 8.17–8.12 (m, 2H), 7.67 (d, J = 7.7 Hz, 1H), 7.59 (d, J = 9.0 Hz, 1H), 7.55 (d, J = 7.9 Hz, 1H), 7.49 (d, J = 8.3 Hz, 2H), 7.30 (dd, J = 9.1, 1.5 Hz, 1H), 7.19 (d, J = 8.2 Hz, 2H), 6.26 (dd, J = 11.6, 3.8 Hz, 1H, CH), 4.15 (dd, J = 17.2, 11.7 Hz, 1H, CH_2), 3.39 (dd, J = 17.2, 3.8 Hz, 1H, CH_2), 2.61 (s, 3H, CH_3), 2.46 (s, 3H, CH_3), 2.34 (s, 3H, CH_3); ^{13}C NMR (100 MHz, $CDCl_3$): δ 173.72, 149.77, 148.71, 147.53, 145.98, 144.00, 135.94, 135.71, 132.21, 130.44, 129.93, 129.41, 125.99, 124.40, 124.11, 122.77, 120.80, 116.30, 112.92, 77.32, 77.00, 76.69, 61.32, 44.62, 21.01, 18.88, 16.74.

3.20. *N-(3,5-Bis(trifluoromethyl)phenyl)-3-(2,6-dimethylimidazo[1,2-a]pyridin-3-yl)-5-(3-nitrophenyl)-4,5-dihydro-1H-pyrazoline-1-carbothioamide* (**4g**)

Yellow solid; ^{1}H NMR (400 MHz, $CDCl_3$): δ 9.31 (s, 1H), 9.13 (s, 1H), 8.26 (s, 2H), 8.15 (s, 1H), 7.66 (s, 1H), 7.64-7.53 (m, 2H), 7.34 (d, J = 9.0 Hz, 1H), 7.26 (t, J = 3.7 Hz, 1H), 7.18 (s, 1H), 6.21 (dd, J = 11.5, 3.8 Hz, 1H, CH), 4.20 (dd, J = 17.3, 11.6 Hz, 1H, CH_2), 3.45 (dd, J = 17.3, 3.8 Hz, 1H, CH_2), 2.63 (s, 3H, CH_3), 2.49 (s, 3H, CH_3); ^{13}C NMR (100 MHz, $CDCl_3$): δ 172.21, 150.70, 148.74, 148.46, 146.34, 143.25, 140.10, 132.24, 131.85, 130.74, 130.06, 129.00, 128.18, 125.98, 125.25, 124.33, 123.03, 122.65, 121.68, 120.79, 118.36, 116.52, 112.64, 61.33(CH), 44.73(CH_2), 18.73(CH_3), 16.83(CH_3); MS = 606.13, *m/z* = 607.19 [M + 1]$^+$

3.21. *N-Butyl-3-(2,6-dimethylimidazo[1,2-a]pyridin-3-yl)-5-(3-nitrophenyl)-4,5-dihydro-1H-pyrazoline-1-carbothioamide* (**4h**)

White solid; ^{1}H NMR (400 MHz, $CDCl_3$): δ 9.09 (s, 1H), 8.16–8.06 (m, 2H), 7.56 (ddd, J = 25.3, 17.7, 7.8 Hz, 3H), 7.28 (dd, J = 9.1, 1.5 Hz, 1H), 7.20 (t, J = 5.1 Hz, 1H), 6.15 (dd, J = 11.7, 3.9 Hz, 1H CH), 4.08 (dd, J = 17.2, 11.8 Hz, 1H, CH_2), 3.69 (qd, J = 13.2, 6.3 Hz, 2H, CH_2), 3.34 (dd, J = 17.2, 4.0 Hz, 1H, CH_2), 2.58 (s, 3H, CH_3), 2.46 (s, 3H, CH_3), 1.70 (dd, J = 8.5, 7.1 Hz, 2H, CH_2), 1.50 (dd, J = 15.1, 7.4 Hz, 2H, CH_2), 1.00 (t, J = 7.3 Hz, 3H, CH_3); ^{13}C NMR (100 MHz, $CDCl_3$): δ 175.43, 149.31, 148.69, 146.88, 145.85, 144.44, 132.15, 130.15, 129.88, 125.94, 123.79, 122.66, 120.69, 116.25, 113.04, 61.30(CH), 44.53(CH_2),

44.50(CH_2), 31.22(CH_2), 20.27(CH_2), 18.70(CH_3), 16.63(CH_3), 13.8(CH_3); MS = 450.1838, m/z = 451.2311 $[M + 1]^+$

3.22. Cell Viability Assay

MCF-7 cells were procured from Procell Life Science and Technology Co. LTD, Wuhan, China. Human breast cancer T47D, BT474, and SK-BR-3 cells were obtained from American Type Culture Collection (Washington, DC, NW, USA). Cells (2000) were cultured in MEM or Leibovitz's L-15 medium enriched with 2% FBS and maintained in a humidified atmosphere of 5% CO_2 at 37 °C. A stock solution of DMSO-dissolved compounds was prepared, and this solution was diluted with culture medium as required. MCF-7 Cells (4×10^3) were cultured in 96-well plates for 12 h and then treated for 72 h with compounds at concentrations of 0, 0.01, 0.1, 10, 100, and 1000 mM. Using Alamar Blue, the compounds were evaluated for their inhibitory effects [44,45]

3.23. Preparation of Whole Cell Lysates

As previously reported, whole-cell lysates from cells treated with compound **3f** were prepared to detect protein expression and phosphorylation [46] using a lysis buffer [Tris (20 mM, pH 7.4), NaCl (250 mM), EDTA (2 mM, pH 8.0), Triton X-100 (0.1%), aprotinin (0.01 mg/mL), leupeptin (0.005 mg/mL), phenylmethane sulfonyl fluoride (0.4 mM), and NaVO4 (4 mM)]. To remove insoluble material, lysates were centrifuged at 13,000 rpm for 15 min.

3.24. Western Blot Analysis

Equal protein concentrations of cell lysate were resolved on sodium dodecyl sulfate-polyacrylamide gel electrophoresis, followed by transfer to nitrocellulose membranes as reported previously [47]. Incubation was carried out overnight at 4 °C with antibodies after treatment with 5% skim milk. Afterward, the membranes were washed, probed with HRP-conjugated secondary antibodies for 2 h, and then visualized using chemiluminescence.

3.25. In Silico DFT Calculations

The theoretical calculations were performed using Gaussian 09 [48] and Gaussview 5 program. The polarized and diffused basis set 6-311+G(d, p) provides accurate values for all theoretical calculations. The computational studies utilized the most useful and precise hybrid method of B3LYP [49]. The structures of the compounds were fully optimized with no constraint. The global chemical reactivity descriptors (GCRD) were evaluated to understand the chemical properties of a molecule, such as ionization potential, electron affinity, chemical hardness (η), softness (S), potential (μ), electronegativity (χ), and electrophilic index (ψ). The global hardness [$\eta = (E_{LUMO} - E_{HOMO})/2$], softness ($S = 1/2\eta$), chemical potential [$\mu = (E_{HOMO} + E_{LUMO})/2$], electronegativity [$\chi = (I+A)/2$], and electrophilic index ($\psi = \mu^2/2\eta$) were calculated by taking the energies of HOMO as ionization potential (I) and LUMO as electron affinity (A). Chemical hardness, softness, and potential were used to understand the chemical reactivity of the molecular system [50].

3.26. Docking Simulation

AUTODOCK4.0 [51] software was employed for molecular docking studies. The docking receptor STAT3 (PDB ID: 1BG1) was retrieved from the RCSB Protein Data Bank. The graphical user interface AUTODOCK TOOLS was utilized to build up the protein molecule. The water molecules were removed from the protein crystal and only polar hydrogens were applied. The predicted gasteigers charge was found to be −25.9962. For both dockings, the grid box size was 127 × 127 × 85 with a grid spacing of 0.55Å. The receptor and the complex were saved in the pdbqt file format. Using Lamarckian genetic algorithm searches, twenty runs were performed. The default parameters were employed, with a maximum of 2.5×10^6 energy assessments and an initial population of 50 randomly placed individuals [52]. The autogrid4.exe and autodock4.exe functions were executed at the end of the docking process to generate glg and dlg files. Maestro (v2020.4) [53] and PyMOL (v2.5.2) [54] were used to generate the interaction pictures and visualization plots.

3.27. Immunocytochemistry Assay

As described earlier, STAT3 phosphorylation in cells was quantified [55]. After compound **3f** treatment (10 μM for 4 h), cells were fixed for 20 min with paraformaldehyde (4%). Thereafter, cells were treated with 0.2% Triton X-100 in phosphate-buffered saline for permeabilization, followed by blocking with 5% bovine serum albumin for 1 h. Then, the preparation was incubated overnight at 4 °C with a rabbit polyclonal anti-human STAT3 antibody (dilution, 1:100). The next day, slides were subjected to washing and incubation with Alexa Fluor 594 (dilution, 1:1000) anti-Rabbit IgG1 for 1 h at room temperature in the dark. In the next step, DAPI (5 μg/mL) was used for counterstaining the nuclei. The slides were mounted and analyzed under an Olympus FluoView FV1000 confocal microscope (Tokyo, Japan).

4. Conclusions

A series of imidazopyridine-tethered-purazoles were synthesized and screened for loss of viability of breast cancer cells. The lead compound **3f** inhibited STAT3 phosphorylation in MCF-7 and T47D cells. The DFT calculations and molecular docking experiments showed a theoretical bioactivity correlation for compound **3f** towards STAT3. In conclusion, compound **3f** effectively inhibited the phosphorylation of STAT3 in MCF-7 and T47D cells, indicating that ITPs may be an alternative method to target STAT3 in BC.

Supplementary Materials: The following are available online at https://www.mdpi.com/article/10.3390/bioengineering10020159/s1. Supplementary data for newly synthesized molecules and their IC50 values determined against human breast cancer cells.

Author Contributions: A.R., R.S., M.H.Y., J.-R.Y., Z.X., A.S. and O.N. performed the experiments; B.B., M.M., P.E.L., K.S.A., A.C., S.A.A. and V.P. designed the experiments; B.B., G.S., K.S.A., M.M. and P.E.L. provided resources; B.B., P.E.L., G.S. and K.S.A. wrote the manuscript. All authors have read and agreed to the published version of the manuscript.

Funding: This work was supported by DBT-NER, and Vision Group on Science and Technology (CESEM), Government of Karnataka. This work was also supported by Singapore MOE Tier 1 grant to GS. This work was supported by a National Research Foundation of Korea (NRF) grant funded by the Korean government (MSIP) (NRF-2021R1I1A2060024). This work was also supported by the Shenzhen Key Laboratory of Innovative Oncotherapeutics (ZDSYS20200820165400003) (Shenzhen Science and Technology Innovation Commission), China; Shenzhen Development and Reform Commission Subject Construction Project ([2017]1434), China; Overseas Research Cooperation Project (HW2020008) (Tsinghua Shenzhen International Graduate School), China; Tsinghua University Stable Funding Key Project (WDZC20200821150704001); the Shenzhen Bay Laboratory (21310031), China and TBSI Faculty Start-up Funds, China. This project was supported by Researchers Supporting Project number (RSP2023R5), King Saud University, Riyadh, Saudi Arabia.

Institutional Review Board Statement: Not applicable.

Informed Consent Statement: Not applicable.

Data Availability Statement: Not applicable.

Conflicts of Interest: The authors declare no conflict of interest.

Sample Availability: The synthesized compounds are available at Basappa Laboratory.

References

1. Ghoncheh, M.; Pournamdar, Z.; Salehiniya, H. Incidence and Mortality and Epidemiology of Breast Cancer in the World. *Asian Pac. J. Cancer Prev.* **2016**, *17*, 43–46. [CrossRef] [PubMed]
2. Bray, F.; Ferlay, J.; Soerjomataram, I.; Siegel, R.L.; Torre, L.A.; Jemal, A. Global cancer statistics 2018: GLOBOCAN estimates of incidence and mortality worldwide for 36 cancers in 185 countries. *CA Cancer J. Clin.* **2018**, *68*, 394–424. [CrossRef] [PubMed]

3. Wenger, C.R.; Clark, G.M. S-phase fraction and breast cancer—A decade of experience. *Breast Cancer Res Treat.* **1998**, *51*, 255–265. [CrossRef] [PubMed]
4. Pandya, V.; Githaka, J.M.; Patel, N.; Veldhoen, R.; Hugh, J.; Damaraju, S.; McMullen, T.; Mackey, J.; Goping, I.S. BIK drives an aggressive breast cancer phenotype through sublethal apoptosis and predicts poor prognosis of ER-positive breast cancer. *Cell Death Dis.* **2020**, *11*, 448. [CrossRef] [PubMed]
5. van Kuilenburg, A.B.; Maring, J.G. Evaluation of 5-fluorouracil pharmacokinetic models and therapeutic drug monitoring in cancer patients. *Pharmacogenomics* **2013**, *14*, 799–811. [CrossRef] [PubMed]
6. Osborne, C.K. Tamoxifen in the treatment of breast cancer. *N. Engl. J. Med.* **1998**, *339*, 1609–1618. [CrossRef] [PubMed]
7. Gourmelon, C.; Bourien, H.; Augereau, P.; Patsouris, A.; Frenel, J.S.; Campone, M. Vinflunine for the treatment of breast cancer. *Expert Opin. Pharmacother.* **2016**, *17*, 1817–1823. [CrossRef] [PubMed]
8. Tutt, A.N.J.; Garber, J.E.; Kaufman, B.; Viale, G.; Fumagalli, D.; Rastogi, P.; Gelber, R.D.; de Azambuja, E.; Fielding, A.; Balmaña, J.; et al. Clinical Trial Steering Committee and Investigators. Adjuvant Olaparib for Patients with BRCA1- or BRCA2-Mutated Breast Cancer. *N. Engl. J. Med.* **2021**, *384*, 2394–2405. [CrossRef] [PubMed]
9. Symmans, F.W. Breast cancer response to paclitaxel in vivo. *Drug Resist. Updat.* **2001**, *4*, 297–302. [CrossRef] [PubMed]
10. Cuzick, J.; Sestak, I.; Forbes, J.F.; Dowsett, M.; Cawthorn, S.; Mansel, R.E.; Loibl, S.; Bonanni, B.; Evans, D.G.; Howell, A. IBIS-II investigators. Use of anastrozole for breast cancer prevention (IBIS-II): Long-term results of a randomised controlled trial. *Lancet* **2020**, *395*, 117–122. [CrossRef] [PubMed]
11. Cuzick, J.; Sestak, I.; Baum, M.; Buzdar, A.; Howell, A.; Dowsett, M.; Forbes, J.F. ATAC/LATTE investigators. Effect of anastrozole and tamoxifen as adjuvant treatment for early-stage breast cancer: 10-year analysis of the ATAC trial. *Lancet Oncol.* **2010**, *11*, 1135–1141. [CrossRef] [PubMed]
12. Hortobagyi, G.N.; Stemmer, S.M.; Burris, H.A.; Yap, Y.S.; Sonke, G.S.; Paluch-Shimon, S.; Campone, M.; Blackwell, K.L.; André, F.; Winer, E.P.; et al. Ribociclib as First-Line Therapy for HR-Positive, Advanced Breast Cancer. *N. Engl. J. Med.* **2016**, *375*, 1738–1748. [CrossRef] [PubMed]
13. Jia, L.Y.; Shanmugam, M.K.; Sethi, G.; Bishayee, A. Potential role of targeted therapies in the treatment of triple-negative breast cancer. *Anticancer. Drugs* **2016**, *27*, 147–155. [CrossRef] [PubMed]
14. Lee, J.H.; Kim, C.; Kim, S.H.; Sethi, G.; Ahn, K.S. Farnesol inhibits tumor growth and enhances the anticancer effects of bortezomib in multiple myeloma xenograft mouse model through the modulation of STAT3 signaling pathway. *Cancer Lett.* **2015**, *360*, 280–293. [CrossRef] [PubMed]
15. Furtek, S.L.; Backos, D.S.; Matheson, C.J.; Reigan, P. Strategies and Approaches of Targeting STAT3 for Cancer Treatment. *ACS Chem. Biol.* **2016**, *11*, 308–318. [CrossRef] [PubMed]
16. Garg, M.; Shanmugam, M.K.; Bhardwaj, V.; Goel, A.; Gupta, R.; Sharma, A.; Baligar, P.; Kumar, A.P.; Goh, B.C.; Wang, L.; et al. The pleiotropic role of transcription factor STAT3 in oncogenesis and its targeting through natural products for cancer prevention and therapy. *Med. Res. Rev.* **2021**, *41*, 1291–1336. [CrossRef] [PubMed]
17. Lee, J.H.; Kim, C.; Baek, S.H.; Ko, J.H.; Lee, S.G.; Yang, W.M.; Um, J.Y.; Sethi, G.; Ahn, K.S. Capsazepine inhibits JAK/STAT3 signaling, tumor growth, and cell survival in prostate cancer. *Oncotarget* **2017**, *8*, 17700–17711. [CrossRef] [PubMed]
18. Baek, S.H.; Ko, J.H.; Lee, H.; Jung, J.; Kong, M.; Lee, J.W.; Lee, J.; Chinnathambi, A.; Zayed, M.E.; Alharbi, S.A.; et al. Resveratrol inhibits STAT3 signaling pathway through the induction of SOCS-1: Role in apoptosis induction and radiosensitization in head and neck tumor cells. *Phytomedicine* **2016**, *23*, 566–577. [CrossRef] [PubMed]
19. Ma, J.H.; Qin, L.; Li, X. Role of STAT3 signaling pathway in breast cancer. *Cell Commun. Signal.* **2020**, *18*, 33. [CrossRef] [PubMed]
20. Hartenfeller, M.; Schneider, G. De novo drug design. *Methods Mol. Biol.* **2011**, *672*, 299–323. [CrossRef] [PubMed]
21. Khatun, S.; Singh, A.; Bader, G.N.; Sofi, F.A. Imidazopyridine, a promising scaffold with potential medicinal applications and structural activity relationship (SAR): Recent advances. *J. Biomol. Struct. Dyn.* **2022**, *40*, 14279–14302. [CrossRef] [PubMed]
22. Brown, D.G.; Wobst, H.J. A Decade of FDA-Approved Drugs (2010–2019): Trends and Future Directions. *J. Med. Chem.* **2021**, *64*, 2312–2338. [CrossRef] [PubMed]
23. He, L.J.; Yang, D.L.; Chen, H.Y.; Huang, J.H.; Zhang, Y.J.; Qin, H.X.; Wang, J.L.; Tang, D.Y.; Chen, Z.Z. A Novel Imidazopyridine Derivative Exhibits Anticancer Activity in Breast Cancer by Inhibiting Wnt/β-catenin Signaling. *Onco. Targets Ther.* **2020**, *13*, 10111–10121. [CrossRef] [PubMed]
24. Su, J.C.; Chang, C.H.; Wu, S.H.; Shiau, C.W. Novel imidazopyridine suppresses STAT3 activation by targeting SHP-1. *J. Enzyme Inhib. Med. Chem.* **2018**, *33*, 1248–1255. [CrossRef] [PubMed]
25. Godse, P.; Kumar, P.; Yewalkar, N.; Deore, V.; Lohar, M.; Mundada, R.; Padgaonkar, A.; Manohar, S.; Joshi, A.; Bhatia, D.; et al. Discovery of P3971 an orally efficacious novel anticancer agent targeting HIF-1α and STAT3 pathways. *Anticancer. Agents Med. Chem.* **2013**, *13*, 1460–1466. [CrossRef] [PubMed]
26. Nichols, W.C.; Kvols, L.K.; Ingle, J.N.; Edmonson, J.H.; Ahmann, D.L.; Rubin, J.; O'Connell, M.J. Phase II study of triazinate and pyrazofurin in patients with advanced breast cancer previously exposed to cytotoxic chemotherapy. *Cancer Treat Rep.* **1978**, *62*, 837–839. [PubMed]
27. Tołoczko-Iwaniuk, N.; Dziemiańczyk-Pakieła, D.; Nowaszewska, B.K.; Celińska-Janowicz, K.; Miltyk, W. Celecoxib in Cancer Therapy and Prevention—Review. *Curr. Drug Targets.* **2019**, *20*, 302–315. [CrossRef] [PubMed]
28. Nehra, B.; Rulhania, S.; Jaswal, S.; Kumar, B.; Singh, G.; Monga, V. Recent advancements in the development of bioactive pyrazoline derivatives. *Eur. J. Med. Chem.* **2020**, *205*, 112666. [CrossRef] [PubMed]

29. Santoro, A.; Pisanti, S.; Grimaldi, C.; Izzo, A.A.; Borrelli, F.; Proto, M.C.; Malfitano, A.M.; Gazzerro, P.; Laezza, C.; Bifulco, M. Rimonabant inhibits human colon cancer cell growth and reduces the formation of precancerous lesions in the mouse colon. *Int. J. Cancer* **2009**, *125*, 996–1003. [CrossRef] [PubMed]
30. Mamytbeková, A.; Hájícek, J.; Grimová, J.; Rezábek, K. Reductive effect of lonazolac on lung metastasis formation in mice. *Neoplasma* **1990**, *37*, 349–355. [PubMed]
31. Srinivasa, V.; Li, F.; Siveen, K.S.; Dai, X.; Swamy, S.N.; Sethi, G.; Mantelingu, K.; Bender, A.; Rangappa, K.S. Synthesis and biological evaluation of tetrahydropyridinepyrazoles ('PFPs') as inhibitors of STAT3 phosphorylation. *Med. Chem. Commun.* **2014**, *5*, 32–40.
32. Zhang, L.; Peterson, T.E.; Lu, V.M.; Parney, I.F.; Daniels, D.J. Antitumor activity of novel pyrazole-based small molecular inhibitors of the STAT3 pathway in patient derived high grade glioma cells. *PLoS ONE* **2019**, *14*, e0220569. [CrossRef] [PubMed]
33. Wang, F.; Feng, K.R.; Zhao, J.Y.; Zhang, J.W.; Shi, X.W.; Zhou, J.; Gao, D.; Lin, G.Q.; Tian, P. Identification of novel STAT3 inhibitors bearing 2-acetyl-7-phenylamino benzofuran scaffold for antitumour study. *Bioorg. Med. Chem.* **2020**, *28*, 115822. [CrossRef] [PubMed]
34. Anilkumar, N.C.; Sundaram, M.S.; Mohan, C.D.; Rangappa, S.; Bulusu, K.C.; Fuchs, J.E.; Girish, K.S.; Bender, A.; Basappa Rangappa, K.S. A One Pot Synthesis of Novel Bioactive Tri-Substitute-Condensed-Imidazopyridines that Targets Snake Venom Phospholipase A2. *PLoS ONE* **2015**, *10*, e0131896. [CrossRef] [PubMed]
35. Kuthyala, S.; Hanumanthappa, M.; Kumar, S.M.; Sheik, S.; Karikannar, N.G.; Prabhu, A. Crystal, Hirshfeld, ADMET, drug-like and anticancer study of some newly synthesized imidazopyridine containing pyrazoline derivatives. *J. Mol. Struct.* **2019**, *1197*, 65–72. [CrossRef]
36. Basappa, B.; Chumadathil Pookunoth, B.; Shinduvalli Kempasiddegowda, M.; Knchugarakoppal Subbegowda, R.; Lobie, P.E.; Pandey, V. Novel Biphenyl Amines Inhibit Oestrogen Receptor (ER)-α in ER-Positive Mammary Carcinoma Cells. *Molecules* **2021**, *26*, 783. [CrossRef] [PubMed]
37. Bharathkumar, H.; Mohan, C.D.; Ananda, H.; Fuchs, J.E.; Li, F.; Rangappa, S.; Surender, M.; Bulusu, K.C.; Girish, K.S.; Sethi, G.; et al. Microwave-assisted synthesis, characterization and cytotoxic studies of novel estrogen receptor α ligands towards human breast cancer cells. *Bioorg. Med. Chem. Lett.* **2015**, *25*, 1804–1807. [CrossRef] [PubMed]
38. Kalakoti, Y.; Yadav, S.; Sundar, D. Deep Neural Network-Assisted Drug Recommendation Systems for Identifying Potential Drug-Target Interactions. *ACS Omega* **2022**, *7*, 12138–12146. [CrossRef] [PubMed]
39. Sebastian, A.; Pandey, V.; Mohan, C.D.; Chia, Y.T.; Rangappa, S.; Mathai, J.; Baburajeev, C.P.; Paricharak, S.; Mervin, L.H.; Bulusu, K.C.; et al. Novel Adamantanyl-Based Thiadiazolyl Pyrazoles Targeting EGFR in Triple-Negative Breast Cancer. *ACS Omega* **2016**, *1*, 1412–1424. [CrossRef] [PubMed]
40. Lee, J.H.; Mohan, C.D.; Deivasigamani, A.; Jung, Y.Y.; Rangappa, S.; Basappa, S.; Chinnathambi, A.; Alahmadi, T.A.; Alharbi, S.A.; Garg, M.; et al. Brusatol suppresses STAT3-driven metastasis by downregulating epithelial-mesenchymal transition in hepatocellular carcinoma. *J. Adv. Res.* **2020**, *26*, 83–94. [CrossRef] [PubMed]
41. Fleming, I. *Molecular Orbitals and Organic Chemical Reactions*; Wiley: Hoboken, NJ, USA, 2010; ISBN 9780470746585.
42. Ansari, M.F.; Siddiqui, S.M.; Ahmad, K.; Avecilla, F.; Dharavath, S.; Gourinath, S.; Azam, A. Synthesis, antiamoebic and molecular docking studies of furan-thiazolidinone hybrids. *Eur. J. Med. Chem.* **2016**, *124*, 393–406. [CrossRef] [PubMed]
43. Becker, S.; Groner, B.; Müller, C.W. Three-dimensional structure of the Stat3beta homodimer bound to DNA. *Nature* **1998**, *394*, 145–151. [CrossRef] [PubMed]
44. Pandey, V.; Wang, B.; Mohan, C.D.; Raquib, A.R.; Rangappa, S.; Srinivasa, V.; Fuchs, J.E.; Girish, K.S.; Zhu, T.; Bender, A.; et al. Discovery of a small-molecule inhibitor of specific serine residue BAD phosphorylation. *Proc. Natl. Acad. Sci. USA* **2018**, *115*, E10505–E10514. [CrossRef] [PubMed]
45. Barash, U.; Rangappa, S.; Mohan, C.D.; Vishwanath, D.; Boyango, I.; Basappa, B.; Vlodavsky, I.; Rangappa, K.S. New Heparanase-Inhibiting Triazolo-Thiadiazoles Attenuate Primary Tumor Growth and Metastasis. *Cancers* **2021**, *13*, 2959. [CrossRef] [PubMed]
46. Zhang, J.; Sikka, S.; Siveen, K.S.; Lee, J.H.; Um, J.Y.; Kumar, A.P.; Chinnathambi, A.; Alharbi, S.A.; Basappa Rangappa, K.S.; Sethi, G.; et al. Cardamonin represses proliferation, invasion, and causes apoptosis through the modulation of signal transducer and activator of transcription 3 pathway in prostate cancer. *Apoptosis* **2017**, *22*, 158–168. [CrossRef] [PubMed]
47. Mohan, C.D.; Bharathkumar, H.; Bulusu, K.C.; Pandey, V.; Rangappa, S.; Fuchs, J.E.; Shanmugam, M.K.; Dai, X.; Li, F.; Deivasigamani, A.; et al. Development of a novel azaspirane that targets the Janus kinase-signal transducer and activator of transcription (STAT) pathway in hepatocellular carcinoma in vitro and in vivo. *J. Biol. Chem.* **2014**, *289*, 34296–34307. [CrossRef] [PubMed]
48. Frisch, M.J.; Trucks, G.W.; Schlegel, H.B.; Scuseria, G.E.; Robb, M.A.; Cheeseman, J.R. *Gaussian 09*; Gaussian Inc.: Wallingford, CT, USA, 2009.
49. Lee, W.C.; Yang, R.G. Parr, Development of the Colle-Salvetti correlation-energy formula into a functional of the electron density. *Phys. Rev. B* **1988**, *37*, 785. [CrossRef]
50. Ananda, S.; Khamees, H.A.; Mahendra, M.; Kumara, C.; Jagadeesh Prasad, D.; Hegde, T.A.; Vinitha, G. Structural, thermal, dielectric, nonlinear optical properties and DFT investigations of a novel material 2-(6-chloropyridin-3-yl)-N'-(2, 3-dihydro-1, 4-benzodioxin-6-ylmethylidene) acetohydrazide for optoelectronic applications. *J. Mater. Sci. Mater. Electron.* **2021**, *32*, 14677–14702. [CrossRef]
51. Schrödinger, L.; DeLano, W. PyMOL. 2020. Available online: http://www.pymol.org/pymol (accessed on 1 October 2022).

52. Sanner, M.F. Python: A programming language for software integration and development. *J Mol Graph Model* **1999**, *17*, 57–61.
53. Abbasi-Radmoghaddam, Z.; Riahi, S.; Gharaghani, S.; Mohammadi-Khanaposhtanai, M. Design of potential anti-tumor PARP-1 inhibitors by QSAR and molecular modeling studies. *Mol. Divers.* **2021**, *25*, 263–277. [CrossRef]
54. Schrödinger, LLC. *Schrödinger Release 2020-1: Maestro*; Schrödinger, LLC: New York, NY, USA, 2020.
55. Kim, J.W.; Gautam, J.; Kim, J.E.; Kim, J.A.; Kang, K.W. Inhibition of tumor growth and angiogenesis of tamoxifen-resistant breast cancer cells by ruxolitinib, a selective JAK2 inhibitor. *Oncol. Lett.* **2019**, *17*, 3981–3989. [CrossRef]

bioengineering

MDPI

Article

Hyper-Branched Cyclodextrin-Based Polymers as Anticoagulant Agents: In Vitro and In Vivo Studies

Yousef Khazaei Monfared [1], Mohammad Mahmoudian [2], Gjylije Hoti [1], Daniel Mihai Bisericaru [1], Fabrizio Caldera [1], Roberta Cavalli [3], Parvin Zakeri-Milani [4,*], Adrián Matencio [1,*,†] and Francesco Trotta [1,*,†]

1 Dipartimento di Chimica and NIS, Università di Torino, Via P. Giuria 7, 10125 Torino, Italy
2 Faculty of Pharmacy, Tabriz University of Medical Sciences, Tabriz 5166-15731, Iran
3 Dipartimento di Scienza e Tecnologia del Farmaco, Università di Torino, Via P. Giuria 9, 10125 Torino, Italy
4 Liver and Gastrointestinal Diseases Research Centre and Faculty of Pharmacy, Tabriz University of Medical Sciences, Tabriz 5166-15731, Iran
* Correspondence: pzakeri@tbzmed.ac.ir (P.Z.-M.); adrian.matencioduran@unito.it or adrian.matencio@um.es (A.M.); francesco.trotta@unito.it (F.T.)
† These authors contributed equally to this work.

Abstract: This study tested the anticoagulant effect of cyclodextrin (CD) hyper-branched-based polymers (HBCD-Pols). These polymers were synthesized and tested for their coagulant characteristics in vitro and in vivo. Due to their polymeric structure and anionic nature, the polymers can chelate Ca^{2+}, reducing the free quantity in blood. HBCD-Pol increased the blood clotting time, PT, and aPTT 3.5 times over the control, showing a better effect than even ethylenediaminetetraacetic acid (EDTA), as occured with recalcification time as well. A titration of HBCD-Pol and EDTA showed exciting differences in the ability to complex Ca^{2+} between both materials. Before executing in vivo studies, a hemocompatibility study was carried out with less than 5% red blood cell hemolysis. The fibrinogen consumption and bleeding time were analyzed in vivo. The fibrinogen was considerably decreased in the presence of HBCD-Pol in a higher grade than EDTA, while the bleeding time was longer with HBCD-Pols. The results demonstrate that the anticoagulant effect of this HBCD-Pol opens novel therapy possibilities due to the possible transport of drugs in this carrier. This would give combinatorial effects and a potential novel anticoagulant therapy with HBCD-Pol per se.

Keywords: cyclodextrin; hyper-branched polymers; coagulation; calcium; EDTA; therapy; in vivo

Citation: Monfared, Y.K.; Mahmoudian, M.; Hoti, G.; Bisericaru, D.M.; Caldera, F.; Cavalli, R.; Zakeri-Milani, P.; Matencio, A.; Trotta, F. Hyper-Branched Cyclodextrin-Based Polymers as Anticoagulant Agents: In Vitro and In Vivo Studies. *Bioengineering* **2022**, 9, 765. https://doi.org/10.3390/bioengineering9120765

Academic Editor: Traian Chirila

Received: 26 October 2022
Accepted: 2 December 2022
Published: 4 December 2022

1. Introduction

Recent reports suggest thrombosis is responsible for one in four deaths worldwide, which is expected to increase due to the aging of the population. There are two types of thrombosis, arterial and venous; they are closely related to the existence of one another, and different factors, such as infections or diet, can influence their appearance. For example, they may be associated with cardiovascular diseases such as hyperlipidemia, smoking, or diabetes, which provoke the rupture of atherosclerotic plaques and promote the formation of a thrombus [1].

Antithrombotic agents are divided into three families: anticoagulants, anticoagulants and fibrinolytic agents. Another type is the anticoagulants that prevent fibrin formation, which predominates in venous thrombi. There are four FDA-approved anticoagulants: (i) heparins, which bind antithrombin [2], (ii) vitamin K antagonists (warfarin, [3]), (iii) direct thrombin inhibitors, and (iv) direct FXa inhibitors (Table 1). Typically, all current agents are associated with different secondary effects, such as bleeding. They can be oral or injected, although oral anticoagulants could present non-significant reductions in overall major bleeding, significantly lower rates of intracranial hemorrhage, and higher rates of gastrointestinal bleeding [4]. Due to all of the above, it has been continuously interesting to discover novel drugs with reduced adverse effects or carriers that increase the activity of the drug while decreasing the adverse effects. The increased bioavailability of drugs is an excellent option to achieve better results and safe profiles of drugs. In this field, the use of a carrier called cyclodextrin (CD) is highly considered.

Table 1. Examples of anticoagulant drugs in comparison with EDTA and HBCD-Pol.

	Anticoagulant Drugs		**Molecule**	**Polymer**
	Enoxaparin (synthetic heparin) ǂ	Warfarin ǂ	EDTA ǂ	HBCD-Pol [5]
Schematic structure	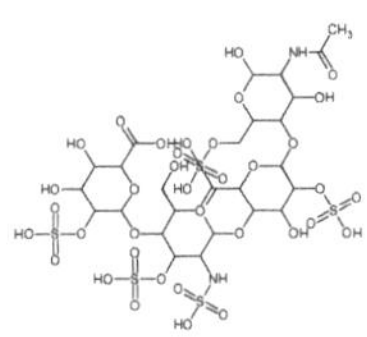	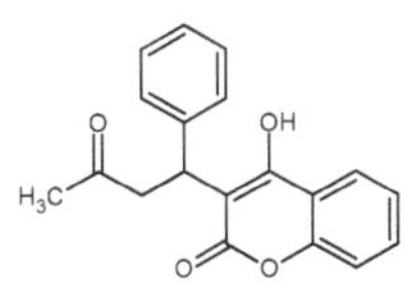	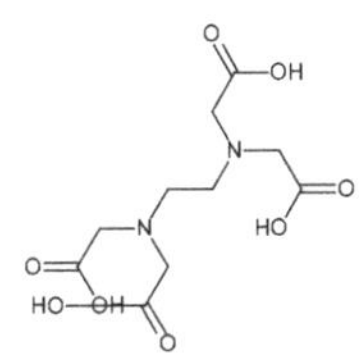	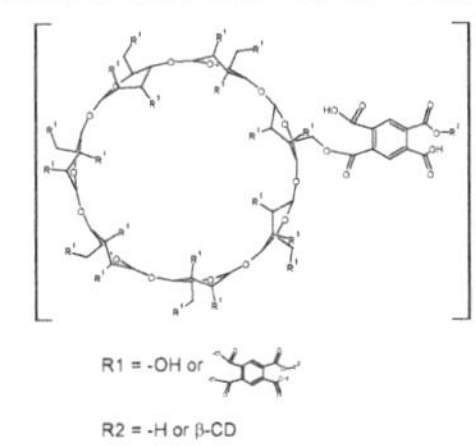
Description	Synthetic heparin, drug	Synthetic anticoagulant	Chelant	Polymer
Molecular formula	C26H42N2O37S5	C19H16O4	C10H16N2O8	(C162H94o107)n
Molecular weight	1134.9	308.3	292.24	37–42Kda
Solubility (mg/mL)	50	17×10^{-3}	1×10^{3}	800
Bioactivity	Drug similar to that of heparin (binds to antithrombin, AT), although it exhibits a higher ratio of anti-Factor Xa to anti-Factor Iia activity	Drug; inhibits the regeneration of vitamin K1 epoxide and thus the synthesis of vitamin K-dependent clotting factors, which include Factors II, VII, IX and X, and the anticoagulant proteins C and S	Chelant; induces anticoagulation by Ca^{2+} binding	Aim of present work
Can vehiculize biomolecules?	-	-	-	Yes

ǂextracted from PUBCHEM (CID 772, 54678486 and 6049 on 23 November 2022).

CDs are well-known to the scientific community for their ability to solubilize poorly soluble drugs [6–8]. Chemically, CDs are cone-shaped oligosaccharides obtained from starch with α-(1,4) linked glucose units. The most common CDs are the natural derivatives with six, seven, and eight glucose units, α-, β- and γ-cyclodextrin (CD), respectively. The CD ring is a conical cylinder of an amphiphilic nature, with a hydrophilic outer layer (formed by the hydroxyl groups) and a lipophilic cavity [9]. Classically, poorly soluble drugs are complexed with CD, creating the so-called "inclusion complex", which increases its solubility, stability, or bioactivity [10–12].

In some cases, their capacity lacks efficiency due to the drug's complex structure or the desire for different profiles (e.g., slower release). Different materials have been prepared to improve their properties. This is the case of the hyper-branched CD-based polymers (HBCD-Pol, [5], Table 1), which are soluble but with a three-dimensional network and tunable structure with little/no toxicity [13]. Curiously, although free CDs did not present a coagulation effect [14], HBCD-Pol demonstrated the capacity to chelate different cationic ions, which could originate a combinatorial effect between the increase in activity of some anticoagulant for the complexation and the interferences with the Ca^{2+} signal [5,15].

Based on this last point, these intrinsic capacities for eventual use as an anticoagulant are presented in this research. This will open a gate for novel therapeutic approaches where HBCD-Pol can be administered with different drugs to control and manage the coagulation from a combinatorial point of view. In particular, different points will be studied in three different blocks:

(I) The capacity of HBCD-Pol to chelate Ca^{2+};
(II) The in vitro anticoagulant activity and its hemocompatibility to determine the blood clotting time and the plasma recalcification time. After these, we study the prothrombin time (PT) and the partial prothrombin time (PTT);
(III) The in vivo PT, PTT, fibrinogen, and bleeding time using rats as a model.

2. Materials and Methods

2.1. Materials

β-cyclodextrin (β-CD) was kindly provided by Roquette Freres (Lestrem–France). Pyromellitic anhydride (PMDA), dimethylsulfoxide (DMSO), triethylamine (Et_3N), ethylenediaminetetraacetic acid (EDTA), thrombin from bovine plasma, partial thromboplastin time (aPTT) and prothrombin time (PT) assay agents were purchased from Sigma-Aldrich (Milan, Italy). Animal food was supplied by Javaneh Khorasan Co. (Iran). All chemicals and reagents used were of analytical grade unless otherwise specified.

2.2. In Vitro Protocols and Experiments

2.2.1. Synthesis of HBCD-Pol

Hyper-branched water-soluble β-CD polymer (HBCD-Pol) was prepared as reported [5]. Briefly, 6 mL of anhydrous DMSO and 1 mL of triethylamine were placed in a glass scintillation vial round-bottom flask and 0.997 g of β-CD was added until complete dissolution. Then, the required quantity of PMDA (pyromellitic dianhydride) was added to achieve a 1:12 (CD:linker) molar ratio, and the solution was allowed to react for 24 h at room temperature. Once the reaction was completed, the product was precipitated by adding ethyl acetate, and an excess of ethyl acetate was added during filtration to remove impurities. After drying, the product was solubilized in deionized water, lyophilized, and finally, residuals and unreacted reagents were completely removed by Soxhlet extraction with acetone for 24 h. Elemental analysis was used to confirm the total removal of DMSO. The white powder was dried and ground in a mortar, then preserved in dark and dry conditions.

2.2.2. Potentiometric Titration

The chelating agents used in potentiometric titration were HBCD-Pol and EDTA, forming chelates through the reaction with Ca^{2+}. The titration was carried out according to the procedure described in the literature [16]. A 0.1 M NaOH solution, pre-standardized with 0.025 M oxalic acid solution, was used as a titrant. Titration was carried out using a volumetric manual burette by adding 0.1 mL of the titrant under gentle stirring of the analyte solution (50 mL) at room temperature. Subsequently, 500 mg of EDTA, HBCD-Pol and their complexes with 20 mg of Ca^{2+} ions were used as analytes. To reach an equilibrium between readings, it was necessary to add the titrant around every 60 s. pH values were continuously measured and recorded after each addition using a pH meter until a pH of 12 was reached. The titration curve of pH versus NaOH titration volume was generated, and the curve's inflection point, using a second derivative method, was found for the indicated transition. Then, the volume of NaOH consumed at the inflection point is applied to the Equation (1) used for the calculation of the milliequivalents of acidity per 100 g of sample (m_{eq}):

$$\frac{m_{eq}\text{of acidity}}{100\text{ g sample}} = \frac{V_s * c_{NaOH} * 100}{m_s} \quad (1)$$

where V_s is the volume of NaOH consumed by the sample, c_{NaOH} is the concentration of NaOH in mol/L and m_s is the mass of the sample [16].

2.2.3. Anticoagulant Activity

Fresh Wistar male rat's blood was used to evaluate the anticoagulant activity of HBCD-Pol compared to EDTA and control at room temperature without citrate sodium addition, according to the literature [17]. Different concentrations of 250, 500, 1000, 1500 and 2000 (μg/mL blood) of HBCD-Pol (5%) and concentrations of 500 and 1000 (μg/mL blood) for EDTA (5%) as a positive control were separately added to glass tubes with 1 mL of fresh rat blood. The blood that was free of additions was drawn to the tube as a control. Almost all blood samples were then checked over the next 1 and 3 h for detectable changes at room temperature, such as the formation of a clot.

2.2.4. In Vitro Clotting Time

Blood clotting time (CBT) is a primary method to identify the prominent hemostatic agent [18]. Blood samples were collected from Wistar male rats and mixed with 3.8% sodium citrate. CBT was started by adding 1500, 1000 and 500 μg/mL of HBCD-Pol (5%) and the same concentrations for EDTA (5%) into a 5 mL glass tube, followed by incubating at 37 °C for 5 min, and the control group added nothing. Then, 1 mL of citrated rat blood was mixed with the samples and continued for incubation at 37 °C for 3 min. After that, 500 μL of 0.025 M $CaCl_2$ aqueous solution was added into the tubes to trigger the coagulation pathway. The tubes were taken out of the water bath and inclined every 30 s until the blood in the tube did not flow. The clotting time was recorded as the result of CBT.

2.2.5. Plasma Recalcification Time (PRT)

Plasma recalcification time (PRT) was determined according to the hook method [18]. Briefly, collected blood was mixed with 3.8% trisodium citrate (1/10 vol). Platelet-poor plasma (PPP) was obtained by centrifuging the fresh blood at 3000 rpm for 15 min at 4 °C. A fresh 300 μL of PPP was incubated at 37 °C with 1500, 1000 and 500 μg/mL blood of HBCD-Pol and the same amount for EDTA at tube tests for 3 min. The recalcification of plasma was checked using 30 μL of 0.5 M $CaCl_2$. Plasma was observed every 1 min to estimate fibrin thread formation in PPP. PRT was designated as the time to recalcification from the time of the addition of $CaCl_2$.

2.2.6. Partial Thromboplastin Time (aPTT) and Prothrombin Time (PT) Assays

All the test samples, test reagents and $CaCl_2$ solution were incubated at 37 °C in advance [18,19]. To test aPTT, 100 μL of PPP and 100 μL of aPTT reagent were incubated at 37 °C for 5 min, and then different concentrations (1500, 1000 and 500 μg/mL blood) of HBCD-Pol 5% and the same volume of EDTA 5% was added. Then, 100 μL of $CaCl_2$ 0.025 M was added to the tubes. The prothrombin time (PT) test was performed by adding 100 μL of PPP, which was mixed with both HBCD-Pol and EDTA 5% as the same used concentrations for aPTT and then 200 μL of PT reagents to the tube successively. The negative controls were considered pure aPTT and PT reagents without samples. The PTT and PT were tested with an automatic coagulometer (ACL 300R, Instrumentation Laboratory).

2.2.7. Hemocompatibility Test

The hemolytic activity of HBCD-Pol was evaluated according to the standard technique [20]. The freshly citrated prepared blood samples were collected and centrifuged at 1000 rpm for 5 min. Then, the plasma was removed, and approximately 5 mL of sterile phosphate-buffered saline (PBS), pH 7.4, was added with repeated centrifugation at 1000 rpm for 5 min to remove the plasma residues. Then, 1000 μL of HBCD-Pol were mixed with blood tubes and incubated at 37 °C and 150 RPM for 90 min. After incubation, the tubes were placed on an ice bath for 5 min and then centrifuged at 1000 rpm for 5 min. Then, 100 μL of supernatant was removed from the tube and then diluted with a refrigerated PBS solution to yield 2% v/v. The same procedure was repeated with PBS as a negative control and Triton X-100 (0.1% by volume) as a positive control. Finally, each sample's absorbance was calculated at a value of 540 nm.

2.3. *In Vivo Experiments*

For in vivo experiments' published protocols with modifications were used [21]. Briefly, BALB/c male mice weighing 20–25 g were used. The HBCD-Pol and EDTA were administered intraperitoneally (i.p.) in a fixed volume of 200 μL in three dosages, 40, 20 and 10 mg/kg, for both agents. In each experimental session, at least five animals per treatment group were tested; control groups were run at the beginning and at the end of every experimental session. Mice were accustomed to handling by the investigators, and the injections were carried out by skilled investigators with minimal disturbance to the animals. The experiments were approved by the Tabriz University of Medical Sciences Pharmacy Department under the Iran Ethical Code: IR.TBZMED.AEC.1401.022.

2.3.1. APPT and PT Assays

Blood was collected from ether-anesthetized mice by cardiac puncture and anticoagulated with 3.8% trisodium citrate (1/10 vol) of mice treated by HBCD-Pol and EDTA at the determined dosage 10 min after injection. Anticoagulated blood was immediately centrifuged for 5 min at 12,000× *g*, and the supernatant platelet-poor plasma (PPP) was separated and transferred onto melting ice until tested (generally within 1 h) or frozen at –80 °C. Both activities were measured by standard assays, as mentioned in Section 2.2.6.

2.3.2. Fibrinogen Measurement

Blood was collected from ether-anesthetized BALB/c mice by cardiac puncture and anticoagulated with trisodium citrate (1/10 vol) of mice treated with HBCD-Pol and EDTA at a determined dosage 2 min before the thrombotic challenge, which was induced by the i.p. injection of bovine thrombin. The dose of thrombin used was selected as 1000 U/kg [22]. Plasma fibrinogen was measured by the Clauss method in a Coagulab MJ coagulometer (Ortho Diagnostics) using bovine thrombin.

2.3.3. Bleeding Time

Bleeding time was assessed by a tail transection method [22,23]. Briefly, BALB/c mice were treated with HBCD-Pol and EDTA at 40 m/kg dosage; this dosage was selected according to the results of our previously mentioned experiments for 10 min. Then, the mice were positioned in a special immobilization cage that kept the tail steady and immersed in saline thermostated at 37 °C. After 2 min, the tip of the tail was transected with a razor blade at approximately 2.5 mm from its end. The tail was immediately reimbursed in warm saline, and the bleeding time was recorded. The endpoint was the arrest of bleeding lasting for more than 30 s.

3. Results and Discussion

3.1. *Potentiometric Titration*

As mentioned above, the capacity of HBCD-Pol to chelate metals was previously described [5]. However, it is crucial to determine the interaction ability of the synthesized polymer. Ethylenediaminetetraacetic acid (EDTA) is a well-known chelating agent used in several industries to determine Ca^{2+}, although its activity can be affected by other ions, such as Na^+ [24,25].

Figure 1 presents the delay on the equivalent point of EDTA compared to HBCD-Pol. The results suggest the more robust binding sites of EDTA that are characterized by 581.6 m_{eq} of acidity, compared to HBCD-Pol with 461.1 m_{eq} of acidity. It is important to mention that near the physiological and blood pH, the volumes of NaOH are ~25 mL for HBCD-Pol and ~31 mL for EDTA. Therefore, the higher the presence of COOH groups in the solution, the higher the interaction, and the volume of NaOH required to neutralize the compound (the equivalence point) is also high.

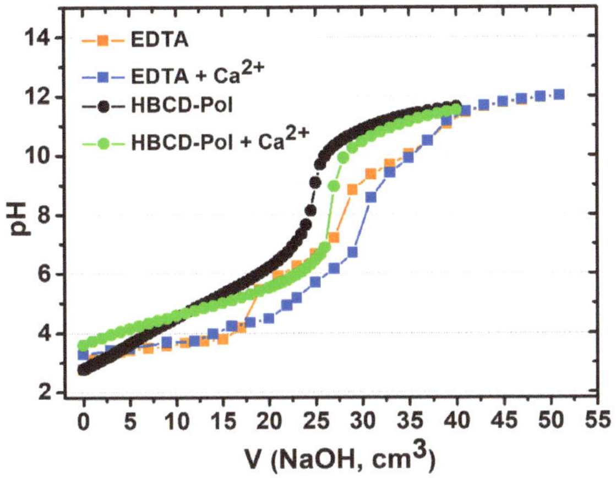

Figure 1. Potentiometric titration curves of EDTA, HBCD-Pol, and their complexes with Ca^{2+}.

The results demonstrated (i) a higher chelating strength of EDTA in comparison with HBCD-Pol, with closer values at physiological pH; and (ii) the presence of Ca^{2+} affects both compounds. It is reasonable to have a lower effect of HBCD-Pol than EDTA against Ca^{2+}. However, the effect on global coagulation is affected by several factors, not only Ca^{2+}, such as the presence of other molecules that can be complexed by HBCD-Pol, affecting the global coagulation profile.

3.2. In-Vitro Anticoagulant Study

The anticoagulant activity was evaluated by adding the HBCD-Pol and EDTA to rat blood samples at different concentrations. Figure 2 shows the results, where the control sample without addition, HBCD-Pol (250 µg/mL) and EDTA (500 µg/mL) initiated coagulation after 15, 20 and 18 min of incubation, respectively. In addition, the thickened blood samples were checked after 60 min to find a large clot in the tubes. Surprisingly, after one hour, there were no signs of clotting in the HBCD-Pol and EDTA at 500 and 1000 µg/mL, but after three hours, only some clotting was observed for both of them, although these clots did not prevent blood flow.

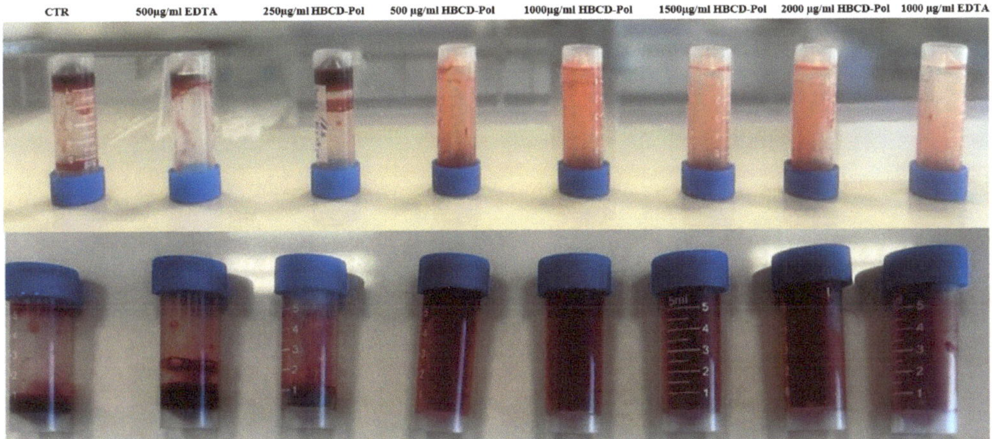

Figure 2. Anticoagulant activities of HBCD-Pol and EDTA after 3 h.

The results suggest that HBCD-Pol could present a higher anticoagulation effect than EDTA. Therefore, to better understand its ability, the in vitro clotting time (CT) using fresh blood clotting time was carried out according to the literature [2]. Blood samples were collected and mixed with 3.8% sodium citrate as an anticoagulant, and then the samples were warmed to 37 °C in a water bath and added to 20 μL of 0.2 M $CaCl_2$. The CT was designated as the time when no flow of samples was observed after the tubes were inverted.

Interestingly, the results showed that the time of clotting was significantly increased by about two times by HBCD-Pol at different concentrations when compared to EDTA-K3 as the positive control ($p < 0.01$). In contrast, EDTA at 500 μg/mL could not show a difference from the control (Figure 3).

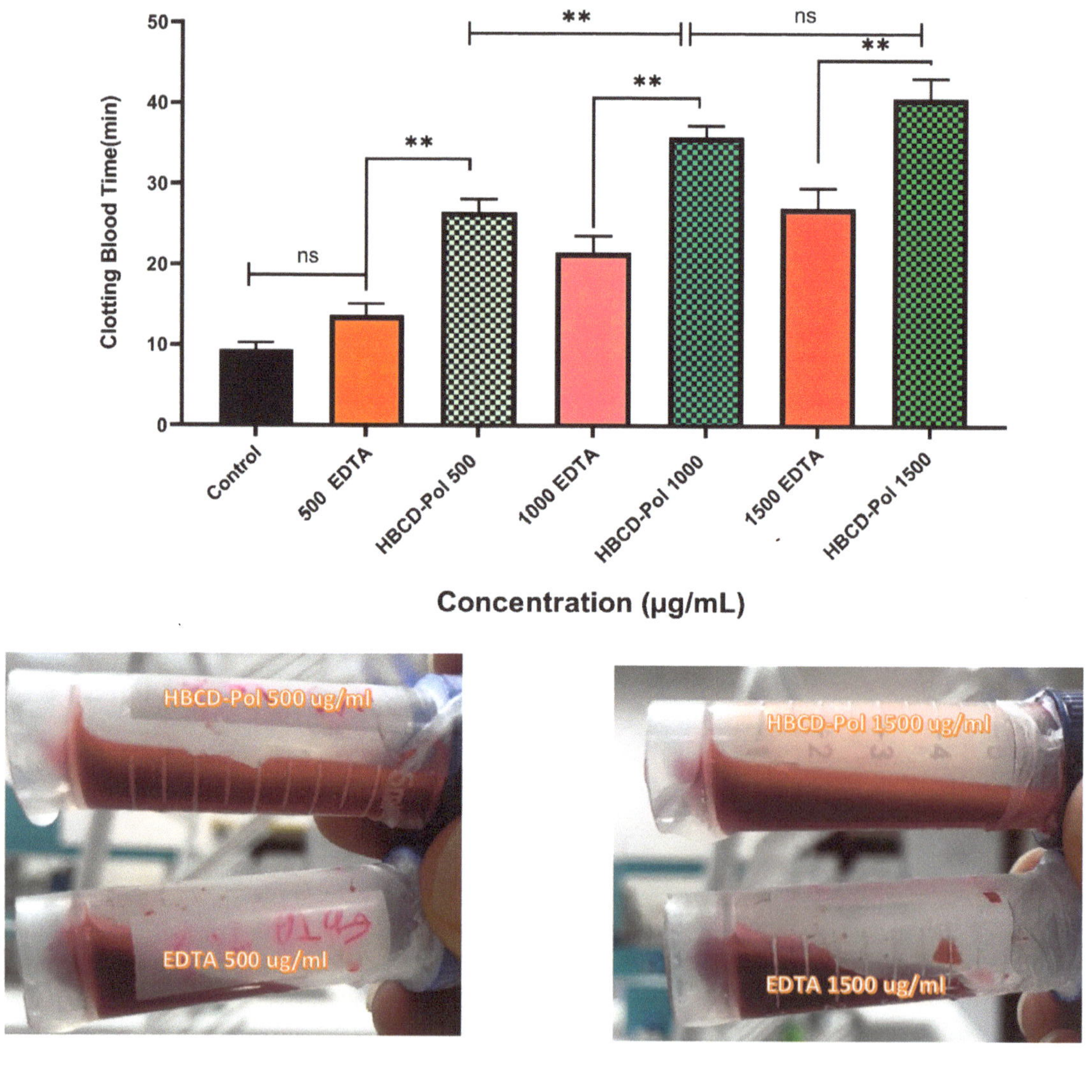

Figure 3. Blood clotting time assessment on different concentrations of HBCD-Pol and EDTA (5%). ** $p < 0.01$ and ns (not significant).

In the next step, the plasma recalcification time parameter was evaluated on platelet-poor plasma (PPP) at three different concentrations for both HBCD-Pol and EDTA. Surprisingly, the results showed that adding HBCD-Pol into plasma at 1500 g/mL led to a considerable increase in clotting time of about 4 and 2 times compared to control and EDTA at that concentration, respectively, while the EDTA at 500 g/mL did not show differences compared to the control. Meanwhile, HBCD-Pol at other concentrations showed a better ability to enhance the clotting time compared to EDTA at the same amount (Figure 4).

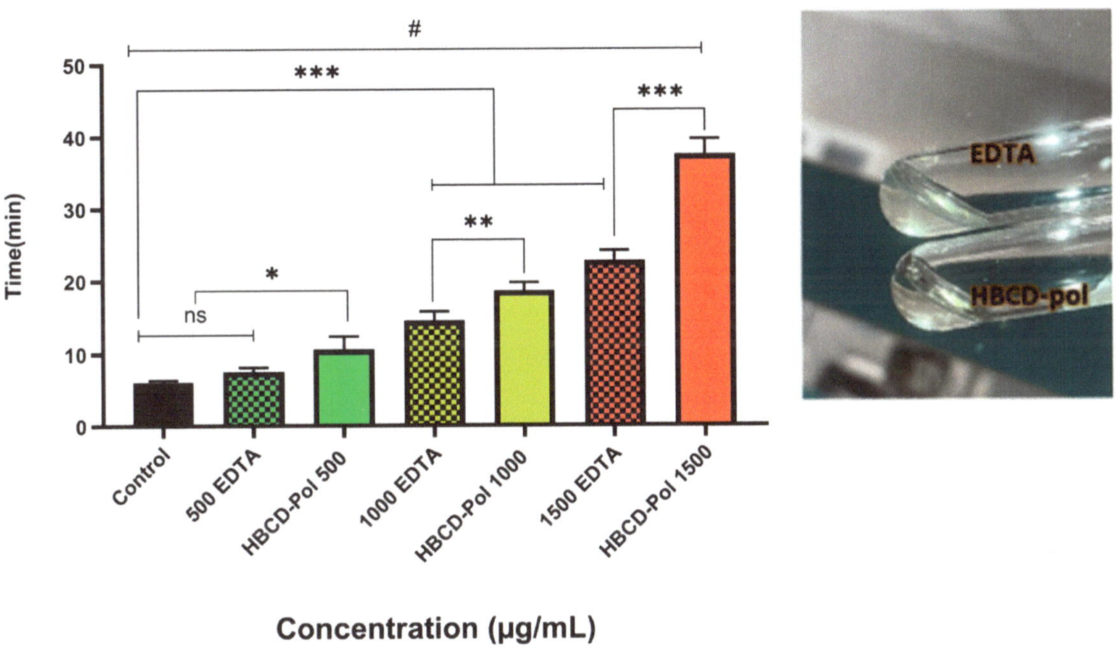

Figure 4. Plasma recalcification time on platelet-poor plasma for different amounts of HBCD-Pol and EDTA. # $p < 0.0001$, *** $p < 0.001$, ** $p < 0.01$, * $p < 0.05$ and ns (not significant).

3.3. *PT and aPTT Tests*

The next step in the study was to evaluate the activated partial thromboplastin time (aPTT) and prothrombin time (PT) [7,8]. The pre-warmed PPP, aPTT and PT reagents were incubated at 37 °C for 5 min with HBCD-Pol 5% and EDTA 5% at various concentrations. Then, the reactions were started by adding the $CaCl_2$ 0.025 M to the tubes. The results showed that HBCD-Pol and EDTA could not affect the aPTT parameter except at 1500 g/mL ($p < 0.01$). HBCD-Pol illustrated a much better effect at all concentrations, especially at 1500 g/mL ($p < 0.001$), regarding increasing the PT parameter compared to EDTA. Indeed, the outcomes of this experiment confirmed that HBCD-Pol may be able to inhibit the extrinsic and common coagulation pathway because of its effect on increasing the PT time (Figure 5). In the bibliography, it is indicated that concentrations of α-, β- and γ-CD of 0.5% did not show any aPTT or PT variation [14]. However, the polymer above 0.05% showed differences, possibly due to the combination of Ca^{+2} chelation and the more complex structure and complexation capacities of the polymer with some coagulation factors.

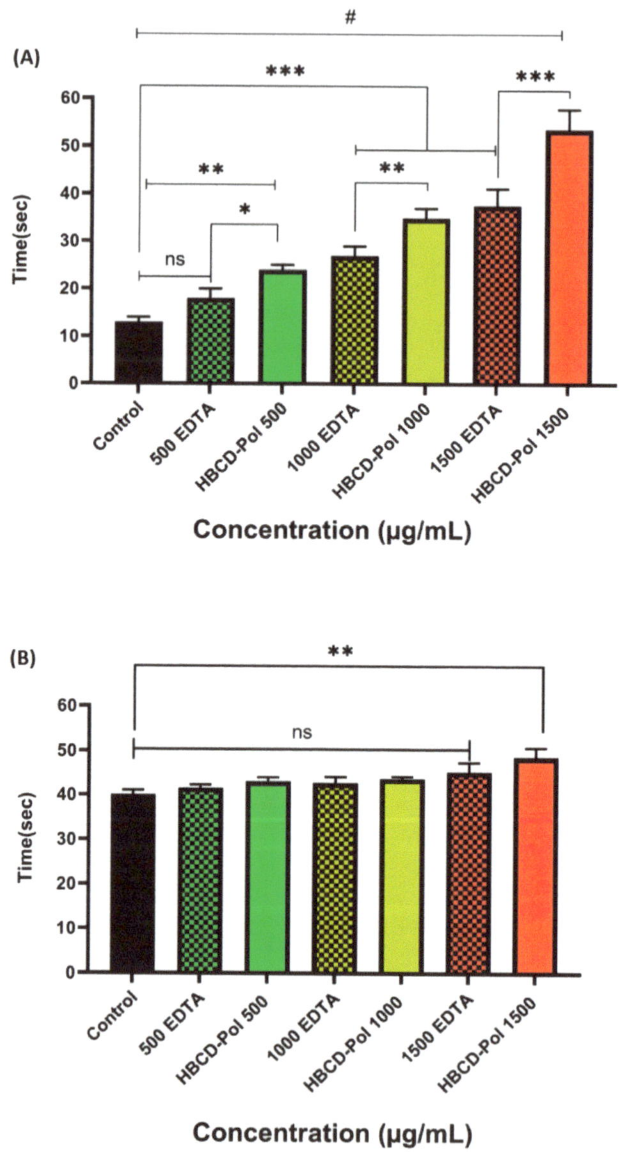

Figure 5. (**A**) aPTT and (**B**) PT assessment at various concentrations. # $p < 0.\ 0001$, *** $p < 0.001$, ** $p < 0.01$, * $p < 0.05$ and ns (not significant).

3.4. Hemocompatibility

The hemocompatibility of the HBCD-Pol was tested at different concentrations as a crucial parameter to use in an in-vivo experiment [9,10]. The hemolysis activity was studied in order to prevent hemolysis, thrombosis and embolization [11]. The hemocompatibility performance was shown in Figure 6, with a value of about 3% for both 2500 and 2000 μg/mL concentrations, while other concentrations revealed less than 2.5% hemolysis. All samples can be characterized as hemocompatible according to ISO 10993 (<5%) [12]. It is important to mention that α- and β-CD showed toxicity (up to 40%) at concentrations around 1% [14], while the polymer around 0.25% (2500 µg/mL) not only obtained no toxicity but also showed an effect on coagulation.

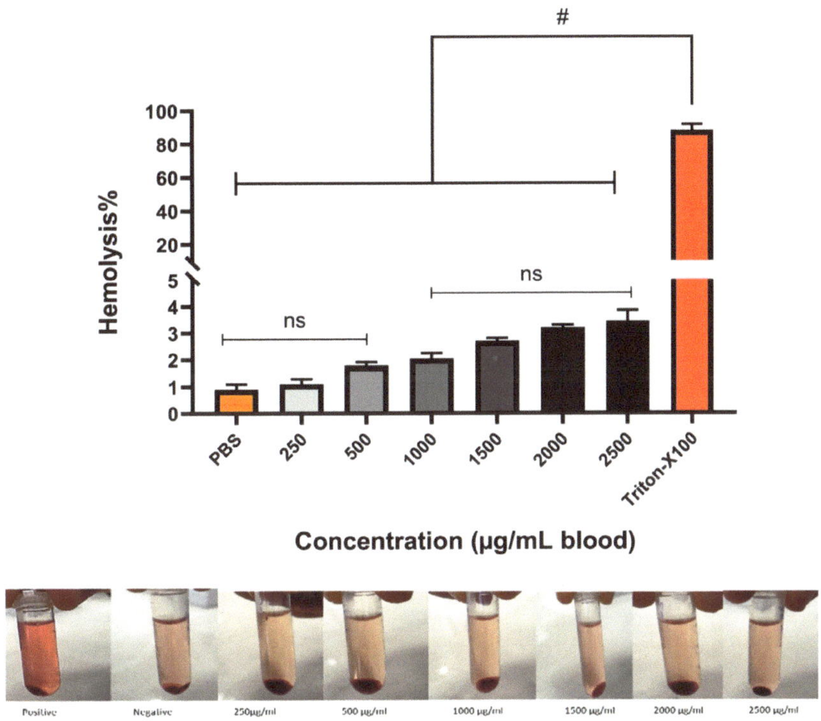

Figure 6. Hemolysis percentages of RBCs at 250–2500 µg/mL of HBCD-Pol free. PBS and Triton X-100 were used as the negative and positive control, respectively. # $p < 0.0001$ and ns (not significant).

3.5. In Vivo Anticoagulant Characterization

3.5.1. Effects of Anticoagulation Agents on Clinical Coagulation Parameters

Using a mouse model, the anticoagulant effects of HBCD-Pol and EDTA with thrombin were assessed. We selected a maximum dose of 40 mg/kg without visual signs of toxicity. aPTT and PT assays on samples collected 10 min after the agents' injection showed a dose-dependent prolongation of aPTT and PT (Figure 7). However, doses corresponding to 40 mg/kg of HBCD-Pol showed the highest level of prolonged aPTT (>3.5 fold than control), while EDTA at that dose (40 mg/kg) caused a prolongation less than HBCD-Pol ($p < 0.05$). In addition, there were no significant differences for other dosages between the HBCD-Pol and EDTA. Interestingly, no increase in PTT in the in vitro experiment was observed (Figure 5). Furthermore, the results of the CT test showed a significant effect of HBCD-Pol to increase the plasma clotting time more than three times compared to control at the highest dosage. Moreover, HBCD-Pol could show a better anticoagulation effect at that dosage (40 mg/kg) when compared to EDTA ($p < 0.01$), while this difference was not seen at lower dosages (Figure 7). These results support the idea of a combinatorial effect due to the complexation of some coagulant agents, which decreased the effectivity of the pathway.

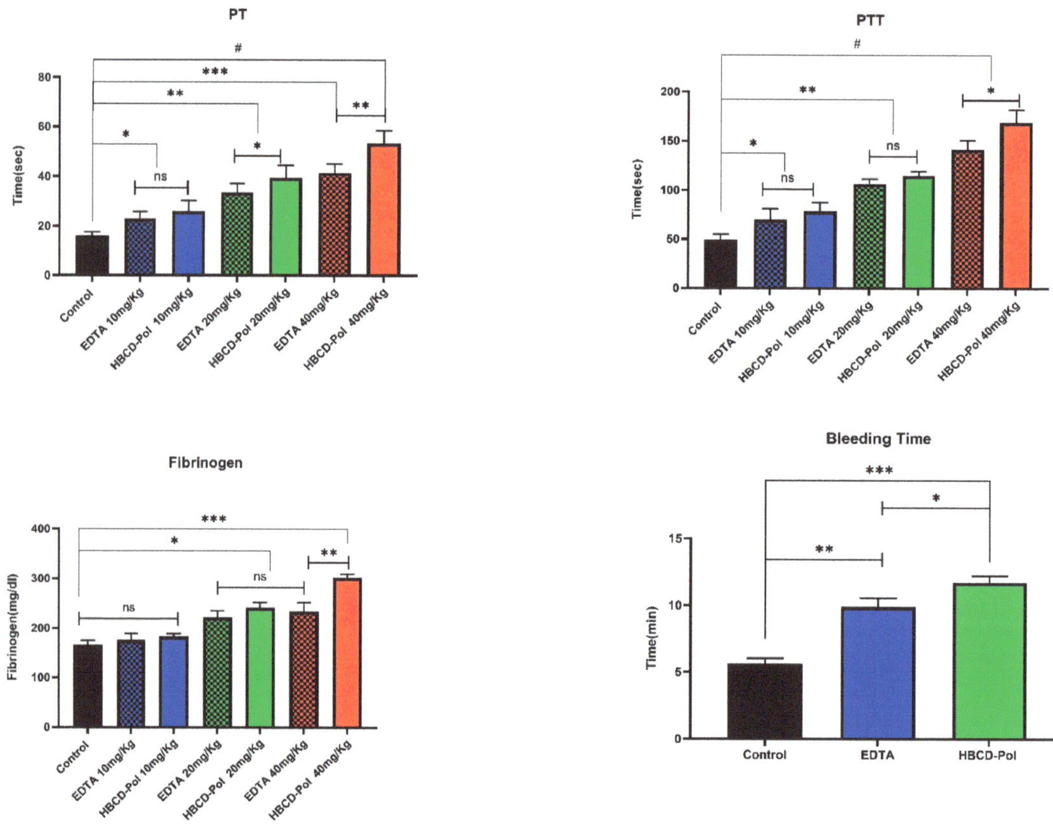

Figure 7. PT and aPTT in mice treated with HBCD-Pol and EDTA. Fibrinogen levels in mice injected with thrombin (1000 U/kg) after pretreatment and the tail transection bleeding time of animals pretreated. Data are the mean ± SEM (five animals for each group). # $p < 0.0001$, *** $p < 0.001$, ** $p < 0.01$, * $p < 0.05$ and ns (not significant).

3.5.2. Fibrinogen Consumption and Bleeding Time Test

CDs have a weak impact on fibrinogen's microenvironment but can slightly affect its activity [14].Therefore, this section evaluated the capability of HBCD-Pol and EDTA to prevent fibrinogen consumption as an in vivo measure of thrombin inhibition. Mice were treated with HBCD-Pol and EDTA at the same dosages that were used for in vivo PT and PTT tests. Then, the plasma fibrinogen was determined on samples taken 2 min after the thrombin challenge. Both HBCD-Pol and EDTA at the highest dosage (40 mg/kg) effectively attenuated fibrinogen consumption by preserving about 45% and 30% of the circulating protein, respectively (Figure 7). On the contrary, the 10 mg/kg dosage showed a clearly weaker effect. The difference in fibrinogen levels between HBCD-Pol and EDTA-treated animals at 40 mg/kg was statistically significant ($p < 0.01$).

The tail transection bleeding time was measured in animals receiving the dosage of 40 mg/kg of each agent 10 min after treatment. In line with the aPTT results, the bleeding time was significantly more prolonged in mice treated with HBCD-Pol and EDTA (2- and 1.7-fold increase, respectively) than in the control group mice (Figures 7 and 8).

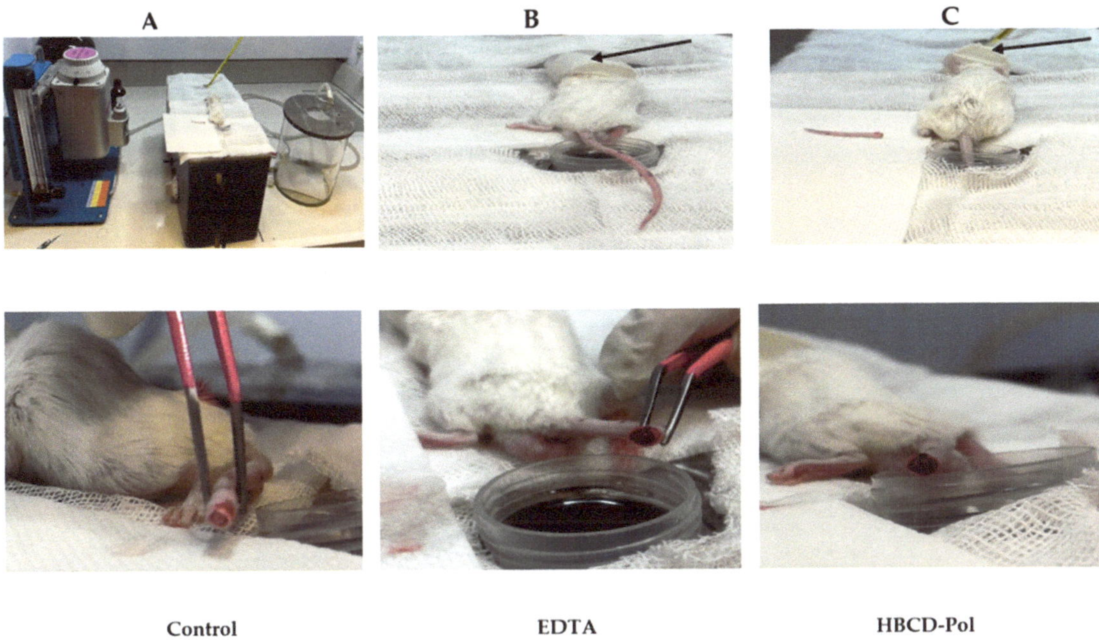

Figure 8. (**A**) including nebulizer machine, water bath, anesthesia induction chamber, left to right, respectively; (**B**,**C**), the tail transection procedure. The results of bleeding time in vivo experiment after 6 min of tail cutting. Small anesthesia chamber indicated with an arrow (↗).

As a whole, these results suggest that HBCD-Pol increased the slight effect on the coagulation of natural CDs. Although it is true that its Ca^{2+} chelation capacity is lower than EDTA at the studied conditions, a higher in vitro and in vivo effect is obtained. Between both intrinsic (aPTT) and extrinsic (PT) pathways, the most affected is the intrinsic pathway. This pathway depends on more coagulation factors than the extrinsic pathway. The importance of Ca^{2+} in both pathways is evident, as suggested in Figure 5A or Figure 7, as in both cases, this is the pathway more affected. However, EDTA does not present a possible mechanism than HBCD-Pol yes: firstly, HBCD-Pol is able to chelate Ca^{2+}, possibly creating an environment where $[Ca^{2+}]_{app}$ is lower, avoiding coagulation. Secondly, the polymeric structure could interact with the different factors of the pathway. Although natural CDs do not affect the aPPT pathway [14], the additional cavities where non-typical cyclodextrin guest can be complexed could increase the interaction and prevent the continuation of the pathway [26]. For example, the low interaction of fibrinogen with CDs could be increased for HBCD-Pol, boosting the effect of the material in coagulation [14].

4. Conclusions

To conclude, in this study, the anticoagulant effect of HBCD-Pol, a CD-based polymer was analyzed in vitro and in vivo. Due to its polymeric structure and anionic nature, the polymer can form a chelate with Ca^{2+} in vitro but with a lower strength than EDTA. However, HBCD-Pol increases the CT, PT and aPTT 3.5 times over the control, showing a better effect than even EDTA, as occurs with recalcification time, as well. The hemocompatibility test showed better results than EDTA and CD values reported in the bibliography. The fibrinogen consumption and bleeding time were analyzed in vivo; in all cases, the effect of HBCD-Pol at the same concentration of EDTA was higher. The results demonstrate that the anticoagulant effect of this HBCD-Pol could be used as a novel therapy from a combinatorial point of view: adding a drug at the same time as the polymer.

Author Contributions: Conceptualization, Y.K.M., A.M., R.C., P.Z.-M. and F.T.; data curation, Y.K.M., A.M., M.M., D.M.B., G.H. and F.T.; funding acquisition, F.T.; investigation, Y.K.M., G.H., D.M.B. and F.C.; methodology, Y.K.M., M.M., R.C. and A.M; project administration, P.Z.-M., A.M. and F.T.; validation, Y.K.M., A.M., F.C., G.H. and F.T.; writing—original draft preparation, Y.K.M., G.H. and A.M.; writing—review and editing, Y.K.M., A.M., M.M., F.C., G.H., P.Z.-M. and F.T. All authors have read and agreed to the published version of the manuscript.

Funding: This research received no external funding.

Institutional Review Board Statement: The experiments were approved by the Tabriz University of Medical Sciences Pharmacy Department under the Iran Ethical Code: IR.TBZMED.AEC.1401.022.

Informed Consent Statement: Not applicable.

Data Availability Statement: Not applicable.

Acknowledgments: This work is the partial result of a contract for the University of Turin (Italy) for Y.K.M. for Ph.D. training, and for A.M., an RTDA contract from the D.M 1062/2021 (Ministero dell'Università e della Ricerca) for the University of Turin.

Conflicts of Interest: The authors declare no conflict of interest.

References

1. Wendelboe, A.M.; Raskob, G.E. Global Burden of Thrombosis: Epidemiologic Aspects. *Circ. Res.* **2016**, *118*, 1340–1347. [CrossRef] [PubMed]
2. Hirsh, J.; Raschke, C.R.; Warkentin, T.E.; Dalen, J.E.; Deykin, D.; Poller, L. Heparin: Mechanism of Action, Pharmacokinetics, Dosing Considerations, Monitoring, Efficacy, and Safety. *CHEST* **1995**, *108*, 258S–275S. [CrossRef] [PubMed]
3. Bell, R.G.; Sadowski, J.A.; Matschiner, J.T. Mechanism of Action of Warfarin. Warfarin and Metabolism of Vitamin K1. *Biochemistry* **1972**, *11*, 1959–1961. [CrossRef] [PubMed]
4. Wiggins, B.S.; Dixon, D.L.; Neyens, R.R.; Page, R.L.; Gluckman, T.J. Select Drug-Drug Interactions With Direct Oral Anticoagulants. *J. Am. Coll. Cardiol.* **2020**, *75*, 1341–1350. [CrossRef]
5. Trotta, F.; Caldera, F.; Cavalli, R.; Mele, A.; Punta, C.; Melone, L.; Castiglione, F.; Rossi, B.; Ferro, M.; Crupi, V.; et al. Synthesis and Characterization of a Hyper-Branched Water-Soluble β-Cyclodextrin Polymer. *Beilstein J. Org. Chem.* **2014**, *10*, 2586–2593. [CrossRef] [PubMed]
6. Jansook, P.; Ogawa, N.; Loftsson, T. Cyclodextrins: Structure, Physicochemical Properties and Pharmaceutical Applications. *Int. J. Pharm.* **2018**, *535*, 272–284. [CrossRef]
7. Brewster, M.E.; Loftsson, T. Cyclodextrins as Pharmaceutical Solubilizers. *Adv. Drug Deliv. Rev.* **2007**, *59*, 645–666. [CrossRef]
8. Santos, C.I.A.V.; Ribeiro, A.C.F.; Esteso, M.A. Drug Delivery Systems: Study of Inclusion Complex Formation between Methylxanthines and Cyclodextrins and Their Thermodynamic and Transport Properties. *Biomolecules* **2019**, *9*, 196. [CrossRef]
9. Kurkov, S.V.; Loftsson, T. Cyclodextrins. *Int. J. Pharm.* **2013**, *453*, 167–180. [CrossRef]
10. Matencio, A.; Bermejo-Gimeno, M.J.; García-Carmona, F.; López-Nicolás, J.M. Separating and Identifying the Four Stereoisomers of Methyl Jasmonate by RP-HPLC and Using Cyclodextrins in a Novel Way. *Phytochem. Anal.* **2017**, *28*, 151–158. [CrossRef]
11. Matencio, A.; Caldera, F.; Rubin Pedrazzo, A.; Khazaei Monfared, Y.K.; Dhakar, N.; Trotta, F. A Physicochemical, Thermodynamical, Structural and Computational Evaluation of Kynurenic Acid/Cyclodextrin Complexes. *Food Chem.* **2021**, *356*, 129639. [CrossRef]
12. López-Nicolás, J.M.; García-Carmona, F. Effect of Hydroxypropyl-β-Cyclodextrin on the Aggregation of (E)-Resveratrol in Different Protonation States of the Guest Molecule. *Food Chem.* **2010**, *118*, 648–655. [CrossRef]
13. Matencio, A.; Guerrero-Rubio, M.A.; Caldera, F.; Cecone, C.; Trotta, F.; García-Carmona, F.; López-Nicolás, J.M. Lifespan Extension in Caenorhabditis Elegans by Oxyresveratrol Supplementation in Hyper-Branched Cyclodextrin-Based Nanosponges. *Int. J. Pharm.* **2020**, *589*, 119862. [CrossRef] [PubMed]
14. Fu, Y.; Wang, X.; Zhang, Y.; Liu, Z.; Xue, W. Effect of Cyclodextrins on the Structure and Functions of Blood Components in Vitro. *J. Bioact. Compat. Polym.* **2015**, *30*, 541–554. [CrossRef]
15. Palta, S.; Saroa, R.; Palta, A. Overview of the Coagulation System. *Indian J. Anaesth.* **2014**, *58*, 515–523. [CrossRef] [PubMed]
16. Francese, R.; Cecone, C.; Costantino, M.; Hoti, G.; Bracco, P.; Lembo, D.; Trotta, F. Identification of a BCD-Based Hyper-Branched Negatively Charged Polymer as HSV-2 and RSV Inhibitor. *Int. J. Mol. Sci.* **2022**, *23*, 8701. [CrossRef]
17. Azeez, M.A.; Durodola, F.A.; Lateef, A.; Yekeen, T.A.; Adubi, A.O.; Oladipo, I.C.; Adebayo, E.A.; Badmus, J.A.; Abawulem, A.O. Green Synthesized Novel Silver Nanoparticles and Their Application as Anticoagulant and Thrombolytic Agents: A Perspective. *IOP Conf. Ser. Mater. Sci. Eng.* **2020**, *805*, 012043. [CrossRef]
18. Janvikul, W.; Uppanan, P.; Thavornyutikarn, B.; Krewraing, J.; Prateepasen, R. In Vitro Comparative Hemostatic Studies of Chitin, Chitosan, and Their Derivatives. *J. Appl. Polym. Sci.* **2006**, *102*, 445–451. [CrossRef]
19. Winter, W.E.; Flax, S.D.; Harris, N.S. Coagulation Testing in the Core Laboratory. *Lab. Med.* **2017**, *48*, 295–313. [CrossRef]

20. Momi, S.; Nasimi, M.; Colucci, M.; Nenci, G.G.; Gresele, P. Low Molecular Weight Heparins Prevent Thrombin-Induced Thrombo-Embolism in Mice despite Low Anti-Thrombin Activity. Evidence That the Inhibition of Feed-Back Activation of Thrombin Generation Confers Safety Advantages over Direct Thrombin Inhibition. *Haematologica* **2001**, *86*, 297–302. [CrossRef]
21. Kopić, A.; Benamara, K.; Schuster, M.; Leidenmühler, P.; Bauer, A.; Glantschnig, H.; Höllriegl, W. Coagulation Phenotype of Wild-Type Mice on Different Genetic Backgrounds. *Lab. Anim.* **2019**, *53*, 43–52. [CrossRef] [PubMed]
22. Gresele, P.; Momi, S.; Berrettini, M.; Nenci, G.G.; Schwarz, H.P.; Semeraro, N.; Colucci, M. Activated Human Protein C Prevents Thrombin-Induced Thromboembolism in Mice. Evidence That Activated Protein c Reduces Intravascular Fibrin Accumulation through the Inhibition of Additional Thrombin Generation. *J. Clin. Investig.* **1998**, *101*, 667–676. [CrossRef] [PubMed]
23. Dejana, E.; Villa, S.; de Gaetano, G. Bleeding Time in Rats: A Comparison of Different Experimental Conditions. *Thromb. Haemost.* **1982**, *48*, 108–111. [CrossRef] [PubMed]
24. Kim, J.; Vipulanandan, C. Effect of PH, Sulfate and Sodium on the EDTA Titration of Calcium. *Cem. Concr. Res.* **2003**, *33*, 621–627. [CrossRef]
25. Tucker, B.B.; Kurtz, L.T. Calcium and Magnesium Determinations by EDTA Titrations. *Soil Sci. Soc. Am. J.* **1961**, *25*, 27–29. [CrossRef]
26. Lucia Appleton, S.; Khazaei Monfared, Y.; Vidal-Sánchez, F.J.; Caldera, F.; Cavalli, R.; Trotta, F.; Matencio, A. Cyclodextrin-Based Nanosponges and Proteins. *Encyclopedia* **2022**, *2*, 752–760. [CrossRef]

bioengineering

MDPI

Article

3D-Printed Piezoelectric Porous Bioactive Scaffolds and Clinical Ultrasonic Stimulation Can Help in Enhanced Bone Regeneration

Prabaha Sikder *, Phaniteja Nagaraju and Harsha P. S. Naganaboyina

Department of Mechanical Engineering, Cleveland State University, Cleveland, OH 44115, USA
* Correspondence: p.sikder@csuohio.edu

Abstract: This paper presents a comprehensive effort to develop and analyze first-of-its-kind design-specific and bioactive piezoelectric scaffolds for treating orthopedic defects. The study has three major highlights. First, this is one of the first studies that utilize extrusion-based 3D printing to develop design-specific macroporous piezoelectric scaffolds for treating bone defects. The scaffolds with controlled pore size and architecture were synthesized based on unique composite formulations containing polycaprolactone (PCL) and micron-sized barium titanate ($BaTiO_3$) particles. Second, the bioactive PCL-$BaTiO_3$ piezoelectric composite formulations were explicitly developed in the form of uniform diameter filaments, which served as feedstock material for the fused filament fabrication (FFF)-based 3D printing. A combined method comprising solvent casting and extrusion (melt-blending) was designed and deemed suitable to develop the high-quality PCL-$BaTiO_3$ bioactive composite filaments for 3D printing. Third, clinical ultrasonic stimulation (US) was used to stimulate the piezoelectric effect, i.e., create stress on the PCL-$BaTiO_3$ scaffolds to generate electrical fields. Subsequently, we analyzed the impact of scaffold-generated piezoelectric stimulation on MC3T3 pre-osteoblast behavior. Our results confirmed that FFF could form high-resolution, macroporous piezoelectric scaffolds, and the poled PCL-$BaTiO_3$ composites resulted in the d_{33} coefficient in the range of 1.2–2.6 pC/N, which is proven suitable for osteogenesis. In vitro results revealed that the scaffolds with a mean pore size of 320 μm resulted in the highest pre-osteoblast growth kinetics. While 1 Hz US resulted in enhanced pre-osteoblast adhesion, proliferation, and spreading, 3 Hz US benefited osteoblast differentiation by upregulating important osteogenic markers. This study proves that 3D-printed bioactive piezoelectric scaffolds coupled with US are promising to expedite bone regeneration in orthopedic defects.

Keywords: piezoelectric; electroactive; bioactive; orthopedics; 3D printing; ultrasonic stimulation

Citation: Sikder, P.; Nagaraju, P.; Naganaboyina, H.P.S. 3D-Printed Piezoelectric Porous Bioactive Scaffolds and Clinical Ultrasonic Stimulation Can Help in Enhanced Bone Regeneration. *Bioengineering* **2022**, *9*, 679. https://doi.org/10.3390/bioengineering9110679

Academic Editors: Gary Chinga Carrasco and Ali Zarrabi

Received: 18 October 2022
Accepted: 8 November 2022
Published: 11 November 2022

1. Introduction

Piezoelectric materials are defined to be "smart" or "electroactive" materials that generate electrical fields (EFs) or electrical stimuli in response to mechanical deformations. The deformations cause an asymmetric shift of ions or charges in the materials, which induces a change in the electrical polarization, thus generating EFs [1]. Piezoelectric materials are widely used in electronic applications such as transducers, sensors, and actuators, but in the past decade, there has been a growing interest in using them as an implant or scaffold for tissue regeneration [2]. One of the primary reasons to drive this interest is the capability of piezoelectric materials to provide therapeutic EFs to cells and tissues without the need for additional power sources and wire connections. The EFs are beneficial for tissue regeneration, as they influence essential cell functions such as cell proliferation and migration [3]. In orthopedics, the applied EFs simulate bioelectrical cues of natural living bone and regulate metabolic activities such as bone growth, structural remodeling, and fracture healing [4]. Moreover, physiological compressive loads have been proven to increase the negative charges in a fractured bone region due to the bone's innate piezoelectric potential, which helps promote matrix mineralization, osteoblast functions, and bone regeneration [5,6]. Hence, sufficient evidence indicates that applied piezoelectric stimulation can help improve bone regeneration.

Previous studies have confirmed that the stimulation or the EF generated from electroactive piezoelectric implants can help expedite bone fracture repair [7,8]. Such electroactive scaffolds develop EFs with the application of physiological loads at the implantation site, which is usually the region of the bone defect. Precisely, the hyperpolarization from the piezoelectric effect activates voltage-gated calcium channels of the cell membrane, which helps in osteoblast proliferation, resulting in new bone formation [7,9]. The piezoelectric scaffolds also develop surface electrical charges under external stress, similar to bone, and the polarized piezoelectric surfaces improve the osteogenic performance of natural bone. For instance, Liu et al. [10] developed electropositive ferroelectric bismuth ferrite ($BiFeO_3$) nanofilms and observed that the films enhanced protein adsorption and mesenchymal stem cell attachment, spreading, and osteogenic differentiation. In the rat femur, the polarized $BiFeO_3$ nanofilms in the presence of built-in electric fields resulted in rapid and superior osseointegration [10]. Similarly, Ribeiro et al. [11] implanted poled and un-poled β-poly(vinylidene fluoride) (β-PVDF) films in bone defects created in rats and observed significantly more defect closure and bone remodeling with poled β-PVDF films as opposed to the non-poled ones. Hence, it is well comprehended that the controlled application of electrical stimuli or EFs from implanted piezoelectric scaffolds can be an effective therapeutic tool in bone regeneration and consequentially treat bone defects expeditedly.

Generally, the piezoelectric scaffolds are developed from biomaterials such as piezo-bioceramics and piezo-biopolymers. Piezoelectric bioceramics have also been designed as in vivo energy harvesters for biosensors. The most well-known piezoelectric bioceramics are barium titanate ($BaTiO_3$), boron nitride (BN), and magnesium silicate ($MgSiO_3$) [12]. The most well-known piezo-biopolymers are PVDF and their copolymers, such as P(VDF-TrFE), and natural polymers, such as collagen and chitosan [12]. Moreover, various composites such as hydroxyapatite–$BaTiO_3$ (HA-$BaTiO_3$), PVDF–$BaTiO_3$, and HA–$MgSiO_3$–Chitosan have been developed to take advantage of the combined properties of the piezo-ceramics and polymers [8,12]. Yet, the majority of the piezoelectric scaffolds were manufactured using conventional techniques such as electrospinning, solvent casting, slip casting, solvent forming-particulate leaching, or freeze casting [12], which do not take into consideration the design-specific needs of the medical implant industry. Especially due to the high clinical demand for developing patient-specific therapies to augment treatment outcomes, of late, there is an imperative need to develop defect- and anatomy-specific regenerative implants.

Three-dimensional printing, or additive manufacturing, is a high-powered fabrication technique that can produce parts of any complex architecture, shape, and dimension layer by layer. In recent years, even though there has been a surge in the application of 3D printing to manufacture biomedical devices, very few reports have focused on developing bioactive and regenerative piezoelectric implants utilizing this process. The majority of the focus was on using 3D printing to develop biosensors, not bioactive and regenerative implants. [13]. For instance, Grinberg et al. [14] developed Polyamide-11 (PA-11) and $BaTiO_3$ 3D-printable composite filaments and used the fused deposition modeling technique to print proof-of-concept models of piezoelectric sensors to monitor the applied force on the knee prosthesis. Furthermore, even though researchers used 3D printing to develop regenerative scaffolds, they utilized complex techniques. For instance, Jiang et al. [15] developed porous $BaTiO_3$ scaffolds with the help of digital light processing (DLP) technology, but it required sintering as an additional processing step to develop the scaffolds. On the other hand, Polley et al. [16] used the binder jet 3D printing technique to develop porous HA-$BaTiO_3$ scaffolds. Similarly, Yang et al. [17] used selective laser sintering (SLS) to create graphene and $BaTiO_3$-based scaffolds. However, both SLS and binder jet involve long wait times after the printing process, processing complexities, and rigorous thermal-post treatment such as sintering.

On the contrary, "fused filament fabrication" (FFF) is a one-step, rapid and easy fabrication method to develop implants with favorable material properties. FFF can directly form robust scaffolds, which might not need sintering as an additional step [18–22]. Importantly, it involves much less material wastage than SLS or binder-jet printing. Our research group used the FFF to develop various design-specific, regenerative implants comprising novel and bespoke polymer–ceramic compositions [23,24].

Not much effort has been made to develop bioactive and biodegradable piezoelectric compositions. For instance, extensive efforts have been made to develop PVDF-based scaffolds [11,25–27], but PVDF is neither a bioactive nor biodegradable piezoelectric biomaterial. Poly (lactic acid-co-glycolic acid) (PLGA)–$BaTiO_3$ composites [28] and PLLA–$BaTiO_3$–graphene [17] have been developed, but PLGA and PLLA are known to degrade into acidic byproducts such as glycolic acid and lactic acid that might be harmful to healthy cells [29]. Graphene is also known to exhibit cytotoxicity in various kinds of cells [30].

Hence, to address some of the literature gaps, in this study, we developed a unique electroactive and bioactive polymer–ceramic composition in the form of 3D-printable filaments such that the filaments could be utilized in a FFF 3D printing setup to manufacture design-specific piezoelectric orthopedic scaffolds. To achieve that, first, we introduced micron-sized piezoelectric $BaTiO_3$ particles in the polycaprolactone (PCL) polymer matrix by solvent-casting and subsequently used a customized extrusion process to form uniform-diameter 3D-printable PCL–$BaTiO_3$ composite filaments. Then, the filaments served as feedstock material for an FFF setup to develop the design-specific, porous scaffolds. We chose PCL because it is an FDA-approved, biodegradable polymer that has been extensively used in implantable biomaterials, and we chose $BaTiO_3$ because of its well-known piezoelectric and biocompatible properties. After the scaffold development, we performed thorough material, mechanical, and in vitro property analyses of the 3D-printed structures. The phase composition and morphology of the implants were determined by X-ray diffraction and scanning electron microscopy (SEM). Mechanical properties of the composites were determined using ASTM standards, and the proliferative and differentiative capacity of the scaffolds was assessed in vitro using MC3T3 pre-osteoblast cells.

We believe that the result of this proof-of-concept study is the first step toward developing an effective orthopedic-based regenerative medicine for the society. EFs have been used in clinics for wound healing for several years. Furthermore, the components/materials of the orthopedic scaffold used in this study are FDA-approved, hopefully making the regulatory route to be 510(k). More effective collaboration between clinician scientists, engineers, orthopedic surgeons, and federal funding agencies is required to conduct clinical trials and translate innovative medical devices such as this piezoelectric polymer–ceramic scaffold for expedited bone regeneration [31,32]. Notably, such scaffold-mediated expedited bone regeneration can be instrumental in the military community, where it is expected for the wounded soldiers to recover quickly and rejoin active duty. It can also benefit patients with diabetes with compromised or delayed wound healing and tissue regeneration rates [33].

2. Materials and Methods

2.1. Raw Materials

Polycaprolactone (PCL, molecular weight = 80,000) pellets, $BaTiO_3$ (average particle size < 2 μm, molecular weight = 233.19) powder, and dichloromethane (CH_2Cl_2, anhydrous, ≥99.8%) were procured from Sigma Aldrich and used as is.

2.2. Composite Development

The entire process of developing the piezoelectric composites and the 3D-printable piezoelectric filaments is shown in Figure 1. First, PCL pellets were dissolved in dichloromethane following 15 % (w/v) concentration. Once the polymer was completely dissolved, a varying amount of $BaTiO_3$ powder (25, 45, and 65 vol.%) was added to it, followed by a 3-step mixing procedure. The specimens are named as PCL-x$BaTiO_3$, where x denotes the vol.% of $BaTiO_3$ in the composites = 25, 45, and 65; we refer to PCL-x$BaTiO_3$ as "PCLxBT" for simplified nomenclature. The steps comprised one hour of stirring followed by another hour of ultrasonic dispersion and another two hours of stirring. Subsequently, the PCL-BT slurry was poured into glass Petri dishes and left in a laminar fume hood overnight for the dichloromethane to evaporate. The resultant was PCL-BT composite films (approximately 120 mm in diameter and 1.5 mm thick). These films were sheared into roughly 10 × 10 mm squares and made ready for filament extrusion (melt-blending). The squares were thoroughly dried in a desiccator before they were used for extrusion.

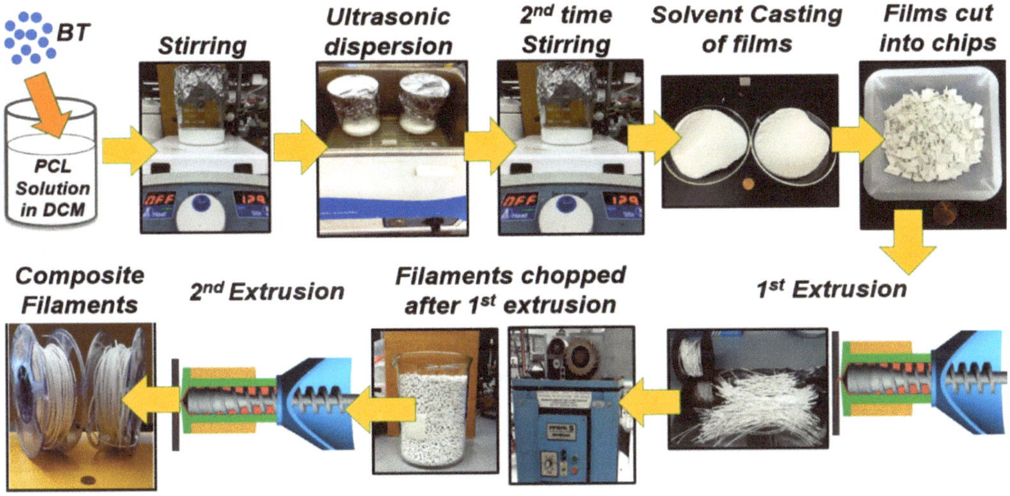

Figure 1. Composite filament development. Schematics show the entire workflow of developing the polycaprolactone–barium titanate (PCL-BT) composites and the 3D-printable PCL-BT filaments.

2.3. 3D-Printable Filament Development

The uniform diameter composite filaments were developed by a customized device assembly comprising (1) a single-screw extruder, (2) a sensor controlling the speed of filament winding, and (3) a motorized filament winding setup, as shown in Figure 2a. All the parts for the extrusion setup were procured from Filastruder (Snellville, GA, USA). The single screw extruder (with a Ø 2.75 mm nozzle) was kept at an elevated position with respect to the sensor and the winder setup. The sheared squares of the composite films were fed into the single-screw extruder, and the extrusion temperature and speed were set to 120 °C and 10 rpm, respectively. We used a fan just next to the nozzle, as shown in Figure 2a, to cool down the filaments. Once a continuous extrusion of uniform diameter filament was achieved, it was guided through the sensor. The sensor detects the filament's position and adjusts the speed of the motorized winder in the filament winding setup. Usually, directly after the sensor, the filament is guided through a polytetrafluoroethylene (PTFE) tube loop onto the filament winding spool. The PTFE tube makes sure that there are no tangles in the filament. The motorized winding spool rotates at a specific speed (depending on the filament's position as detected by the sensor), provides tension to the continuously extruded filaments, and helps form the uniform diameter 3D-printable composite filaments.

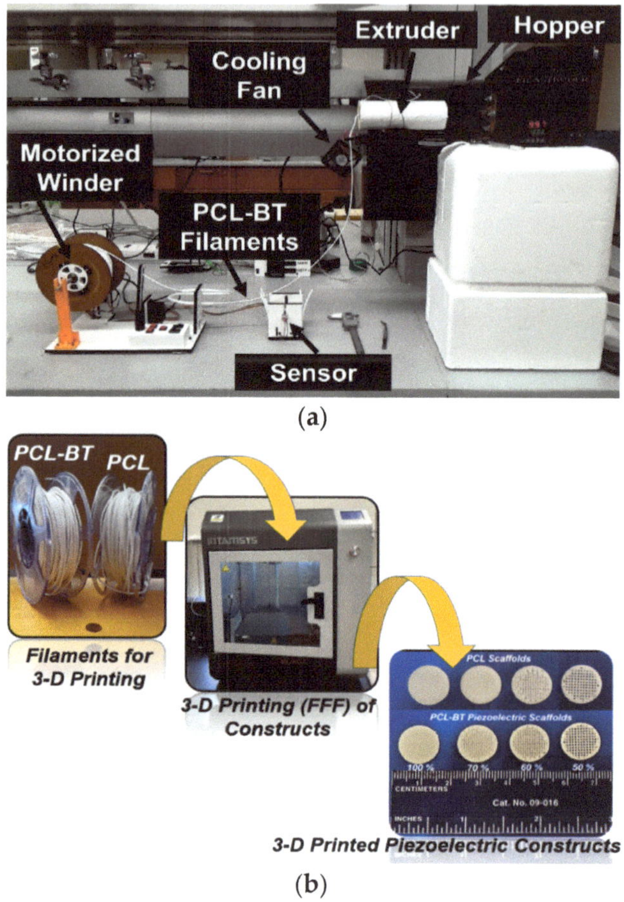

Figure 2. Setup for filament extrusion and 3D printing. (**a**) The assembled Filastruder extruder and filament winder setup that was used to extrude the 3D-printable filaments. (**b**) The workflow shows that the filaments were used in a 3D printer setup to develop the piezoelectric PCL- BT composite scaffolds.

2.4. 3D Printing of Constructs

The custom-made filaments were loaded onto an FDM 3D printer (Funmat HT Enhanced, Intamsys) for developing the macroporous piezoelectric scaffolds, as shown in Figure 2b. The pore sizes of these microporous scaffolds were controlled by changing the infill density from 50–100%. The Simplify 3D software was used for slicing. For all the prints, a 0.4 mm nozzle was used. The print and bed temperatures were set at 130 and 30 °C, respectively. The layer height was maintained at 0.2 mm, and the print speed was 30 mm/s.

2.5. Material Characterization

Phase compositions of the specimens were detected using X-ray diffraction (XRD, Ultima III, RI, USA) with monochromated Cu Kα radiation (44 kV, 40 mA), focused beam mode over a 2θ range of 10–60°. Step width and count time were fixed to be 0.05° and 8 s during the analysis. A whole pattern fitting (WPF) analysis and Rietveld refinement were conducted using the MDI Jade software 2010 (Materials Data Inc., Livermore, CA, USA) to determine the goodness-of-fit. The morphology of the 3D-printed scaffolds was analyzed by a secondary electron detector in a Field Emission Scanning Electron Microscope (FESEM,

Inspect F50, FEI, OR, USA). The cross-sections of the various 3D-printable PCL-BT filaments were also analyzed by FESEM.

2.6. Corona Poling and Electroactive Properties

The 3D-printed specimens were poled by the Corona Poling process to align the $BaTiO_3$ dipoles in the composites (Figure 3a). We developed a customized setup, as shown in Figure 3b, to perform the corona poling process. Briefly, the specimens were kept on an aluminum plate that was grounded. The specimen was contained in an enclosed glass chamber, and a corona needle connected to the positive terminal of a high-voltage DC source was placed 3 mm away from the specimen surface. The DC voltage (8–12 kV) and the time (5, 15, and 30 min) were varied to identify the most optimum conditions for corona poling. Right after the poling, specimens were tested on a wide-range d_{33} meter (APC International, Ltd, Mackeyville, PA, USA) to measure the d_{33} coefficient and analyze the effect of corona poling on the electroactive properties of the samples. Specimens were also immersed in Phosphate Buffered Saline (PBS) for 14 days and then analyzed in a d_{33} meter.

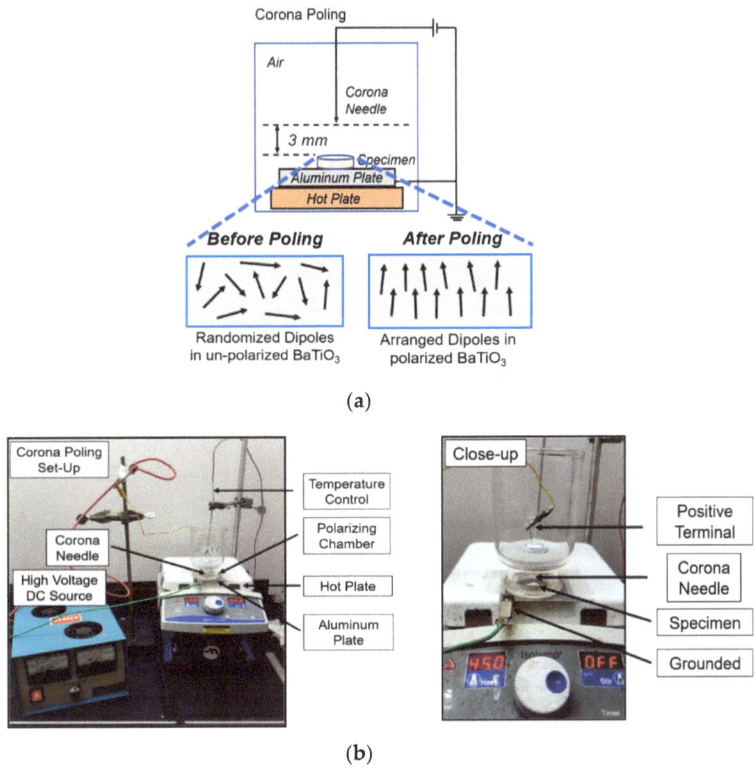

Figure 3. Polarization of the PCL-BT scaffolds. (**a**) Schematics of the Corona Poling setup. (**b**) Digital image of the coronal poling setup.

2.7. Mechanical Properties

The mechanical testing specimens were 3D printed according to the dimensions mentioned in ASTM D638 (Figure 4). The same 3D printing parameters were followed as mentioned in Section 2.4 to develop/print the dog-bone-shaped tensile specimens. Subsequently, the specimens were mounted on a Universal Testing Machine (UTM). An Instron 3369 UTM with a 50 kN load cell was used for the tensile tests. Then, 5 mm/min strain deformation was used to measure the specimens' tensile strength, and 1 mm/min strain deformation was used to measure the modulus, as per ASTM standards.

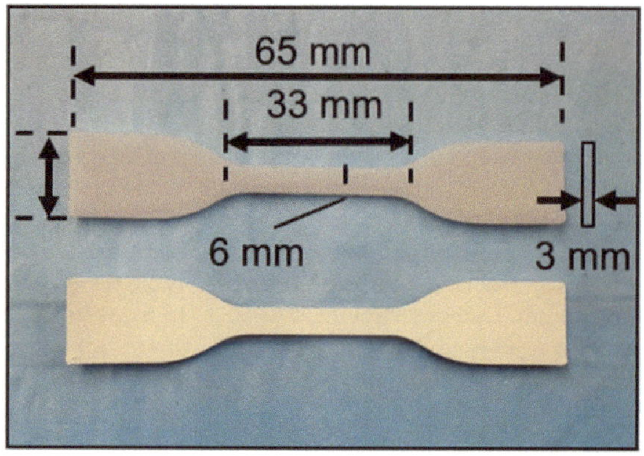

Figure 4. Digital image of the 3D-printed PCL and PCL-45BT scaffolds.

2.8. In Vitro Properties

Various PCL-BT were 3D printed into disc-shaped specimens (8 mm Ø × 2 mm height) for all the in vitro studies. Specimens were disinfected by immersing them in 90% ethyl alcohol for 30 min and then applying UV for another 10 min. Preosteoblast MC3T3 mouse cells (CRL-2593™, ATCC, Manassas, VA, USA) were used for the cell studies. The cells were cultured in complete minimum essential medium alpha (α-MEM, Thermo Scientific), supplemented with 10% fetal bovine serum (FBS, HyClone) and 1% penicillin/streptomycin (0.2 g/mL) at 5% CO_2 and 37 °C.

2.8.1. Routine Ultrasonic Stimulation (US) Application

For this assay, only PCL-65BT was chosen, as it was hypothesized that the latter would exhibit the highest electroactive properties. This assay was specifically designed to validate the effect of US stimulation on the piezoelectric scaffolds. The disc-shaped specimens were placed in 24-well plates and soaked with culture media. Then, ~2.3 × 10^4 cells were directly seeded onto the specimens and allowed to attach to the scaffolds for 4 h. Subsequently, US was applied with the help of a clinical US setup (ComboCare Clinical Electrotherapy & Ultrasound Combo Unit). The probe of the US stimulator was attached to the bottom of the well plates, and the US was applied following a specific routine as shown in Figure 5. For validation purposes, US (1 or 3 MHz) was applied for 1 min continuously, every 3 h, two times/day for 72 h (3 days). At the end of 48 and 72 h, thiazolyl blue tetrazolium bromide (MTT) was used to determine the effect of US on cell proliferation if any. The cultured specimens were retrieved, rinsed three times in PBS, and prepared for the assay. The MTT solution was added to the specimens and control, followed by 4 h of incubation at 37 °C. Dimethyl sulfoxide was used to dissolve formazan, and finally, OD570 readings were recorded using a spectrophotometer.

2.8.2. Cell Proliferation

The same US application procedure was followed for 3 or 7 days. At the end of 3 or 7 days, the specimens were retrieved, and the cell growth kinetics on those 3D printed piezoelectric specimens were determined using MTT assay as mentioned in Section 2.8.1.

2.8.3. Cell Adhesion and Morphology

The 3D-printed specimens were soaked in α-MEM for 4 h and were seeded with ~6.5 × 10^4 cells. The cell-seeded specimens were cultured for 24 h or 7 days, and at the end of the period, samples were retrieved, washed three times in PBS, and immersed in a

3% glutaraldehyde fixing solution for 1 h at room temperature. After fixation, they were dried sequentially in 30, 50, 70, 90, 95, and 100% ethyl alcohol, followed by a combined Hexamethyldisilazane (HMDS) and ethyl alcohol treatment. HMDS helps in chemically drying the specimens. Further, HMDS drying helps preserve the adhered cells' structure on the specimens. Finally, they were dried overnight in a fume hood, plated with gold palladium for 60 secs, and imaged under FESEM.

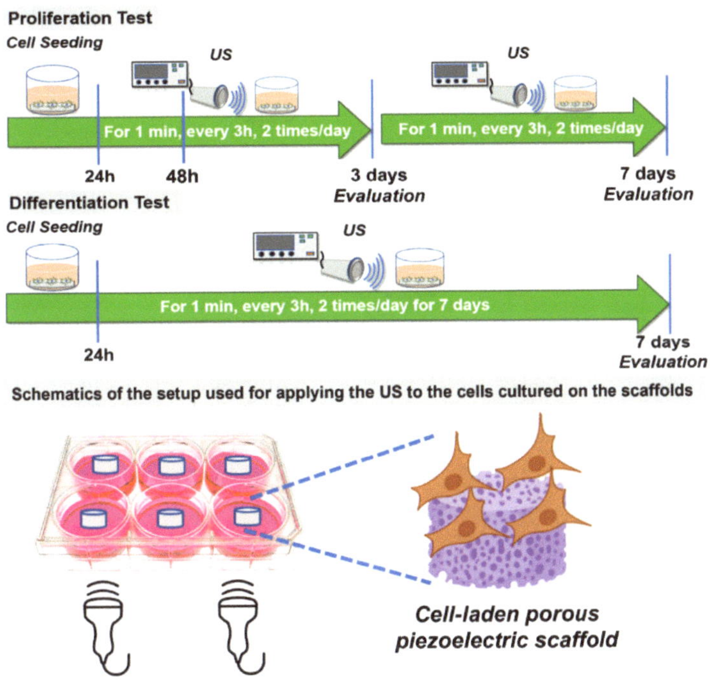

Figure 5. Schematics showing the routine for applying the US on the PCL-BT piezoelectric scaffolds.

2.8.4. Cell Differentiation

In this assay, MC3T3 cells were seeded on multiple (more than the number needed for statistical analysis) specimens of the same composition. Then, 3 Mhz US was applied to the cells cultured on the scaffolds as shown in Figure 5. This is to maximize the amount of RNA collection. The cell-seeded specimens were cultured for 3 or 7 days, and at the end of specific times, RNA from these cells was extracted by TRIzol reagent (Invitrogen, Carlsbad, CA, USA). Complementary DNA was produced by reverse transcription using M-MLV reverse transcriptase (Promega, Madison, WI, USA), and quantitative Real-Time (qRT-PCR) was performed using SsoFast EvaGreen Supermix (Bio-Rad, Hercules, CA, USA) by a 2-step amplification program (30 s at 95 °C and 60 s at 62 °C) on a thermal cycler (Eppendorf, Hamburg, Germany). The relative quantification of mRNA of the gene of interest was determined by the $2^{-\Delta\Delta CT}$ method and presented as fold change compared to the control sample (PCL). The forward and reverse primers for targeted genes are listed as follows:

Osteocalcin (OCN; forward 5′-GCAATAAGGTAGTGAACAGACTCC-3′ and reverse 5′-CTTTGTAGGCGGTCTTCAAGC-3′),

Osteopontin (OPN; forward 5′-CTTTCACTCCAATCGTCCCTAC-3′ and reverse 5′-GCTCTCTTTGGAATGCTCAAGT-3′),

Alkaline Phosphatase (ALP; forward 5′-ATCTTTGGTCTGGCTCCCATG-3′ and reverse 5′-TTTCCCGTTCACCGTCCAC-3′)

Type 1 Collagen (Col-1; forward 5′-GAGCGGAGTACTGGATCG-3′ and reverse 5′-GCTTCTTTTCCTTGGGGTT-3′), and

Runt-related transcription factor 2 (Runx2, forward 5′- CCAGATGGGACTGTGGTTA-CC3′ and reverse 5′-ACTTGGTGCAGAGTTCAGGG3′)

2.9. Statistical Analysis

All test results represented means ± standard deviation in triplicates. One-way and two-way analysis of variance (ANOVA) with the Tukey test was conducted to determine the statistical difference between groups, and $p < 0.05$ was considered significant.

3. Results and Discussion

3.1. Physical Characterization Results

Figure 6 shows the XRD plots of the various PCL-BT composite specimens. The unmodified PCL specimen exhibits sharp and well-defined peaks corresponding to the polymer. Specifically, two major peaks at the 2θ angles of 21.4° and 23.8° can be observed. These two peaks correspond to the (110) and (200) crystallographic planes of the semi-crystalline nature of PCL biopolymer, respectively [34,35]. However, incorporating $BaTiO_3$ in the PCL matrix markedly reduced the intensity of the PCL diffraction peaks, and instead, new high-intensity and well-defined diffraction peaks were spotted. These peaks correspond to $BaTiO_3$ as per JCPDS: #34-0129 [36,37]. The highest diffraction peak corresponding to $BaTiO_3$ can be spotted at the 2θ angle of 30°, which corresponds to the (101) lattice plane. Furthermore, the other low-intensity yet sharp and well-defined peaks of $BaTiO_3$ correspond to (111), (002), (112), (202), and (103) lattice planes [38]. Furthermore, the well-defined nature of the $BaTiO_3$ diffraction peaks indicates the high crystallinity of the $BaTiO_3$ particles. These findings are in accordance with previous studies that dealt with PCL- $BaTiO_3$ composites, as mentioned in Ref [39].

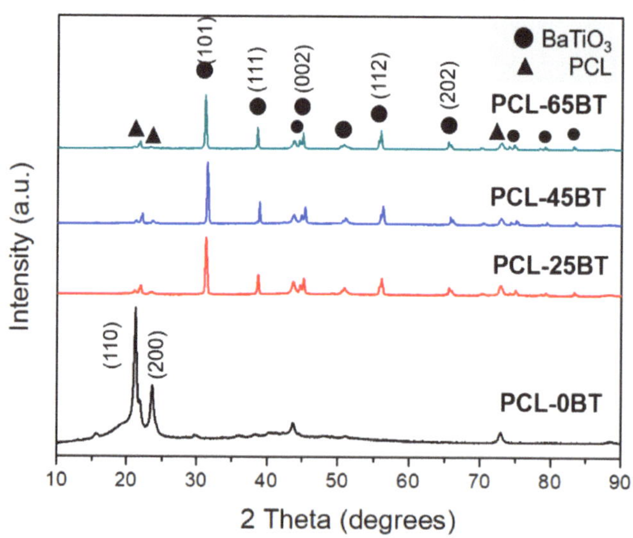

Figure 6. X-ray diffraction (XRD) of the PCL-BT composite scaffolds with various concentrations of $BaTiO_3$.

The SEM micrographs in Figure 7a indicate the presence of uniform-sized and interconnected pores in the PCL-BT composite scaffolds. The average pore sizes of the scaffolds are mentioned in the image headers. Similar to our previous efforts in developing scaffolds with highly consistent geometry and uniform-sized pores, FFF stands out to be a highly efficient yet sustainable manufacturing technique for developing scaffolds with controlled pore sizes and geometry [19,40]. Previous studies focused on developing porous $BaTiO_3$-based piezoelectric scaffolds using conventional methods such as freeze casting [36] or particulate (salt)-leaching [28]. However, even though such techniques can form porous scaffolds, there is little to no control over the pore size and geometry. Moreover, the probability of creating interconnected pores critical for cell nutrient flow might be less when using such conventional methods. As opposed, FFF is a powerful yet simple technique to develop scaffolds with highly consistent interconnected pores that should be leveraged to develop porous piezoelectric scaffolds [41].

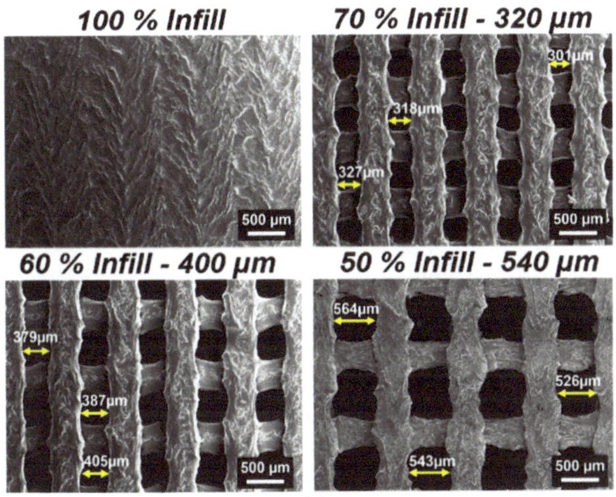

(a)

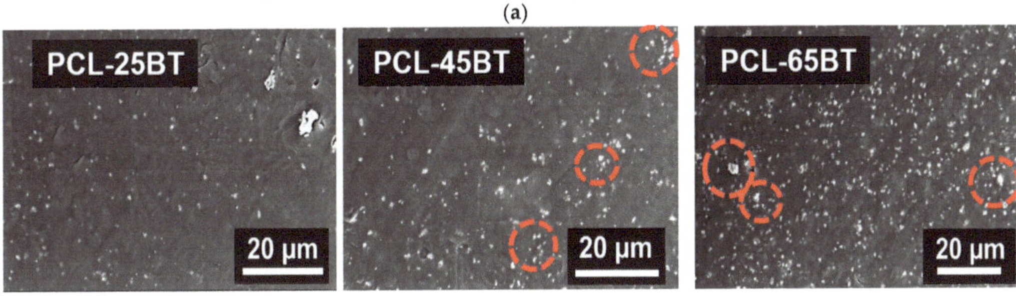

(b)

Figure 7. SEM analysis of the scaffolds using secondary electron detector. (**a**) SEM of the non-porous and porous 3D-printed PCL-45BT scaffolds. The pore sizes of the scaffolds were varied by changing the infill percentage in the slicing software. (**b**) SEM of various PCL-BT filament cross-sections showing the uniform dispersion of the white $BaTiO_3$ particles in the PCL matrix.

The cross-sections of the various composite filaments are shown in Figure 7b. For composite materials, obtaining a homogenous dispersion of the secondary phase in the polymer matrix is critical to decreasing the chances of mechanical property degradation. The white dots in Figure 7b represent the $BaTiO_3$ particles in the polymer matrix, which is seen as dark background, similar to previous reports [39]. The micrographs further indicate the homogenous distribution of the $BaTiO_3$ microparticles in the PCL matrix, and no evident agglomerations were observed. As expected, the concentration of the white particles corresponding to the $BaTiO_3$ particles increased with the higher incorporation percentage of the $BaTiO_3$ in PCL. Importantly, even at 65 vol.% of $BaTiO_3$ incorporation, there were no signs of $BaTiO_3$ particle agglomeration, confirming that the combined steps of solvent casting and the two-step extrusion process are suitable for homogenous mixing of the $BaTiO_3$ ceramic particles in the PCL polymer matrix. However, some $BaTiO_3$ particle clusters were visible in the PCL-45BT and PCL-65BT composites, as identified by the red dotted circles in Figure 7b. The $BaTiO_3$ cluster sizes were less than the agglomeration, similar to previous studies [42].

3.2. Piezoelectrical Properties

We poled the specimens to align the dipoles and to permanently achieve a prominent piezoelectric response. We specifically chose corona poling, as we wanted the specimens to be poled without immersing them in an oil bath, which might lead to eventual cytotoxicity in cells due to the remanent oil. We checked both the samples' dry and wet piezoelectric responses, as the specimens are expected to be implanted in vivo, exposing them to bodily fluids. Hence, we immersed the specimens in PBS for prolonged periods for the wet piezoelectric response. We chose the PCL-25BT specimen to optimize our poling regime, hypothesizing that PCL-25BT would exhibit the lowest piezoelectric response among all the specimens and would be indicative of the piezoelectric effect. Figure 8a shows the dry piezoelectric response of the specimens. Poling at 8 and 10 kV for 5 and 15 min did not result in any piezoelectric response from the PCL-25BT, indicating that those voltages and times were insufficient to pole the specimens. However, the specimen yielded a piezoelectric response of 1.2 pC/N after poling at 10 kV for 30 min. We wanted to avoid prolonged poling times as we noticed the PCL-BT samples partially melted and deformed at the end of 30 min in cases of 12 and 15 kV due to the heat generated from the high voltages. Instead, the higher voltages of 12 and 15 kV for 5 and 10 min were sufficient to pole the samples with no deformation. The increase in poling voltage and time resulted in a gradual increase in the piezoelectric response of the specimens, but there were no significant differences between the specimen groups poled at voltages 12 and 15 kV. Furthermore, the specimens' dry and wet piezoelectric responses are similar (Figure 8a,b). Overall, the specimens exhibited piezoelectric responses in the range of 1.2–2.25 pC/N with respect to different poling voltages and time. We chose our poling regime to be 15 kV for 15 min for further studies, as the PCL-25BT specimen exhibited a piezoelectric coefficient of 1.25 pC/N, similar to previous studies [42,43].

Figure 8c shows the piezoelectric responses of various PCL-BT specimens. The d_{33} piezoelectric coefficient of the PCL-BT composites increased with the increase in $BaTiO_3$ content. Notably, the $BaTiO_3$ inclusion up to 25 vol.% enhanced the piezoelectric response gradually to 1.2 pC/N compared to the unmodified PCL specimen; however, the piezoelectric response increased significantly when the $BaTiO_3$ inclusion was above 25 wt.%. Specifically, the piezoelectric response increased to 2.4 and 2.6 pC/N for the PCL-45BT and PCL-65BT specimens. Several previous reports also observed similar results that dealt with $BaTiO_3$ [36,42,44]. For instance, Liu et al. [42] observed a drastic rise in the d_{33} values (up to 3.9 pC/N) when the $BaTiO_3$ incorporation extent increased above 35 vol.% in the PCL matrix. The drastic rise in the d_{33} coefficient is due to the dense distribution of the $BaTiO_3$ particles in the PCL-45BT and PCL-65BT specimens as shown in Figure 7b. The electric field developed by the closely packed $BaTiO_3$ particles forms a network interacting with each other, resulting in a more robust electroactive response. On the contrary, the $BaTiO_3$ particles are sparsely distributed in the PCL-25BT specimens (Figure 7b), resulting in reduced electroactive responses.

The d_{33} piezoelectric coefficient of the PCL-BT specimens in the present study is similar to the results of the other $BaTiO_3$-based composites previously reported in the literature. For instance, porous HA-$BaTiO_3$ composites explored by Zhang et al. [36] exhibited d_{33} values in the range of 0.3–2.8 pC/N. Tang et al. [44] recorded d_{33} values of polarized HA-$BaTiO_3$ piezoelectric ceramics in the range of 1.3–6.8 pC/N with $BaTiO_3$ content ranging from 80% to 100%. Some of the HA-$BaTiO_3$ composites with less than 80 vol.% $BaTiO_3$ did not exhibit any piezoelectric effect, highlighting the need for high-volume $BaTiO_3$ contents in the composites [45,46]. On the other hand, some of the bulk-sintered HA-$BaTiO_3$ composites exhibited outstandingly high d_{33} values (>50 pC/N) because of the close packing density of the piezoelectric $BaTiO_3$ particles in a sintered sample [47]. Polymer–ceramic composites such as the PCL-BT can never exhibit such a high piezoelectric response because the $BaTiO_3$

particles are not densely packed in a polymer matrix. Moreover, such a high scaffold-mediated piezoelectric response is not needed for bone regeneration, as the bone itself exhibits a piezoelectric response in the range of 0.7–2.3 pC/N [48]. The 3D-printed PCL-BT composites in the present study exhibited d_{33} values that are close to the piezoelectric response of bone and were also on par with HA-$BaTiO_3$ composites developed by various conventional techniques [36,43,44]. Hence, the observed piezoelectric response of the PCL- $BaTiO_3$ composites in the present study highlights their great potential in the bone remodeling process and osteogenesis and expediting treatment of orthopedic defects.

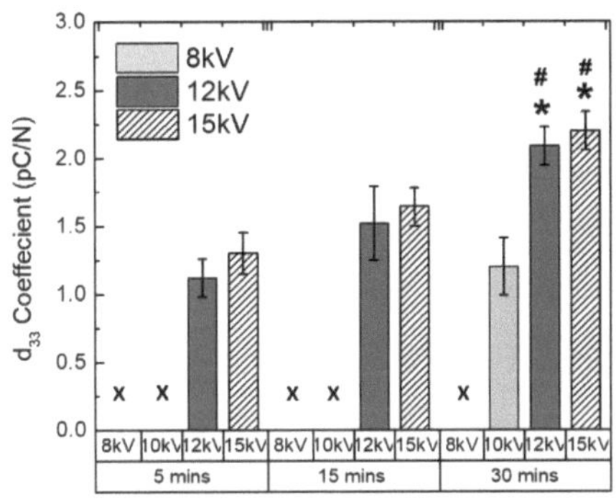

(a)

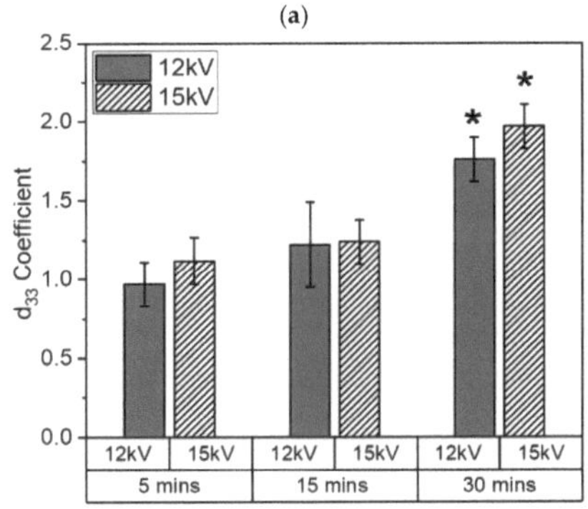

(b)

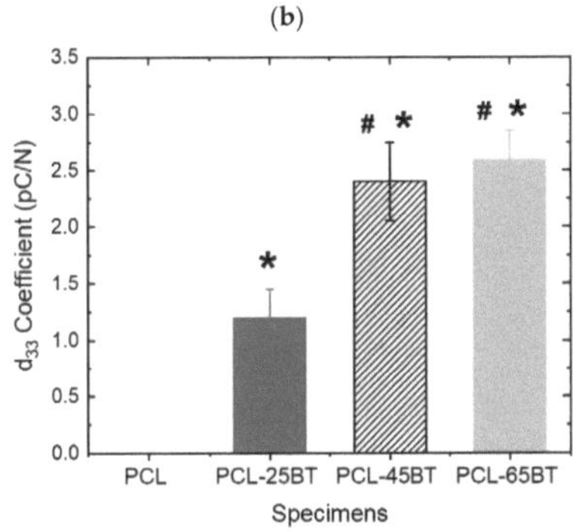

(c)

Figure 8. d_{33} coefficient (picoColoumbs/Newton, pC/N) of (**a**) dry and (**b**) wet PCL-BT scaffolds with respect to the poling voltage and time. # means statistically significant ($p < 0.05$) with respect to the specimen poled at 10 kV in the same group (30 min). * means statistically significant ($p < 0.05$) with respect to the specimens poled for 5 and 15 min. (**c**) d_{33} coefficient of various wet PCL-BT scaffolds. * means statistically significant ($p < 0.05$) with respect to PCL. # means statistically significant ($p < 0.05$) with respect to PCL-25BT.

3.3. Mechanical Properties

Figure 9a,b shows the mechanical properties of the piezoelectric PCL-BT specimens. It is evident that the inclusion of 25 vol.% $BaTiO_3$ particles increased the modulus by 35% (Figure 9a) and the tensile strength by 14% (Figure 9b). Similar results were observed by Bagchi et al. [39] when the authors incorporated various kinds of perovskite ceramic nanoparticles, including $BaTiO_3$, in the PCL matrix. However, in the latter study, the authors observed an increase in the yield strength after adding 20 wt% $BaTiO_3$ in PCL, even though the increase was the lowest (about 10%) when compared to the other ceramic additives explored in that study. As opposed, in the present study, we observed a decrease in the modulus and the tensile strength when the $BaTiO_3$ inclusion extent was increased above 25 wt.%. For instance, increasing the $BaTiO_3$ incorporation percentage to 45 and 65 vol.% reduced the modulus by 9% and 15%, respectively, with respect to unmodified PCL. Furthermore, the tensile strength decreased from 14 MPa (for unmodified PCL) to almost 12 MPa for PCL-45BT and 11 MPa for PCL-65BT specimens (Figure 9b) However, the tensile strain % of the PCL-BT composites was not influenced by the content of the BT particles. All the composites exhibited a high tensile strain % with an average value of 550%, similar to PCL specimens, which exhibited a strain of 590%.

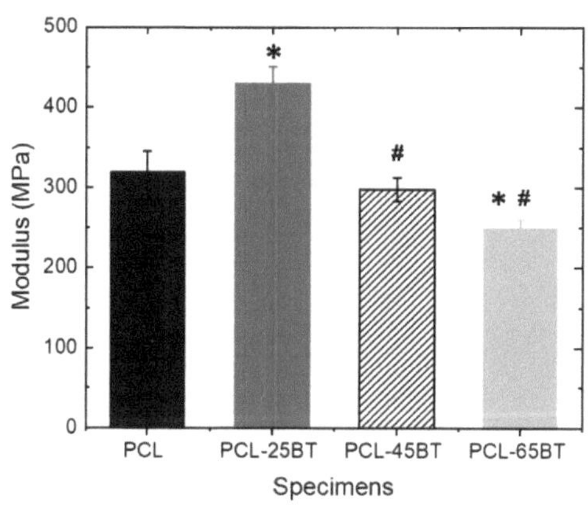

(**a**)

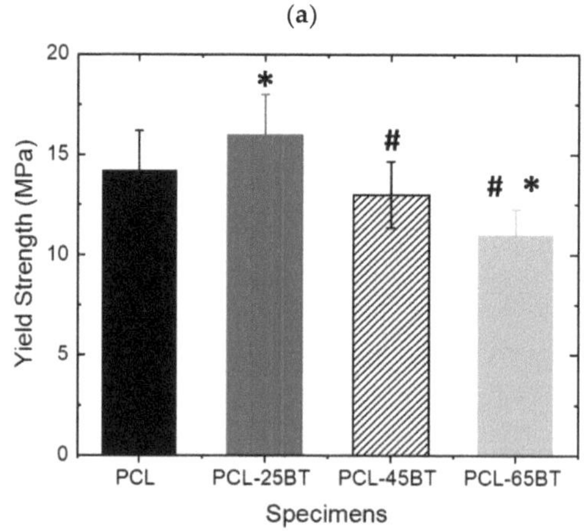

(**b**)

Figure 9. Mechanical testing of the composite specimens. (**a**) Young's modulus and (**b**) yield tensile strength of the 3D-printed PCL and PCL-BT scaffolds. * means statistically significant ($p < 0.05$) with respect to PCL. # means statistically significant ($p < 0.05$) with respect to PCL-25BT.

It is well known that the incorporation of excess secondary phases in polymer composites can hamper the polymer's inherent mechanical properties. Moreover, including ceramic particles in polymers can increase the brittleness of the composites [23,24]. The excess secondary materials can form agglomerations in the matrix, disrupting the efficient load/stress transfer between the ceramic particles and the matrix. However, we did not notice a massive decrease in the mechanical properties with higher $BaTiO_3$ contents (>25 vol.%), indicating that the mixing and distribution of the $BaTiO_3$ particles were favorable. However, the decrease in the mechanical properties of the PCL-45BT and PCL-65BT specimens was due to the clustering of the $BaTiO_3$ particles, as shown in Figure 7b.

All the PCL-BT specimens in the present study were developed by FFF-based 3D printing. In FFF, the printed layers are glued to each other layer by layer to develop the final material; this process of product manufacturing can significantly influence the mechanical properties of the manufactured parts, especially when the print parameters and orientations are incorrect. Even though optimizing the 3D-printing parameters to achieve a robust 3D-printed part was not the scope of this study, our extensive experience in 3D printing of various polymers and composites [18–20,23,24,40] helped us achieve mechanically robust 3D-printed parts. To highlight, the 3D-printed PCL-25BT specimen's mechanical properties were similar to the PCL-BT specimen (with 20 wt.% $BaTiO_3$) that was developed by conventional compression molding [39]. Overall, the mechanical properties of the PCL-BT scaffolds qualify them as mechanically robust scaffolds for orthopedic applications.

3.4. Piezoelectric Effect on the in Vitro Properties of Pre-Osteoblasts

While performing a literature review for this study, we observed that most of the studies did not apply any stimuli to the piezoelectric scaffolds during the in vitro cell studies, such that the scaffolds would exhibit the piezoelectric response during the cell studies. In most cases, the studies used poled samples for cell studies, which raises the question, *"Whether it is the piezoelectric effect from the scaffolds or is it the surface charge on the scaffolds that is helping in cellular activities?"*. We were specifically interested in analyzing the scaffold-mediated piezoelectric effect on the cells; hence, we chose ultrasonic stimulation (US) as the stimuli to create the deformation on the scaffolds and exhibit the piezoelectric effect. Several studies in the literature used US to stimulate $BATiO_3$ 3=-based piezoelectric scaffolds and confirmed that US is a highly effective stimulus to enhance the piezoelectric effect in electroactive scaffolds [17,49]. Moreover, we chose US, as it does not physically come in touch with the cell-laden scaffolds avoiding contamination issues.

3.4.1. Pre-Osteoblast Proliferation

Effect of US and Piezoelectricity on Pre-Osteoblast Proliferation

Figure 10a,b shows the in vitro cell proliferation properties of the MC3T3 cells cultured on the PCL-65BT piezoelectric scaffolds. We chose PCL-65BT as our test specimen, as the latter specimen exhibited the highest piezoelectric effect, and we hypothesized it would have a prominent effect on the pre-osteoblast cells. First, we wanted to analyze the short-term impact of the US and scaffolds' piezoelectric stimulation on the pre-osteoblasts proliferation behavior. As shown in Figure 10a, we observed that US alone had no positive influence on cell proliferation. However, 1 MHz US combined with the piezoelectric scaffolds significantly increased cell proliferation over 48 and 72 h with respect to the only US treatment. The 3 MHz US treatment, combined with the piezoelectric scaffolds, was also beneficial, but the treatment did not result in a significant increase in cell proliferation. The results were similar when the US regime continued for up to 7 days, as shown in Figure 10b. However, after the 7-day treatment and culture, the 3 MHz US treatment combined with the piezoelectric scaffolds yielded nearly similar results to the 1 MHz US, indicating that over long treatment times, the 1 and 3 MHz US application regimens can deliver similar outcomes in terms of pre-osteoblast proliferation. Notably, the cells proliferated significantly over 7 days, irrespective of the specimen and the treatment. Nevertheless, we determined that the 1 MHz US treatment and the piezoelectric stimulation from the PCL-65BT scaffolds are optimum for pre-osteoblast proliferation. Once we determined the optimum US regimen for cell proliferation, we next analyzed the effect of $BaTiO_3$ on cell proliferation.

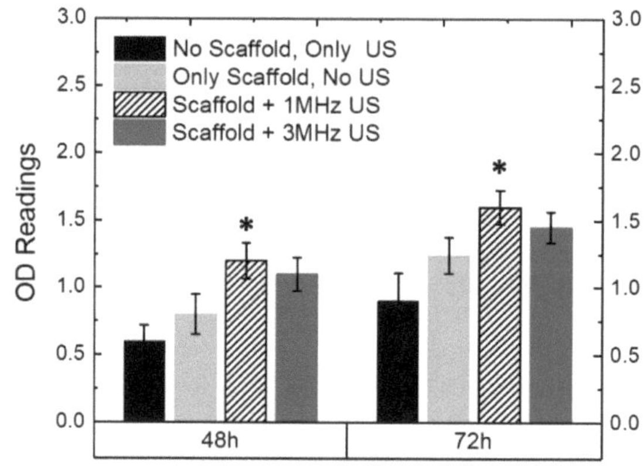

(a)

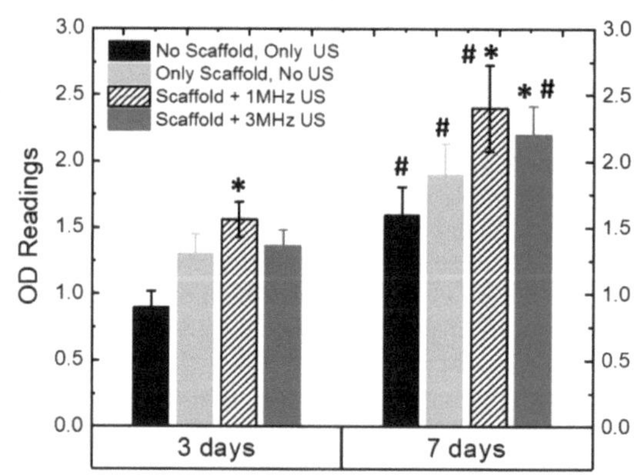

(b)

Figure 10. Ultrasonic stimulation (US) and in vitro properties of the scaffolds. OD readings of MC3T3 cells as cultured on the piezoelectric scaffolds (PCL-65BT) with respect to different intensities of US over (**a**) 72 h and (**b**) 7 days. * means statistically significant ($p < 0.05$) with respect to "no scaffold, only US". # means statistically significant ($p < 0.05$) with respect to the specimens in the "3 days" group.

Effect of Piezoelectric $BaTiO_3$ on Pre-Osteoblast Proliferation

Figure 11a confirms that the $BaTiO_3$ presence in all the composite scaffolds significantly increased the cell growth kinetics over time as compared to the non-piezoelectric PCL scaffolds. This is due to the increased piezoelectric response resulting from the $BaTiO_3$ content in the PCL-BT composites. However, even though there was a significant difference in the extent of cell proliferation as compared to the PCL scaffolds, there was no significant difference in the extent of cell proliferation among the various PCL-BT specimens. The SEM images (Figure 11a,b) of the adhered pre-osteoblast cells on various 3D-printed scaffolds corroborate the proliferation results. Very few cells adhered to the scaffolds after day 1 of the culture (Figure 11a). In addition, many circular and spindled cells were seen, indicating that these cells did not have time to spread over the surface with their filopodia. However, after day 7 of the culture, it is evident that the cells proliferated notably and spread over the scaffolds with extending filopodia, as shown in Figure 11b. Furthermore, it is evident that the increasing $BaTiO_3$ concentration in the composites results in enhanced cell adhesion and spreading. In all the cases, the cells approvingly adhere and spread on the surface with flattened morphology, signifying that the PCL-BT scaffolds exhibit a bioactive and biocompatible surface. The increasing cell adherence and spread over the composites with increasing $BaTiO_3$ concentration demonstrate that the scaffold-generated piezoelectric effect played a major role in cell spreading [50]. The increasing $BaTiO_3$ particles enhance the piezoelectric effect, which helps the cells to proliferate and spread more rapidly on the scaffold surfaces. However, even though more cells proliferated and were seen to adhere on the PCL-45BT and PCL-65BT specimens as compared to PCL-25BT specimens, there was not a significant difference in the amount of adhered cells on all the composites. Similar results were obtained by Tang et al. [44], where the authors observed no significant difference in the osteoblast proliferation on the composite scaffolds containing different contents of $BaTiO_3$ (80 and 90 wt% $BaTiO_3$). Similarly, Tavangar et al. [46] reported no significant difference in cell adhesion and proliferation between the composite scaffolds with varying $BaTiO_3$ (40, 50, and 60% wt $BaTiO_3$). This could be possible because the $BaTiO_3$ particles are embedded in the polymer or ceramic matrix, and in most cases, they are not directly interacting with the cells; hence, the variation in $BaTiO_3$ presence does not play a significant role in cell proliferation [51,52]. Instead, we believe that the extent of the piezoelectric effect generated from the $BaTiO_3$ particles influences cellular activities more prominently than the biocompatibility nature of the $BaTiO_3$.

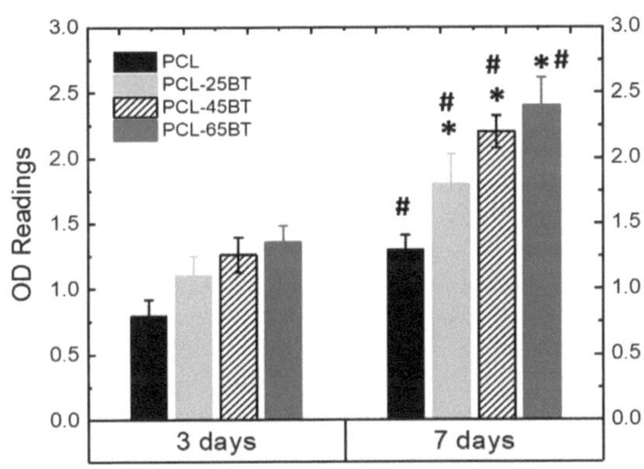

(a)

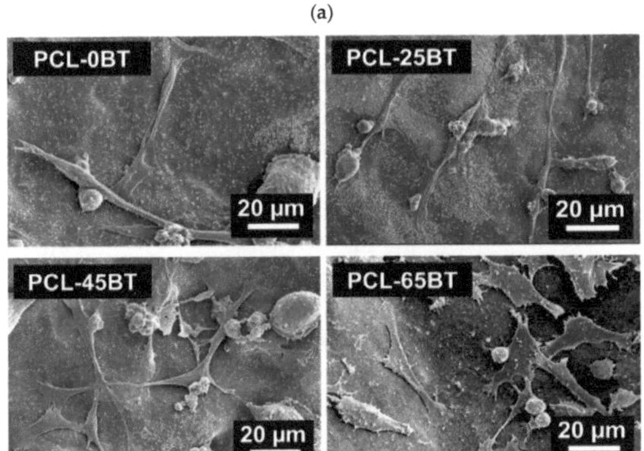

(b)

Figure 11. *Cont.*

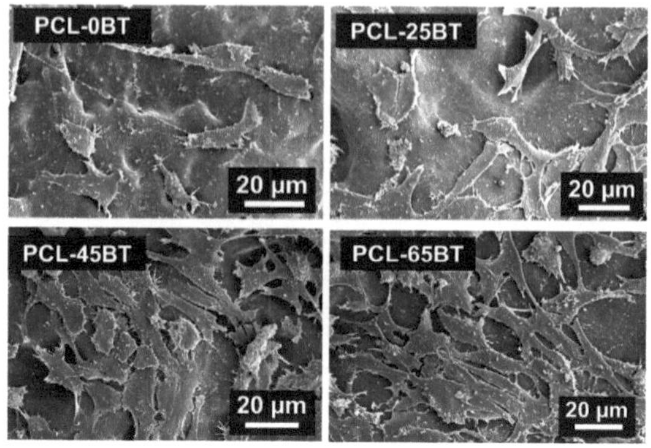

(c)

Figure 11. Effect of $BaTiO_3$ content on the in vitro properties of the piezoelectric scaffolds. (**a**) OD readings of the MC3T3 cells cultured on the PCL and various piezoelectric PCL-BT scaffolds; 1 MHz US were used per the routine for all the proliferation test. * means statistically significant ($p < 0.05$) with respect to PCL specimens. # means statistically significant ($p < 0.05$) with respect to the specimens in the "3 days" group. SEM images of the adhered MC3T3 cells on various scaffolds (**b**) after day 1 and (**c**) after day 7 of culture.

Clearly, US alone does not affect cellular activities in the present case. However, US helped in remarkable cell proliferation when the piezoelectric scaffolds were present, indicating that the US invigorated the scaffolds and generated the piezoelectric effect, increasing the cell growth kinetics. Furthermore, the increase in $BaTiO_3$ content resulting in the enhancement of pre-osteoblast proliferation is a direct indication that US can interact with the $BaTiO_3$ particles and generate the piezoelectric effect. The effect increases when US interacts with more $BaTiO_3$ particles in the composites, notably surging the cell growth kinetics for the PCL-65BT specimens. Shuai et al. [53] developed PVDF-$BaTiO_3$ scaffolds and demonstrated that the electric cues generated by the piezoelectric scaffolds under US efficiently promoted MG-63 cell proliferation. Yang et al. [17] observed similar results when they used US on PLLA- $BaTiO_3$ and PLLA-$BaTiO_3$-graphene scaffolds and deduced that the US was responsible for yielding effective piezoelectric response from the scaffolds. Even $BaTiO_3$ coatings on metallic scaffolds and US have remarkably increased bone cell activity. For instance, Fan et al. [49] coated $BaTiO_3$ onto the porous titanium (Ti) scaffold and applied low-intensity pulsed US (LIPUS), similar to the clinical US used in the present study, and observed that LIPUS was highly influential in activating the piezoelectric effect of $BaTiO_3$. The latter effect significantly enhanced cell viability in vitro, promoted osteogenesis and osseointegration in vivo, and effectively treated large bone defects in a rabbit model. The same research group performed another study in a large animal model (sheep) and confirmed that LIPUS induces a piezoelectric response from $BaTiO_3$ particles, leading to an increased cellular response in vitro and treating sizeable segmental bone defects in vivo [54]. Recently, Chen et al. [50] also used LIPUS on the $BaTiO_3$-coated Ti scaffold and coined the new term "piezodynamic therapy" for such kinds of treatment. The authors observed that the piezodynamic effect of US and piezoelectric $BaTiO_3$ coating activated more mitochondria at the initial stages of cell culture that intervened in the cell culture cycle by promoting cell proliferation and weakening the apoptotic damage. These studies reinforce that US combined with piezoelectric PCL-BT scaffolds has a strong potential to enhance bone cell activities and help in new bone formation and defect regeneration in vivo.

Effect of Pore Size on Pre-Osteoblast Proliferation

We also observed that the pore sizes significantly impacted the cell proliferation kinetics. For instance, as shown in Figure 12, 3D-printed PCL-BT scaffolds with 70% infill (avg pore size: 320 μm) resulted in the highest cell growth compared to the other macro-porous scaffolds and the non-porous scaffolds. This is because the 320 μm pore size provides an optimum distance for cell-cell communication as opposed to other pore sizes. Similarly, Lee et al. [55] reported 350 μm pore sizes to be the optimum for cell proliferation. However, porous scaffolds can compromise the mechanical properties of load-bearing orthopedic scaffolds. In one of our recent studies, we observed that 3D-printed scaffolds with 300 μm pore size exhibit the highest yield compressive strength, and increasing the pore size beyond that would decrease the specimen's yield strength [19]. The 320 μm pore-sized scaffolds in the present study resulted in the highest cell proliferation, indicating that such PCL-BT scaffolds would be beneficial for increasing cell growth kinetics and exhibit favorable mechanical properties.

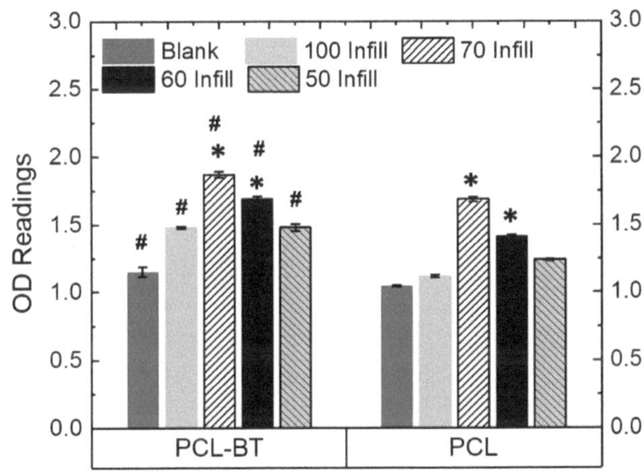

Figure 12. OD readings of MC3T3 cells cultured on the piezoelectric scaffolds (PCL-65BT) with different infill (pore sizes); 1 MHz US was used per the routine for all the proliferation tests. * means statistically significant ($p < 0.05$) with respect to the controls (Blank and 100 Infill specimens) in the same group. # means statistically significant ($p < 0.05$) with respect to the PCL specimens with different infills.

Bone tissue engineering has extensively utilized porous scaffolds to enhance implant–tissue interaction. The osteogenic capability of the orthopedic scaffold can be enhanced significantly by interconnected pores, as they facilitate cell distribution, nutrients, and blood flow. Furthermore, porous scaffolds help in robust anchorage, scaffold-tissue osseointegration, and vascularization [56,57]. However, conventional manufacturing techniques such as salt leaching, gas foaming, phase separation, and freeze-drying do not control pore sizes and numbers. In addition, the formed pores lack interconnectivity, which is critical for nutrient flow, cell migration, vascularization, and tissue ingrowth. Furthermore, scaffolds with random pores and wide variations in pore sizes weaken the scaffold's mechanical strength or load-bearing capacity, thus leading to implant failure. In contrast, 3D printing is a powerful manufacturing technique that can form uniform-sized interconnected pores in scaffolds, facilitating osteogenesis and osseointegration. Thus, the 3D-printed PCL-BT scaffolds developed in the present study will not only help bone cell proliferation via piezoelectricity but will also help expedite new bone formation with the help of the highly uniform pores.

3.4.2. Osteoblast Differentiation

Figure 13a–e show the differentiation behavior of the MC3T3 pre-osteoblasts over time when exposed to the US treatment and the piezoelectric scaffolds. As per previous reports, we selected the 3 MHz US for the differentiation studies and observed that the 3 MHz treatment effectively differentiates pre-osteoblast cells into osteoblasts. All the piezoelectric PCL-BT scaffolds upon US application increased the mRNA expressions notably of all the osteogenic gene markers not after 3 days, but after 7 days of culture. This is usual, as the gene markers explored in this present case are observed in stages of osteoblastic differentiation, and it usually takes longer than 3 days for the MC3T3 pore-osteoblasts to differentiate [58–61]. Notably, the PCL-45BT and PCL-65BT specimens markedly increased the ALP, Col-1, OPN, OCN, and Runx-2 expressions, which are critical in osteogenesis.

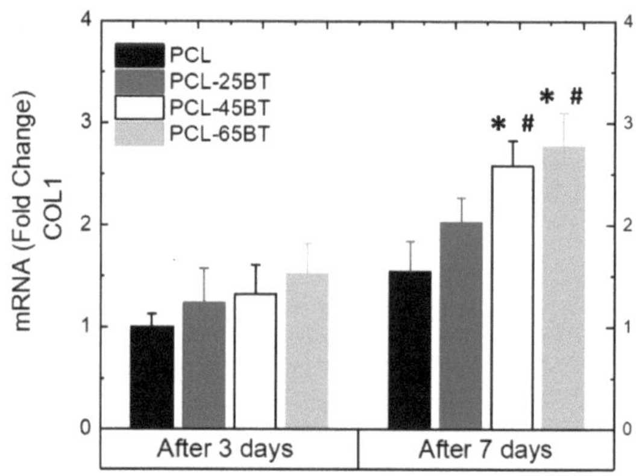

(**a**)

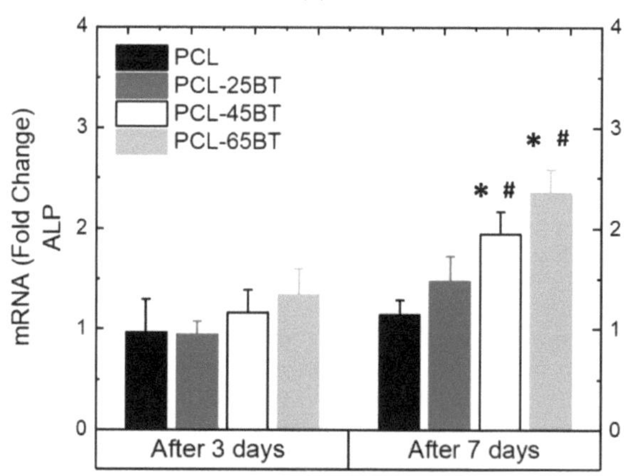

(**b**)

Figure 13. *Cont.*

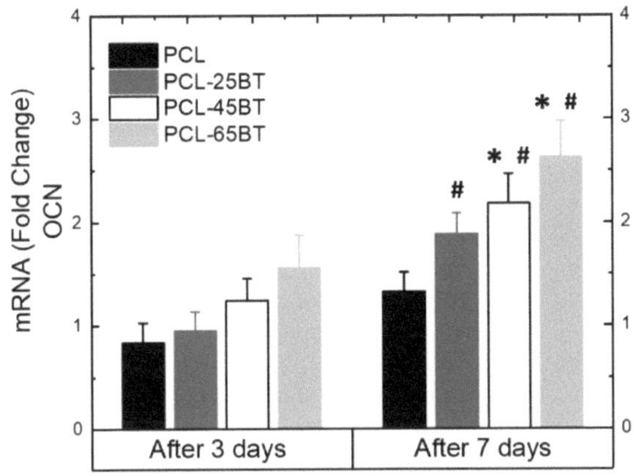

(**c**)

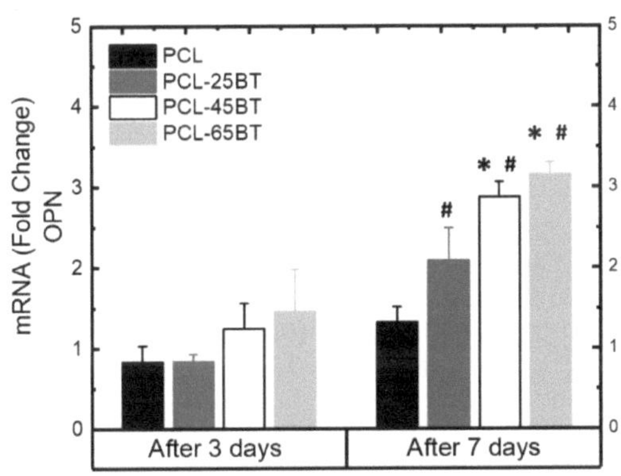

(**d**)

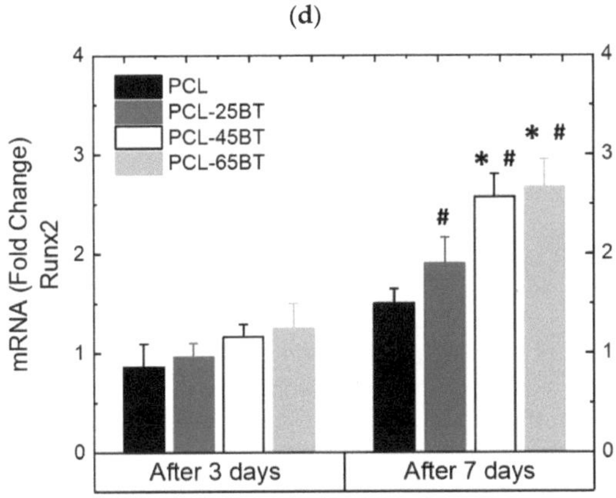

(**e**)

Figure 13. Differentiation properties of MC3T3 cells as cultured on the piezoelectric scaffolds. The following osteogenic gene markers were quantified: (**a**) Col 1 (**b**) ALP (**c**) OCN, (**d**) OPN, and (**e**) Runx 2 and expressed as fold change with respect to the control; 3 MHz US was used per the routine for all the differentiation assays. * means statistically significant ($p < 0.05$) with respect to the control (PCL) in the same group. # means statistically significant ($p < 0.05$) with respect to the fold changes of the same specimen in the "after 3 days" group.

The results in the present study are very similar to previous studies. For instance, Shuai et al. [53] observed no effect of only US on the ALP expressions and osteogenic differentiation of the MG-63 cells. However, ALP activities of the cells significantly increased when US was applied to the piezoelectric PVDF- $BaTiO_3$ scaffolds. Similarly, Fan et al. [49] observed significantly higher mRNA gene expression of Col-1, ALP, OPN, and Runx2 when $BaTiO_3$ coatings and LIPUS were used complementarily after 7 and 14 days. In the present study, we observed upregulation of all the gene markers only after 7 days. As mentioned before, the gene expression of ALP, Col-1, and Runx2 usually appears in the early stages of osteogenic differentiation of bone cells, while OPN and OCN expression occurs late in the process of osteogenic differentiation. Moreover, the US enhanced the mechanical–electrical conversion capability (piezoelectric response) of the scaffolds and yielded more electrical charges. It is proven that electrical charges activate the electrically sensitive calcium ions (Ca^{2+}) signal transduction [10], and Ca^{2+} plays a critical role in regulating osteoblast differentiation via either activation of the calcium-sensing receptors or increasing the Ca^{2+} influx into the osteoblast cells [62]. Thus, the electrical stimulation (due to the piezoelectric effect) from the PCL-BT scaffolds activated the electrically sensitive Ca^{2+} signal transduction, activating the calcium-sensing receptors and increasing the Ca^{2+} influx into MC3T3-E1 cells. The latter series of events resulted in a notable increase in osteogenic gene expression, specifically ALP [63,64]. The remarkable rise in ALP level in MC3T3 cells will result in enhanced bone matrix and mature extracellular matrix synthesis, as observed by Tian et al. [63]. Moreover, in an important recent study, Cai et al. [65] observed a remarkable increase in ALP, OCN, Col-1, and Runx-2 expression when they applied US to $BaTiO_3$-coated Ti scaffolds. The authors proved that the piezoelectric effect from the scaffolds specifically upregulated the mRNA and protein expressions of the calcium channel protein $Ca_V1.2$ in MC3T3 cells, which was responsible for the notable osteogenic differentiation.

Furthermore, Zhou et al. [66] indicated that the charged substrate surface helps in adsorbing proteins and various biological constituents from the culture medium through electrostatic interaction, which might also be another reason for promoting cell attachment, proliferation, and differentiation. Thus, the US successfully triggered an adequate piezoelectric response from the PCL-BT scaffolds, resulting in the upregulation of prominent osteogenic-related gene expressions.

On a separate note, electrical fields are instrumental in cell migration during wound healing [3,67]. Even though analyzing cell migration under the piezoelectric effect was not the scope of this study, we believe that the adhered cells on the scaffolds might have migrated toward the surface regions, which were devoid of cells (similar to a wound), eventually covering the entire scaffold as confirmed by the SEM images in Figure 12b,c. We draw a resemblance that the blank regions on the scaffold at the beginning of the culture (i.e., after day 1) are similar to a wound that is usually devoid of cells. Eventually, by the end of 7 days, the cells approvingly proliferate and spread over the blank surfaces on the scaffolds with the generation of the piezoelectric effect from the scaffolds. Morimoto et al. [68] observed that the migration distance was significantly longer, and the area of the transplanted bone cells was markedly wider in an ischemic stroke model of rats when they applied electrical stimulation. Such studies confirm the various effects of scaffold-mediated electrical fields/electrical charges/electrical stimulation on bone cell activities.

Finally, the combination of the US and PCL-BT scaffolds enhanced the bone cell activities, highlighting that the US could interact with the embedded $BaTiO_3$ particles in the 3D-printed composite scaffolds. Our results strongly confirm that US can be an effective stimulus to enhance the piezoelectric effect in the PCL-BT scaffolds, indicating that the US-activated piezoelectric response from electroactive PCL-BT scaffolds can be a powerful therapeutic tool for enhanced osteogenesis and expedited bone defect or fracture treatment.

4. Conclusions

Overall, this study has three significant findings to highlight. First, this study calls attention to the strong utilization of polymer–ceramic multi-functional 3D-printable filaments. We successfully formed PCL-BT 3D-printable filaments that can be utilized in an FFF setup to seamlessly develop design-specific scaffolds that can lead to the clinical translation of various customized, multi-functional implants for reconstructive applications in the field of orthopedics and beyond. Even though it is beneficial to develop multi-functional implants, few efforts have been made to form polymer–ceramic composite filaments for FFF. This study provides evidence that composite filaments can be leveraged to create multi-functional medical implants and devices. Second, this study highlights the utilization of FFF-based 3D printing to develop design-specific piezoelectric and regenerative implants. Few efforts have been made to 3D print piezoelectric implants focused on regenerative medicine; most efforts are concentrated on piezoelectric sensors. Instead, this study provides evidence that 3D printing and innovative biomaterial compositions can be leveraged to develop regenerative piezoelectric scaffolds not only for orthopedic applications but also for maxillofacial, cranial, and dental applications. Such scaffolds can also be considered in a digital workflow for implantology [69]. Finally, this is one of the few studies to prove that clinical US treatment combined with piezoelectric scaffolds can be instrumental in expediting bone regeneration and cure and help minimize the treatment time in patients with severe orthopedic injuries. However, follow-up in vivo studies is mandatory to further validate the efficacy of the piezoelectric implants in terms of EF generation at the implantation site and the extent of bone regeneration.

Author Contributions: Conceptualization, P.S.; methodology, P.S., P.N. and H.P.S.N.; software, P.N.; validation, P.S., P.N. and H.P.S.N.; formal analysis, P.S.; investigation, P.S., P.N. and H.P.S.N.; resources, P.S.; data curation, P.S., P.N. and H.P.S.N.; writing—original draft preparation, P.S.; writing—review and editing, P.S.; visualization, P.S.; supervision, P.S.; project administration, P.S.; funding acquisition, P.S. All authors have read and agreed to the published version of the manuscript.

Funding: We sincerely thank Sikder's start-up grant STARTUP06 at Cleveland State University for conducting this work.

Institutional Review Board Statement: Not applicable.

Informed Consent Statement: Not applicable.

Data Availability Statement: Data will be made available on request.

Conflicts of Interest: The authors declare no conflict of interest.

References

1. Uchino, K. The Development of Piezoelectric Materials and the New Perspective. In *Advanced Piezoelectric Materials*, 2nd ed.; Elsevier: Amsterdam, The Netherlands, 2017; pp. 1–92.
2. Rajabi, A.H.; Jaffe, M.; Arinzeh, T.L. Piezoelectric materials for tissue regeneration: A review. *Acta Biomater.* **2015**, *24*, 12–23. [CrossRef] [PubMed]
3. Zhao, M. Electrical fields in wound healing—An overriding signal that directs cell migration. *Semin. Cell Dev. Biol.* **2009**, *20*, 674–682. [CrossRef] [PubMed]
4. Baxter, F.R.; Bowen, C.; Turner, I.G.; Dent, A.C.E. Electrically Active Bioceramics: A Review of Interfacial Responses. *Ann. Biomed. Eng.* **2010**, *38*, 2079–2092. [CrossRef]
5. Hastings, G.W.; Mahmud, F.A. Electrical effects in bone. *J. Biomed. Eng.* **1988**, *10*, 515–521. [CrossRef]
6. Bassett, C.A.L.; Becker, R.O. Generation of Electric Potentials by Bone in Response to Mechanical Stress. *Science* **1962**, *137*, 1063–1064. [CrossRef] [PubMed]
7. More, N.; Kapusetti, G. Piezoelectric material—A promising approach for bone and cartilage regeneration. *Med. Hypotheses* **2017**, *108*, 10–16. [CrossRef]
8. Kapat, K.; Shubhra, Q.T.H.; Zhou, M.; Leeuwenburgh, S. Piezoelectric Nano-Biomaterials for Biomedicine and Tissue Regeneration. *Adv. Funct. Mater.* **2020**, *30*, 1909045. [CrossRef]
9. Murillo, G.; Blanquer, A.; Vargas-Estevez, C.; Barrios, L.; Ibáñez, E.; Nogués, C.; Esteve, J. Electromechanical nanogenerator–cell interaction modulates cell activity. *Adv. Mater.* **2017**, *29*, 1605048. [CrossRef]

10. Liu, Y.; Zhang, X.; Cao, C.; Zhang, Y.; Wei, J.; Li, Y.J.; Liang, W.; Hu, Z.; Zhang, J.; Wei, Y.; et al. Built-In Electric Fields Dramatically Induce Enhancement of Osseointegration. *Adv. Funct. Mater.* **2017**, *27*, 1703771. [CrossRef]
11. Ribeiro, C.; Correia, D.; Rodrigues, I.; Guardão, L.; Guimarães, S.; Soares, R.; Lanceros-Méndez, S. In vivo demonstration of the suitability of piezoelectric stimuli for bone reparation. *Mater. Lett.* **2017**, *209*, 118–121. [CrossRef]
12. Khare, D.; Basu, B.; Dubey, A.K. Electrical stimulation and piezoelectric biomaterials for bone tissue engineering applications. *Biomaterials* **2020**, *258*, 120280. [PubMed]
13. Cholleti, E.R. A Review on 3D printing of piezoelectric materials. *IOP Conf. Ser. Mater. Sci. Eng.* **2018**, *455*, 012046. [CrossRef]
14. Grinberg, D.; Siddique, S.; Le, M.-Q.; Liang, R.; Capsal, J.-F.; Cottinet, P.-J. 4D Printing based piezoelectric composite for medical applications. *J. Polym. Sci. Part B Polym. Phys.* **2019**, *57*, 109–115. [CrossRef]
15. Jiang, Z.; Cheng, L.; Zeng, Y.; Zhang, Z.; Zhao, Y.; Dong, P.; Chen, J. 3D printing of porous scaffolds BaTiO3 piezoelectric ceramics and regulation of their mechanical and electrical properties. *Ceram. Int.* **2022**, *48*, 6477–6487. [CrossRef]
16. Polley, C.; Distler, T.; Detsch, R.; Lund, H.; Springer, A.; Boccaccini, A.R.; Seitz, H. 3D Printing of Piezoelectric Barium Titanate-Hydroxyapatite Scaffolds with Interconnected Porosity for Bone Tissue Engineering. *Materials* **2020**, *13*, 1773. [CrossRef]
17. Yang, Y.; Peng, S.; Qi, F.; Zan, J.; Liu, G.; Zhao, Z.; Shuai, C. Graphene-assisted barium titanate improves piezoelectric performance of biopolymer scaffold. *Mater. Sci. Eng. C* **2020**, *116*, 111195. [CrossRef]
18. Challa, B.T.; Gummadi, S.K.; Elhattab, K.; Ahlstrom, J.; Sikder, P. In-house processing of 3D printable polyetheretherketone (PEEK) filaments and the effect of fused deposition modeling parameters on 3D-printed PEEK structures. *Int. J. Adv. Manuf. Technol.* **2022**, *121*, 1675–1688. [CrossRef]
19. Gummadi, S.K.; Saini, A.; Owusu-Danquah, J.S.; Sikder, P. Mechanical Properties of 3D-Printed Porous Poly-ether-ether-ketone (PEEK) Orthopedic Scaffolds. *JOM* **2022**, *74*, 3379–3391. [CrossRef]
20. Sikder, P.; Challa, B.T.; Gummadi, S.K. A comprehensive analysis on the processing-structure-property relationships of FDM-based 3-D printed polyetheretherketone (PEEK) structures. *Materialia* **2022**, *22*, 101427. [CrossRef]
21. Sonaye, S.Y.; Bokam, V.K.; Saini, A.; Nayak, V.V.; Witek, L.; Coelho, P.G.; Bhaduri, S.B.; Bottino, M.C.; Sikder, P. Patient-specific 3D printed Poly-ether-ether-ketone (PEEK) dental implant system. *J. Mech. Behav. Biomed. Mater.* **2022**, *136*, 105510. [CrossRef]
22. Saini, A.; Elhattab, K.; Gummadi, S.; Nadkarni, G.; Sikder, P. Fused Filament Fabrication-3D Printing of Poly-ether-ether-ketone (PEEK) Spinal Fusion Cages. *Mater. Lett.* **2022**, *328*, 133206. [CrossRef]
23. Elhattab, K.; Bhaduri, S.B.; Lawrence, J.G.; Sikder, P. Fused Filament Fabrication (Three-Dimensional Printing) of Amorphous Magnesium Phosphate/Polylactic Acid Macroporous Biocomposite Scaffolds. *ACS Appl. Bio Mater.* **2021**, *4*, 3276–3286. [CrossRef] [PubMed]
24. Elhattab, K.; Bhaduri, S.B.; Sikder, P. Influence of fused deposition modelling nozzle temperature on the rheology and mechanical properties of 3d printed β-tricalcium phosphate (tcp)/polylactic acid (pla) composite. *Polymers* **2022**, *14*, 1222. [CrossRef] [PubMed]
25. Li, Y.; Sun, L.; Webster, T.J. The investigation of zno/poly (vinylidene fluoride) nanocomposites with improved mechanical, piezoelectric, and antimicrobial properties for orthopedic applications. *J. Biomed. Nanotechnol.* **2018**, *14*, 536–545. [CrossRef] [PubMed]
26. Kitsara, M.; Blanquer, A.; Murillo, G.; Humblot, V.; Vieira, S.D.B.; Nogués, C.; Ibáñez, E.; Esteve, J.; Barrios, L. Permanently hydrophilic, piezoelectric pvdf nanofibrous scaffolds promoting unaided electromechanical stimulation on osteoblasts. *Nanoscale* **2019**, *11*, 8906–8917. [CrossRef] [PubMed]
27. Damaraju, S.M.; Shen, Y.; Elele, E.; Khusid, B.; Eshghinejad, A.; Li, J.; Jaffe, M.; Arinzeh, T.L. Three-dimensional piezoelectric fibrous scaffolds selectively promote mesenchymal stem cell differentiation. *Biomaterials* **2017**, *149*, 51–62. [CrossRef]
28. Ciofani, G.; Ricotti, L.; Mattoli, V. Preparation, characterization and in vitro testing of poly(lactic-co-glycolic) acid/barium titanate nanoparticle composites for enhanced cellular proliferation. *Biomed. Microdevices* **2010**, *13*, 255–266. [CrossRef]
29. Lee, H.W.; Seo, S.H.; Kum, C.H.; Park, B.J.; Joung, Y.K.; Son, T.I.; Han, D.K. Fabrication and characteristics of anti-inflammatory magnesium hydroxide incorporated PLGA scaffolds formed with various porogen materials. *Macromol. Res.* **2014**, *22*, 210–218. [CrossRef]
30. Liao, C.; Li, Y.; Tjong, S.C. Graphene nanomaterials: Synthesis, biocompatibility, and cytotoxicity. *Int. J. Mol. Sci.* **2018**, *19*, 3564. [CrossRef]
31. Mediouni, M.; Schlatterer, D.R.; Madry, H.; Cucchiarini, M.; Rai, B. A review of translational medicine. The future paradigm: How can we connect the orthopedic dots better? *Curr. Med. Res. Opin.* **2018**, *34*, 1217–1229. [CrossRef]
32. Mediouni, M.; Madiouni, R.; Gardner, M.; Vaughan, N. Translational medicine: Challenges and new orthopaedic vision (Mediouni-Model). *Curr. Orthop. Pract.* **2020**, *31*, 196–200. [CrossRef]
33. D'Orto, B.; Polizzi, E.; Nagni, M.; Tetè, G.; Capparè, P. Full Arch Implant-Prosthetic Rehabilitation in Patients with Type I Diabetes Mellitus: Retrospective Clinical Study with 10 Year Follow-Up. *Int. J. Environ. Res. Public Health* **2022**, *19*, 11735. [CrossRef]
34. Abdelrazek, E.M.; Hezma, A.M.; El-Khodary, A.; Elzayat, A.M. Spectroscopic studies and thermal properties of PCL/PMMA biopolymer blend. *Egypt. J. Basic Appl. Sci.* **2016**, *3*, 10–15. [CrossRef]
35. Chen, E.-C.; Wu, T.-M. Isothermal crystallization kinetics and thermal behavior of poly (ε-caprolactone)/multi-walled carbon nanotube composites. *Polym. Degrad. Stab.* **2007**, *92*, 1009–1015. [CrossRef]
36. Zhang, Y.; Chen, L.; Zeng, J.; Zhou, K.; Zhang, D. Aligned porous barium titanate/hydroxyapatite composites with high piezoelectric coefficients for bone tissue engineering. *Mater. Sci. Eng. C* **2014**, *39*, 143–149. [CrossRef] [PubMed]

37. Biglar, M.; Gromada, M.; Stachowicz, F.; Trzepieciński, T. Synthesis of Barium Titanate Piezoelectric Ceramics for Multilayer Actuators (MLAs). *Acta Mech. Autom.* **2017**, *11*, 275–279. [CrossRef]
38. Demirors, A.F.; Imhof, A. Batio3, srtio3, catio3, and ba x sr1− x tio3 particles: A general approach for monodisperse colloidal perovskites. *Chem. Mater.* **2009**, *21*, 3002–3007. [CrossRef]
39. Bagchi, A.; Meka, S.R.K.; Rao, B.N.; Chatterjee, K. Perovskite ceramic nanoparticles in polymer composites for augmenting bone tissue regeneration. *Nanotechnology* **2014**, *25*, 485101. [CrossRef]
40. Elhattab, K.; Sikder, P.; Walker, J.M.; Bottino, M.; Bhaduri, S. Fabrication and evaluation of 3-D printed PEEK scaffolds containing Macropores by design. *Mater. Lett.* **2020**, *263*, 127227. [CrossRef]
41. Buj-Corral, I.; Bagheri, A.; Petit-Rojo, O. 3D Printing of Porous Scaffolds with Controlled Porosity and Pore Size Values. *Materials* **2018**, *11*, 1532. [CrossRef]
42. Liu, J.; Gu, H.; Liu, Q.; Ren, L.; Li, G. An intelligent material for tissue reconstruction: The piezoelectric property of polycaprolactone/barium titanate composites. *Mater. Lett.* **2019**, *236*, 686–689. [CrossRef]
43. Dubey, A.K.; Basu, B.; Balani, K.; Guo, R.; Bhalla, A.S. Dielectric and pyroelectric properties of hap-batio3 composites. *Ferroelectrics* **2011**, *423*, 63–76. [CrossRef]
44. Tang, Y.; Wu, C.; Wu, Z.; Hu, L.; Zhang, W.; Zhao, K. Fabrication and in vitro biological properties of piezoelectric bioceramics for bone regeneration. *Sci. Rep.* **2017**, *7*, srep43360. [CrossRef] [PubMed]
45. Bowen, C.R.; Gittings, J.; Turner, I.G.; Baxter, F.; Chaudhuri, J.B. Dielectric and piezoelectric properties of hydroxyapatite-BaTiO3 composites. *Appl. Phys. Lett.* **2006**, *89*, 132906. [CrossRef]
46. Tavangar, M.; Heidari, F.; Hayati, R.; Tabatabaei, F.; Vashaee, D.; Tayebi, L. Manufacturing and characterization of mechanical, biological and dielectric properties of hydroxyapatite-barium titanate nanocomposite scaffolds. *Ceram. Int.* **2020**, *46*, 9086–9095. [CrossRef]
47. Baxter, F.R.; Turner, I.G.; Bowen, C.R.; Gittings, J.P.; Chaudhuri, J.B. An in vitro study of electrically active hydroxyapatite-barium titanate ceramics using Saos-2 cells. *J. Mater. Sci. Mater. Electron.* **2009**, *20*, 1697–1708. [CrossRef] [PubMed]
48. Park, J.; Lakes, R.S. *Biomaterials: An Introduction*; Springer Science & Business Media: Cham, Switzerland, 2007.
49. Fan, B.; Guo, Z.; Li, X.; Li, S.; Gao, P.; Xiao, X.; Wu, J.; Shen, C.; Jiao, Y.; Hou, W. Electroactive barium titanate coated titanium scaffold improves osteogenesis and osseointegration with low-intensity pulsed ultrasound for large segmental bone defects. *Bioact. Mater.* **2020**, *5*, 1087–1101. [CrossRef]
50. Chen, J.; Li, S.; Jiao, Y.; Li, J.; Li, Y.; Hao, Y.-L.; Zuo, Y. In Vitro Study on the Piezodynamic Therapy with a $BaTiO_3$-Coating Titanium Scaffold under Low-Intensity Pulsed Ultrasound Stimulation. *ACS Appl. Mater. Interfaces* **2021**, *13*, 49542–49555. [CrossRef]
51. Koju, N.; Sikder, P.; Gaihre, B.; Bhaduri, S.B. Smart Injectable Self-Setting Monetite Based Bioceramics for Orthopedic Applications. *Materials* **2018**, *11*, 1258. [CrossRef]
52. Sikder, P.; Koju, N.; Lin, B.; Bhaduri, S.B. Conventionally Sintered Hydroxyapatite–Barium Titanate Piezo-Biocomposites. *Trans. Indian Inst. Met.* **2019**, *72*, 2011–2018. [CrossRef]
53. Shuai, C.; Liu, G.; Yang, Y.; Yang, W.; He, C.; Wang, G.; Liu, Z.; Qi, F.; Peng, S. Functionalized BaTiO3 enhances piezoelectric effect towards cell response of bone scaffold. *Colloids Surf. B Biointerfaces* **2020**, *185*, 110587. [CrossRef] [PubMed]
54. Liu, W.; Yang, D.; Wei, X.; Guo, S.; Wang, N.; Tang, Z.; Lu, Y.; Shen, S.; Shi, L.; Li, X.; et al. Fabrication of piezoelectric porous $BaTiO_3$ scaffold to repair large segmental bone defect in sheep. *J. Biomater. Appl.* **2020**, *35*, 544–552. [CrossRef] [PubMed]
55. Lee, J.W.; Ahn, G.; Kim, J.Y.; Cho, D.-W. Evaluating cell proliferation based on internal pore size and 3D scaffold architecture fabricated using solid freeform fabrication technology. *J. Mater. Sci. Mater. Electron.* **2010**, *21*, 3195–3205. [CrossRef]
56. Abbasi, N.; Hamlet, S.; Love, R.M.; Nguyen, N.-T. Porous scaffolds for bone regeneration. *J. Sci. Adv. Mater. Devices* **2020**, *5*, 1–9. [CrossRef]
57. Karageorgiou, V.; Kaplan, D. Porosity of 3D biomaterial scaffolds and osteogenesis. *Biomaterials* **2005**, *26*, 5474–5491. [CrossRef] [PubMed]
58. Sikder, P.; Grice, C.R.; Lin, B.; Goel, V.K.; Bhaduri, S.B. Single-Phase, Antibacterial Trimagnesium Phosphate Hydrate Coatings on Polyetheretherketone (PEEK) Implants by Rapid Microwave Irradiation Technique. *ACS Biomater. Sci. Eng.* **2018**, *4*, 2767–2783. [CrossRef] [PubMed]
59. Sikder, P.; Grice, C.R.; Bhaduri, S.B. Processing-structure-property correlations of crystalline antibacterial magnesium phosphate (newberyite) coatings and their in vitro effect. *Surf. Coat. Technol.* **2019**, *374*, 276–290. [CrossRef]
60. Sikder, P.; Ferreira, J.A.; Fakhrabadi, E.A.; Kantorski, K.Z.; Liberatore, M.W.; Bottino, M.C.; Bhaduri, S.B. Bioactive amorphous magnesium phosphate-polyetheretherketone composite filaments for 3D printing. *Dent. Mater.* **2020**, *36*, 865–883. [CrossRef]
61. Sikder, P.; Bhaduri, S.B.; Ong, J.L.; Guda, T. Silver (Ag) doped magnesium phosphate microplatelets as next-generation antibacterial orthopedic biomaterials. *J. Biomed. Mater. Res. Part B Appl. Biomater.* **2020**, *108*, 976–989. [CrossRef]
62. Zayzafoon, M. Calcium/calmodulin signaling controls osteoblast growth and differentiation. *J. Cell. Biochem.* **2006**, *97*, 56–70. [CrossRef]
63. Tian, J.; Shi, R.; Liu, Z.; Ouyang, H.; Yu, M.; Zhao, C.; Zou, Y.; Jiang, D.; Zhang, J.; Li, Z. Self-powered implantable electrical stimulator for osteoblasts' proliferation and differentiation. *Nano Energy* **2019**, *59*, 705–714. [CrossRef]

64. Ma, B.; Liu, F.; Li, Z.; Duan, J.; Kong, Y.; Hao, M.; Ge, S.; Jiang, H.; Liu, H. Piezoelectric nylon-11 nanoparticles with ultrasound assistance for high-efficiency promotion of stem cell osteogenic differentiation. *J. Mater. Chem. B* **2019**, *7*, 1847–1854. [CrossRef] [PubMed]
65. Cai, K.; Jiao, Y.; Quan, Q.; Hao, Y.; Liu, J.; Wu, L. Improved activity of mc3t3-e1 cells by the exciting piezoelectric batio3/tc4 using low-intensity pulsed ultrasound. *Bioact. Mater.* **2021**, *6*, 4073–4082. [CrossRef] [PubMed]
66. Zhou, Z.; Li, W.; Zhengnan, Z.; Qian, L.; Tan, G.; Ning, C. Polarization of an electroactive functional film on titanium for inducing osteogenic differentiation. *Sci. Rep.* **2016**, *6*, 35512. [CrossRef]
67. Reid, B.; Song, B.; McCaig, C.D.; Zhao, M. Wound healing in rat cornea: The role of electric currents. *FASEB J.* **2005**, *19*, 379–386. [CrossRef]
68. Morimoto, J.; Yasuhara, T.; Kameda, M.; Umakoshi, M.; Kin, I.; Kuwahara, K.; Kin, K.; Okazaki, M.; Takeuchi, H.; Sasaki, T.; et al. Electrical Stimulation Enhances Migratory Ability of Transplanted Bone Marrow Stromal Cells in a Rodent Ischemic Stroke Model. *Cell. Physiol. Biochem.* **2018**, *46*, 57–68. [CrossRef] [PubMed]
69. Manazza, F.; La Rocca, S.; Nagni, M.; Chirico, L.; Cattoni, F. A simplified digital workflow for the prosthetic finishing of implant rehabilitations: A case report. *J. Biol. Regul. Homeost. Agents* **2021**, *35*, 87–97.

bioengineering

MDPI

Brief Report

Verification of the Efficacy of Mexiletine Treatment for the A1656D Mutation on Downgrading Reentrant Tachycardia Using a 3D Cardiac Electrophysiological Model

Ali Ikhsanul Qauli [1,2], Yedam Yoo [1], Aroli Marcellinus [1] and Ki Moo Lim [1,3,*]

1 Department of IT Convergence Engineering, Kumoh National Institute of Technology, Gumi 39177, Korea
2 Robotics and Artificial Intelligence Engineering, Faculty of Advanced Technology and Multidiscipline, Universitas Airlangga, Surabaya 60115, Jawa Timur, Indonesia
3 Department of Medical IT Convergence Engineering, Kumoh National Institute of Technology, Gumi 39253, Korea
* Correspondence: kmlim@kumoh.ac.kr

Abstract: The SCN5A mutations have been long associated with long QT variant 3 (LQT3). Recent experimental and computation studies have reported that mexiletine effectively treats LQT3 patients associated with the A1656D mutation. However, they have primarily focused on cellular level evaluations and have only looked at the effects of mexiletine on action potential duration (APD) or QT interval reduction. We further investigated mexiletine's effects on cardiac cells through simulations of single-cell (behavior of alternant occurrence) and 3D (with and without mexiletine). We discovered that mexiletine could shorten the cell's APD and change the alternant's occurrence to a shorter basic cycle length (BCL) between 350 and 420 ms. The alternant also appeared at a normal heart rate under the A1656D mutation. Furthermore, the 3D ventricle simulations revealed that mexiletine could reduce the likelihood of a greater spiral wave breakup in the A1656D mutant condition by minimizing the appearance of rotors. In conclusion, we found that mexiletine could provide extra safety features during therapy for LQT3 patients because it can change the alternant occurrence from a normal to a faster heart rate, and it reduces the chance of a spiral wave breakup. Therefore, these findings emphasize the promising efficacy of mexiletine in treating LQT3 patients under the A1656D mutation.

Keywords: LQT3; mutation; reentry; APD; alternant

check for updates

Citation: Qauli, A.I.; Yoo, Y.; Marcellinus, A.; Lim, K.M. Verification of the Efficacy of Mexiletine Treatment for the A1656D Mutation on Downgrading Reentrant Tachycardia Using a 3D Cardiac Electrophysiological Model. *Bioengineering* **2022**, *9*, 531. https://doi.org/10.3390/bioengineering9100531

Academic Editor: Ali Zarrabi

Received: 15 September 2022
Accepted: 5 October 2022
Published: 7 October 2022

1. Introduction

Until now, cardiac arrhythmia has been recognized as a global health problem [1]. Sudden cardiac death (SCD) is responsible for 250,000–300,000 fatalities in the United States annually, according to [2]. They also expected that in industrialized nations, an annual incidence of 50–100/100,000 people would acquire SCD. In 2018, one study found that the incidence of fatal cardiovascular disease (CVD) in Asians was 3.68 (95 percent confidence interval: 2.84–4.53) events per 1000 person-years [3]. Long QT syndrome (LQTS), caused by a mutation in an ion channel, is one of the leading causes of inherited CVD and SCD [4]. Previous research [5] also found a 1:2000 chance of developing a genetic disease among healthy live babies.

SCN5A mutations have been linked to LQTS 3, which codes for the α-subunit of the Na+ channel (Nav1.5), according to [6]. A review report [7] found that the LQT3–caused mutations in SCN5A extend the QT interval by minor net increases in the INa current. These are frequently due to an aberrant persistent or sustained late sodium current or to defective a Nav1.5 inactivation observed across the full voltage range and time-course-of-action potential plateau that disrupts the delicate balance between inward and outward currents [8,9]. Furthermore, the QT interval of LQT3 patients is significantly prolonged, especially during rest [10,11], and shortened during a fast heart rate. It has long been hypothesized [12]—and recently proved in a murine model [13]—that diurnal variation in cardiac repolarization patterns may be responsible for the temporally dependent vulnerability to ventricular arrhythmias seen in LQT3 and other heritable cardiac arrhythmia syndromes. However, the precise electrophysiological mechanism underlying this phenomenon is still not fully understood.

In a newborn baby with QT prolongation and non-sustained ventricular tachycardia (NSVT), a unique SCN5A variation known as A1656D was recently discovered [14]. Kim et al. reported that under the influence of mexiletine, A1656D exhibited a distinct response that resulted in the suppression of NSVT and the appearance of premature atrial contraction. Furthermore, another study [15] found that sodium channel mutations N1325S and R1623Q had an unusual reaction to mexiletine, where they could completely restore the electrocardiogram (ECG) and eliminate arrhythmias.

As the sodium blocker, mexiletine has been known to selectively inhibit the slowly–inactivating component of the inward tetrodotoxin–sensitive sodium current (I_{NaL}) without affecting the speed of action potential upstroke [16,17]. The mexiletine can also abbreviate the QT interval shown on ECG without widening the QRS duration [16,18]. In LQT3 patients, the I_{NaL} is greatly enhanced in a slower heart rate because of a slower inactivation of the I_{NaL} in bradycardia conditions [19]. Some researchers found that a faster heart rate can reduce the I_{NaL}, shorten APD and QT interval, thus diminishing the I_{NaL} becomes the basis for LQT3 treatment [20,21].

However, some questions remain about the sodium blocker's effectiveness in LQT3 patients. The significance of sodium channel blockers for LQT3 individuals has been mixed in several trials [22,23]. In this context, a better knowledge of the mechanism of sodium blockers for treating LQT3 patients has become a top goal. The efficacy of mexiletine for treating LQT3 patients has recently been demonstrated in a previous study [14,23,24]. Despite the extensive experimental evaluation of mexiletine for treating LQT3 patients, their findings require more investigation into the effects of mexiletine on a 3D ventricular model of the heart. The single-cell experimental studies may not anticipate additional impacts that a computer simulation at the single-cell level and a three-dimensional model can disclose.

Some researchers have adapted the combined experimental work and in silico simulation to assess the ion channel mutation. A study by Yu et al. [25] assessed the arrhythmia mechanism of the infarcted ventricles following remuscularization with pluripotent stem cell-derived cardiomyocytes (PSC-CMs). They concluded that the main ventricular tachycardia (VT) mechanism involved in patch remuscularization would be reentrant VT. Furthermore, Paci et al. [26] emphasized the ability of populations of in silico simulation to describe the experimental findings of the prolonged action potential as a product of LQT3 hiPSC-CMs mutation and predict differences in drug response of the ion channel mutation. Deo et al. [27] combined an experimental study of the mutation (E299V) in the KCNJ2 gene that encodes the strong inward rectifier K+ channel protein (Kir2.1) and 3D computational modeling of the heart incorporating the His-Purkinje network to predict the 20% reduction of sodium current that can cause reduction of ventricular excitability and increase the vulnerability of ventricular tachyarrhythmia. Kim et al. [14] also produced a computational model of the A1656D mutation, which demonstrated how various medications affect the AP shape of the ventricular cell. One of their significant discoveries is that mexiletine can reduce the APD of the A1656D mutation by restoring late sodium current inactivation kinetics.

The effects of mexiletine on A1656D mutations on single-cell ventricular cells (myocardial, endocardial, and epicardial) were further computationally elaborated in this paper. The study was also extended to a 3D ventricular model. We focused on reentrant arrhythmias that might arise due to alternants in single-cell simulation [28] and spiral waves in a 3D ventricular model.

2. Materials and Methods

In this work, the cell model incorporated in simulations was initially based on the ventricular cell model proposed by [29] with some modifications to resemble the A1656D mutation and the A1656D mutation under mexiletine as proposed by [14]. As shown in Figure 1, the main simulations consisted of single-cell simulation and 3D ventricular simulation. The single-cell simulation was used to study the tachycardia condition by the APD restitution graph showing the BCL region with alternant. Furthermore, the cell models were embedded into the 3D model of the ventricle so that it could allow us to observe the reentry and compare the shape of action potential among different conditions.

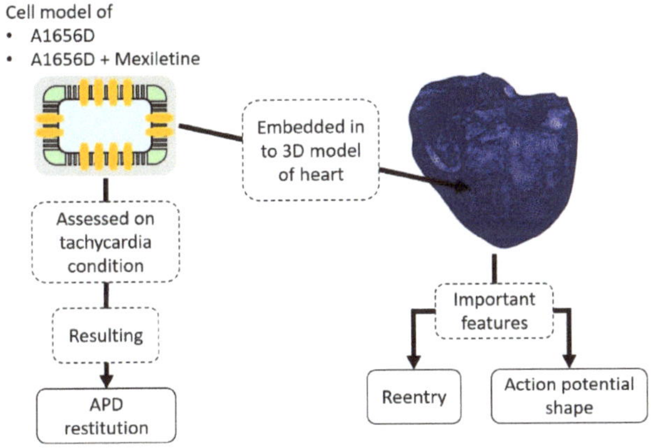

Figure 1. Diagram of the simulation protocol of single cell and 3D model of the heart.

2.1. Model of Ventricular Cell under A1656D Mutation with Incorporated Drug Effects

A recent study by Kim et al. [14] examined and simulated a myocardial cell model based on the work of [29]. The differential equation that characterized the cell is as follows:

$$C_m \frac{dV_m}{dt} = -(I_{ion} + I_{stim}), \tag{1}$$

where C_m is the total membrane capacitance, I_{stim} is the stimulus current, and I_{ion} is the sum of ionic transmembrane currents. In addition, ten Tusscher et al. [30] proposed that the transmembrane currents consist of sodium current (I_{Na}), inward rectifier potassium current (I_{K1}), transient outward potassium current (I_{to}), rapid delayed rectifier potassium current (I_{Kr}), slow delayed rectifier potassium current (I_{Ks}), L-type calcium current (I_{CaL}), sodium/calcium exchange current (I_{NaCa}), pump current (I_{NaK}), plateau Ca^{2+} current (I_{pCa}), plateau K^+ current (I_{pK}), Ca^{2+} background current (I_{bCa}), and K^+ background current (I_{bNa}) [29].

Furthermore, Kim et al. [14] proposed some changes to the (I_{Na}) modeling that alters the channel's activation (m) and inactivation (h) information. Fast (h_f) and slow (h_s) kinetics were used to explain the h-gate [14]. The sodium current can be stated in the following way:

$$m_\infty = \sqrt[3]{\frac{1}{1 + e^{(V - V_{h(m_\infty)})/k_m}}}, \tag{2}$$

$$m_\tau = \frac{a_m}{e^{b_m(V - V_{hm_\tau})/RT} + e^{-c_m(V - V_{hm_\tau})/RT}} + d_m, \tag{3}$$

$$h_{f_\infty} = h_{s_\infty} = \frac{1}{1 + e^{(V - V_{h(h_\infty)})/k_h}}, \tag{4}$$

$$h_{f_\tau} = \frac{a_{h_f}}{e^{b_{h_f}(V - V_{h_{f_\tau}})/RT} + e^{-c_{h_f}(V - V_{h_{f_\tau}})/RT}} + d_{h_f}, \tag{5}$$

$$h_{s_\tau} = \frac{a_{h_s}}{e^{b_{h_s}(V - V_{h_{s_\tau}})/RT} + e^{-c_{h_s}(V - V_{h_{s_\tau}})/RT}} + d_{h_s}, \tag{6}$$

where m_∞ is the steady-state activation, h_∞ is the steady-state inactivation, V is the membrane potential (millivolt), R is the gas constant $\left(8.31\,\mathrm{j\,mol^{-1}K^{-1}}\right)$, T is the absolute temperature in Kelvin. The forward (α) and backward (β) rate constants of each gate can be calculated as follows:

$$\alpha_m = \frac{m}{\tau_m}, \tag{7}$$

$$\beta_m = \frac{1 - m}{\tau_m}, \tag{8}$$

$$\alpha_{h_f} = \frac{m_f}{\tau_{h_f}}, \tag{9}$$

$$\beta_{h_f} = \frac{1 - h_f}{\tau_{h_f}}, \tag{10}$$

$$\alpha_{h_s} = \frac{m_s}{\tau_{h_s}}, \tag{11}$$

$$\beta_{h_s} = \frac{1 - h_s}{\tau_{h_s}}, \tag{12}$$

the relative contribution of fast and slow groups of h-gate to total open probability was r_f and r_s $(= 1 - r_f)$. Finally, the sodium current was calculated using the following formula:

$$I_{Na} = G_{Na_{max}} \times m^3 \times (r_f \times h_f + r_s \times h_s) \times (V - E_{rev}), \tag{13}$$

where G_{Na} is the maximum sodium channel conductance assumed to be 1125.0 for WT, 213.66 for A1656D, and 231.18 for A1656D under mexiletine. Furthermore, Tables 1–4 in [14] provide detailed values for the parameter set of m, h_f, and h_s gates in WT, the A1656D mutation, and the A1656D mutation under mexiletine. Other associated constants for the I_{Na} modeling can be found in [14] and references therein.

For 3D ventricular simulation, by ignoring microscopic cardiac cell structure, the cardiac cells can be modeled as follows [30]:

$$C_m \frac{dV}{dt} = -(I_{ion} + I_{stim}) + \frac{1}{\rho_x S_x} \frac{\partial^2 V}{\partial x^2} + \frac{1}{\rho_y S_y} \frac{\partial^2 V}{\partial y^2} + \frac{1}{\rho_z S_z} \frac{\partial^2 V}{\partial z^2}, \tag{14}$$

where ρ_x, ρ_y, and ρ_z are the cell's resistivity in x, y, and z directions; and S_x, S_y, and S_z are the surface-to-volume ratio in x, y, and z directions, respectively. The 3D mesh incorporated in the electrical simulation consisted of 241,725 nodes and 1,298,751 tetrahedral elements.

2.2. Simulation Protocol

There were three primary simulations in the single-cell electrical simulation: wildtype (WT), the A1656D mutant, and A1656D under mexiletine. The tachycardia state was analyzed using an APD restitution (APDR) graph by varying the BCL from 1400 ms to 300 ms with a step of 10 ms and 20 pacings for each BCL. We applied current stimulation for every pace with a stimulus duration of 1.0 ms and stimulus amplitude of 52 pA/pF.

The A1656D mutant and A1656D under mexiletine were the two primary sections of the 3D ventricular simulations. The BCL of 1000 ms was used in all scenarios. Furthermore, we used the S1 protocol to execute a homogenous simulation in which we stimulated a node from the apex of the heart five times. With the standard S2 procedure, we additionally included reentry/arrhythmia simulation. The electrical stimulation was administered to the heart's apex five times, and half of the stimulated nodes were reset right before the fifth planar wavefront. The stimulus might result in the reentry spreading to the resting nodes. Each 3D simulation time was set to 10 s for both the A1656D mutation and the A1656D mutation with mexiletine.

3. Results

Figure 2 depicts the APDR graph produced by a single-cell simulation with varied BCL values. The branches resulted in Figure 2 represent the alternants. The alternant occurs approximately at BCL between 240–260 ms for the WT in panel A. Among them, the myocardial cell-type results in the alternant with a relatively longer BCL of around 260 ms compared to both epicardial and endocardial cells that result in the alternant at BCL of approximately 240 ms. Furthermore, we can clearly see in panel B for A1656D mutation simulation that the A1656D mutation can cause APD prolongation and a shift of the occurrence of alternant from short BCL to longer BCL compared to WT of around 760 ms for myocardial cell-type, 700 ms for epicardial cell, and 680 ms for endocardial cell. On panel C, we can observe the effect of mexiletine on the A1656D mutation. Under mexiletine, myocardial and epicardial cell types converge to the same BCL of roughly 420 ms for the alternant to develop. However, with a shorter BCL of approximately 350 ms, the endocardial cell-type results in the alternant. We could also observe the progression of the APD values when the alternant occurs among all cell types from panels A to C. The difference in the APD values in WT is more negligible than in A1656D and A1656D under mexiletine. The highest difference in the APD values is observed between the myocardial–epicardial and endocardial cells at around 70 ms in panel C.

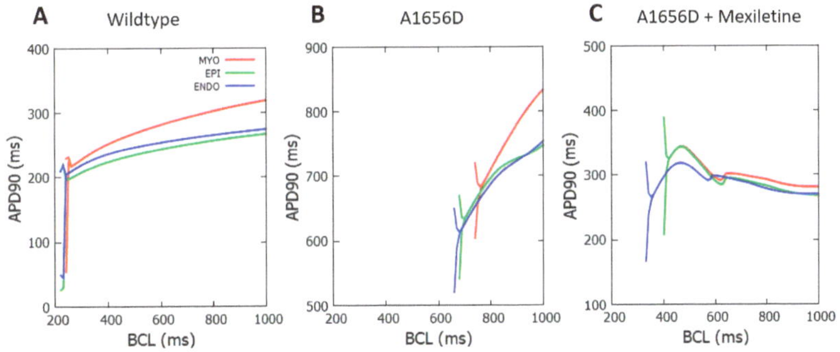

Figure 2. The APD as a function of BCL for WT, the A1656D mutant and A1656D under mexiletine.

Figure 3 depicts the results of a 3D ventricular simulation. In panel A, we can see a comparison of electrical activity from the A1656D mutation and A1656D when treated with mexiletine. At 989 ms, we labeled the rotors formed upon reentry with a black star. We can see that the A1656D mutation has around five rotors in total. They are visible from the "front", "back", "left", and "top" sides of the ventricular mesh. The A1656D mutation, on the other hand, has fewer rotors, with only four, one at the "front", one at the "back", and two at the "top" of the mesh model. The conduction velocity (CV) detected from the simulation was 19 cm/s for A1656D mutation, and 17.3 cm/s for A1656D mutation under mexiletine. Panel B compares the action potential shape of the epicardial cells with the effect of A1656D mutation and A1656D under mexiletine. The A1656D mutation resulted in a more prolonged APD as compared to the A1656D mutation when mexiletine was used.

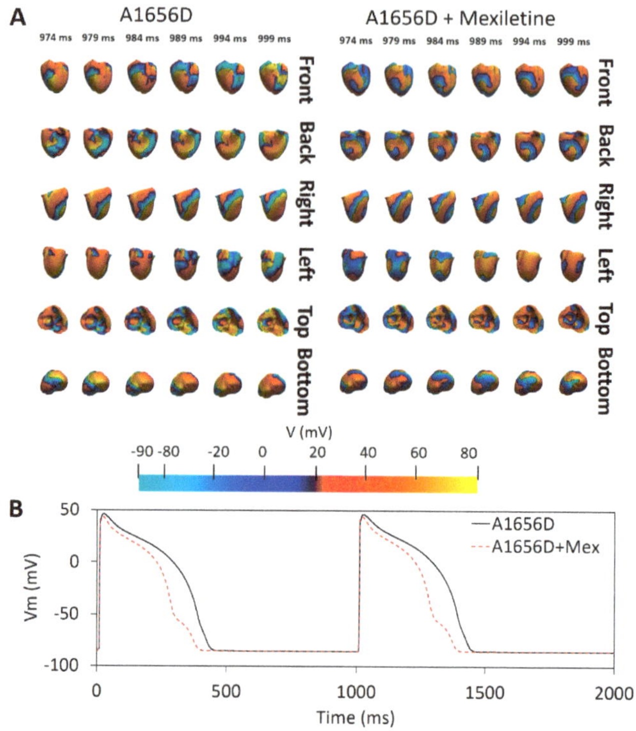

Figure 3. The outcomes of a 3D ventricular simulation. Panel (**A**) depicts the reentry dynamic from various angles, and panel (**B**) defines the action potential shape for A1656D and A1656D under mexiletine. The black star in Panel (**A**) shows the rotor. The cell type depicted in this illustration is the epicardia cell.

4. Discussion

Figure 2 shows that three conditions (WT, A1656D mutation, and A1656D under mexiletine) can produce different APD profiles in the cell. In panel A, the three kinds of cells make APDs greater than 200 ms for instances before the alternant. In panel B, the APD of the cells varies significantly, with the shortest occurring more than 600 ms before the endocardial cell alternant. In panel C, mexiletine can shorten the APD from the A1656D mutation by more than 250 ms before the endocardial cell alternant. The APD shortening by mexiletine demonstrated in our data is consistent with earlier work indicating that the late sodium channel alteration generated by the A1656D mutation is controlled by mexiletine (known to be a late sodium blocker [16,17]). The blockage of late sodium channels can induce repolarization to occur faster, lowering APD. In the model used in this work, the late sodium current is represented by slow kinetics in the formulation of the I_{Na} as proposed by [14].

A high APD difference between two successive action potentials has a high chance of causing discordant alternants [31]. The high APD difference (indicated by the difference between the branch line) shown in Figure 2, panels B and C, may inform us that A1656D and A1656D under mexiletine could cause a discordant alternant more strongly than WT. However, as shown in Figure 2, the A1656D mutation under mexiletine exhibits a change in the alternant incidence towards shorter BCL or a higher heart rate compared to the A1656D mutation. Figure 2, panel B, shows that the alternant occurs between 680 and 760 ms of BCL in the A1656D mutation. The typical adult human heart rate is between 600 ms and 1000 ms of BCL [32], implying that the A1656D mutation might raise the risk of alternant within the normal heart rhythm. Meanwhile, under mexiletine conditions, the alternant occurs at around 350 ms to 420 ms of BCL for the A1656D mutation. Instead, the alternant might appear at a quicker heart rate of 350 ms BCL for endocardial cells and 420 ms BCL for epicardial and myocardial cells. Therefore, mexiletine–treated A1656D show a safer result by lowering the possibility of the alternant occurring at a normal heart rate compared to the pure A1656D mutation. Our findings are consistent with previous research [27], which also reported the occurrence of QT prolongation during rest condition of LQT3 patients. In addition, a recent study indicated that a higher heart rate might reduce the I_{NaL} and shorten APD [33]. As a result, the reduction in the I_{NaL} paired with the decline in APD substantially demonstrates the effectiveness of mexiletine in lowering the risk of the alternant in LQT3 patients during therapy. In fact, the I_{NaL} lowering has been utilized as the foundation of LQT3 therapy using pacemakers [20,21].

The results of the 3D ventricular simulation, as shown in Figure 3, reveal that mexiletine can reduce the number of rotors (from four to two) in the A1656D mutant scenario. More rotors in the A1656D mutant condition may result in greater chaotic electrical activity, which may cause more spiral wave breakup and ventricular fibrillation. According to [34], cardiac electrical restitution properties influence the breakup of reentrant wavefronts during cardiac fibrillation, known as the concordant alternant. The wavelength of the successive wave oscillates between short and long during the concordant alternant. Furthermore, [35] found that rotors may trigger the ventricles at such high frequencies that the wavefronts propagating from it break apart across a wide range of distances, resulting in ventricular fibrillation. As a result, the reduced number of rotors caused by mexiletine therapy for the A1656D mutation may reflect its efficacy in decreasing the risk of spiral wave breakup, which can lead to ventricular fibrillation.

We have limited our 3D ventricular reentry simulation analysis to the epicardial cell type only. Results may be different when simulated using other cell types. However, recent studies [36–38] showed that the epicardial could exhibit more alternants than other regions (myocardial and endocardial), which may be related to the buildup of fat on the epicardial surface. Therefore, we may assume that the results from the epicardial cell are sufficient to describe the occurrence of a spiral breakup.

Furthermore, incorporating the iPSC-CMs into the existing simulation protocol may be beneficial for further study as iPSC-CMs could also mimic the specific genetic description of the LQT3 patients. However, the limitations of iPSC-CMs need to be considered. Cardiomyocytes generated from human iPSCs lack T-tubules and exhibit poor colocalization of calcium channels and ryanodine receptors [39]. As a result, special care should be taken when modeling cardiomyopathies caused by gene mutations that impact calcium transients, such as DMD [40]. Additionally, the immature nature of iPSC-CMs may lead to different electrophysiology, cell morphology, and metabolism [41,42].

In summary, based on our findings, we estimated that mexiletine would be effective in treating LQT3 patients with the A1656D mutation for several reasons:

- Mexiletine could reduce APD.
- Mexiletine could shift the alternant occurrence in the cell from a normal to a quicker heart rate, offering extra safety standards during treatment.
- During reentry, mexiletine could reduce the possibility of a spiral wave breakup, which can contribute to ventricular fibrillation.

Author Contributions: Writing—original draft, A.I.Q.; resource, Y.Y., software, A.M.; supervision, K.M.L.; writing—review & editing, A.I.Q., Y.Y., A.M., K.M.L. All authors have read and agreed to the published version of the manuscript.

Funding: This research was partially supported by the Ministry of Food and Drug Safety (22213MFDS3922), the NRF (National Research Foundation of Korea) under the Basic Science Research Program (2022R1A2C2006326), and the MSIT (Ministry of Science and ICT), Korea, under the Grand Information Technology Research Center support program (IITP-2022-2020-0-01612) supervised by the IITP (Institute for Information & communications Technology Planning & Evaluation).

Institutional Review Board Statement: Not applicable.

Informed Consent Statement: Not applicable.

Data Availability Statement: The data presented in this study are available on request from the corresponding author.

Conflicts of Interest: The authors declare no conflict of interest. The funders had no role: in the design of the study, in the collection, analyses, or interpretation of data, in the writing of the manuscript, or in the decision to publish the results.

References

1. Virani, S.S.; Alonso, A.; Aparicio, H.J.; Benjamin, E.J.; Bittencourt, M.S.; Callaway, C.W.; Carson, A.P.; Chamberlain, A.M.; Cheng, S.; Delling, F.N.; et al. Heart Disease and Stroke Statistics—2021 Update: A Report From the American Heart Association. *Circulation* **2021**, *143*, E254–E743. [CrossRef] [PubMed]
2. Fishman, G.I.; Chugh, S.S.; Dimarco, J.P.; Albert, C.M.; Anderson, M.E.; Bonow, R.O.; Buxton, A.E.; Chen, P.S.; Estes, M.; Jouven, X.; et al. Sudden Cardiac Death Prediction and Prevention Report From a National Heart, Lung, and Blood Institute and Heart Rhythm Society Workshop. *Circulation* **2010**, *122*, 2335. [CrossRef] [PubMed]
3. Irawati, S.; Wasir, R.; Floriaan Schmidt, A.; Islam, A.; Feenstra, T.; Buskens, E.; Wilffert, B.; Hak, E. Long-Term Incidence and Risk Factors of Cardiovascular Events in Asian Populations: Systematic Review and Meta-Analysis of Population-Based Cohort Studies. *Curr. Med. Res. Opin.* **2018**, *35*, 291–299. [CrossRef]
4. Tosaki, A. ArrhythmoGenoPharmacoTherapy. *Front. Pharmacol.* **2020**, *11*, 616. [CrossRef] [PubMed]
5. Garcia-Elias, A.; Benito, B. Ion Channel Disorders and Sudden Cardiac Death. *Int. J. Mol. Sci.* **2018**, *19*, 692. [CrossRef] [PubMed]
6. Wang, Q.; Shen, J.; Splawski, I.; Atkinson, D.; Li, Z.; Robinson, J.L.; Moss, A.J.; Towbin, J.A.; Keating, M.T. SCN5A Mutations Associated with an Inherited Cardiac Arrhythmia, Long QT Syndrome. *Cell* **1995**, *80*, 805–811. [CrossRef]
7. Giudicessi, J.R.; Ackerman, M.J. Genotype- and Phenotype-Guided Management of Congenital Long QT Syndrome. *Curr. Probl. Cardiol.* **2013**, *38*, 417–455. [CrossRef]
8. Ruan, Y.; Liu, N.; Priori, S.G. Sodium Channel Mutations and Arrhythmias. *Nat. Rev. Cardiol.* **2009**, *6*, 337–348. [CrossRef]
9. Bennett, P.B.; Yazawa, K.; Makita, N.; George, A.L. Molecular Mechanism for an Inherited Cardiac Arrhythmia. *Nature* **1995**, *376*, 683–685. [CrossRef]
10. Schwartz, P.J.; Priori, S.G.; Locati, E.H.; Napolitano, C.; Cantù, F.; Towbin, J.A.; Keating, M.T.; Hammoude, H.; Brown, A.M.; Chen, L.-S.K.; et al. Long QT Syndrome Patients With Mutations of the SCN5A and HERG Genes Have Differential Responses to Na+ Channel Blockade and to Increases in Heart Rate. *Circulation* **1995**, *92*, 3381–3386. [CrossRef]
11. Schwartz, P.J.; Priori, S.G.; Spazzolini, C.; Moss, A.J.; Vincent, G.M.; Napolitano, C.; Denjoy, I.; Guicheney, P.; Breithardt, G.; Keating, M.T.; et al. Genotype-Phenotype Correlation in the Long-QT Syndrome. *Circulation* **2001**, *103*, 89–95. [CrossRef] [PubMed]
12. Stramba-Badiale, M.; Priori, S.G.; Napolitano, C.; Locati, E.H.; Viñolas, X.; Haverkamp, W.; Schulze-Bahr, E.; Goulene, K.; Schwartz, P.J. Gene-Specific Differences in the Circadian Variation of Ventricular Repolarization in the Long QT Syndrome: A Key to Sudden Death during Sleep? *Ital. Heart J.* **2000**, *1*, 323–328. [PubMed]
13. Jeyaraj, D.; Haldar, S.M.; Wan, X.; McCauley, M.D.; Ripperger, J.A.; Hu, K.; Lu, Y.; Eapen, B.L.; Sharma, N.; Ficker, E.; et al. Circadian Rhythms Govern Cardiac Repolarization and Arrhythmogenesis. *Nature* **2012**, *483*, 96–99. [CrossRef]
14. Kim, H.J.; Kim, B.G.; Park, J.E.; Ki, C.S.; Huh, J.; Youm, J.B.; Kang, J.S.; Cho, H. Characterization of a Novel LQT3 Variant with a Selective Efficacy of Mexiletine Treatment. *Sci. Rep.* **2019**, *9*, 12997. [CrossRef] [PubMed]
15. Li, G.; Woltz, R.L.; Wang, C.Y.; Ren, L.; He, P.X.; Yu, S.D.; Liu, X.Q.; Yarov-Yarovoy, V.; Hu, D.; Chiamvimonvat, N.; et al. Gating Properties of Mutant Sodium Channels and Responses to Sodium Current Inhibitors Predict Mexiletine-Sensitive Mutations of Long QT Syndrome 3. *Front. Pharmacol.* **2020**, *11*, 1182. [CrossRef] [PubMed]
16. Gao, Y.; Xue, X.; Hu, D.; Liu, W.; Yuan, Y.; Sun, H.; Li, L.; Timothy, K.W.; Zhang, L.; Li, C.; et al. Inhibition of Late Sodium Current by Mexiletine: A Novel Pharmotherapeutical Approach in Timothy Syndrome. *Circ. Arrhythm. Electrophysiol.* **2013**, *6*, 614–622. [CrossRef]
17. Qi, D.; Yang, Z.; Robinson, V.M.; Li, J.; Gao, C.; Guo, D.; Kowey, P.R.; Yan, G.X. Heterogeneous Distribution of INa-L Determines Interregional Differences in Rate Adaptation of Repolarization. *Heart Rhythm.* **2015**, *12*, 1295–1303. [CrossRef]
18. Badri, M.; Patel, A.; Patel, C.; Liu, G.; Goldstein, M.; Robinson, V.M.; Xue, X.; Yang, L.; Kowey, P.R.; Yan, G.X. Mexiletine Prevents Recurrent Torsades de Pointes in Acquired Long QT Syndrome Refractory to Conventional Measures. *JACC Clin. Electrophysiol.* **2015**, *1*, 315–322. [CrossRef]
19. Wu, L.; Ma, J.; Li, H.; Wang, C.; Grandi, E.; Zhang, P.; Luo, A.; Bers, D.M.; Shryock, J.C.; Belardinelli, L. Late Sodium Current Contributes to the Reverse Rate-Dependent Effect of IKr Inhibition on Ventricular Repolarization. *Circulation* **2011**, *123*, 1713. [CrossRef]
20. Benditt, D.; Woodrow Benson, D.; Abbott, J.A.; Herre, J.M.; Scheinman, M.M. Permanent Cardiac Pacing in Patients with the Long QT Syndrome. *J. Am. Coll. Cardiol.* **1987**, *10*, 600–607. [CrossRef]

21. Dorostkar, P.C.; Eldar, M.; Belhassen, B.; Scheinman, M.M. Long-Term Follow-Up of Patients With Long-QT Syndrome Treated With β-Blockers and Continuous Pacing. *Circulation* **1999**, *100*, 2431–2436. [CrossRef] [PubMed]
22. Ruan, Y.; Liu, N.; Bloise, R.; Napolitano, C.; Priori, S.G. Gating Properties of SCN5A Mutations and the Response to Mexiletine in Long-QT Syndrome Type 3 Patients. *Circulation* **2007**, *116*, 1137–1144. [CrossRef] [PubMed]
23. Mazzanti, A.; Maragna, R.; Faragli, A.; Monteforte, N.; Bloise, R.; Memmi, M.; Novelli, V.; Baiardi, P.; Bagnardi, V.; Etheridge, S.P.; et al. Gene-Specific Therapy With Mexiletine Reduces Arrhythmic Events in Patients With Long QT Syndrome Type 3. *J. Am. Coll. Cardiol.* **2016**, *67*, 1053–1058. [CrossRef] [PubMed]
24. Wilde, A.A.M.; Amin, A.S. Clinical Spectrum of SCN5A Mutations: Long QT Syndrome, Brugada Syndrome, and Cardiomyopathy. *JACC Clin. Electrophysiol.* **2018**, *4*, 569–579. [CrossRef] [PubMed]
25. Yu, J.K.; Liang, J.A.; Franceschi, W.H.; Huang, Q.; Pashakhanloo, F.; Sung, E.; Boyle, P.M.; Trayanova, N.A. Assessment of Arrhythmia Mechanism and Burden of the Infarcted Ventricles Following Remuscularization with Pluripotent Stem Cell-Derived Cardiomyocyte Patches Using Patient-Derived Models. *Cardiovasc. Res.* **2022**, *118*, 1247–1261. [CrossRef] [PubMed]
26. Paci, M.; Passini, E.; Severi, S.; Hyttinen, J.; Rodriguez, B. Phenotypic Variability in LQT3 Human Induced Pluripotent Stem Cell-Derived Cardiomyocytes and Their Response to Antiarrhythmic Pharmacologic Therapy: An in Silico Approach. *Heart Rhythm.* **2017**, *14*, 1704–1712. [CrossRef] [PubMed]
27. Deo, M.; Ruan, Y.; Pandit, S.V.; Shah, K.; Berenfeld, O.; Blaufox, A.; Cerrone, M.; Noujaim, S.F.; Denegri, M.; Jalife, J.; et al. KCNJ2 Mutation in Short QT Syndrome 3 Results in Atrial Fibrillation and Ventricular Proarrhythmia. *Proc. Natl. Acad. Sci. USA* **2013**, *110*, 4291–4296. [CrossRef]
28. Qu, Z.; Garfinkel, A.; Chen, P.S.; Weiss, J.N. Mechanisms of Discordant Alternans and Induction of Reentry in Simulated Cardiac Tissue. *Circulation* **2000**, *102*, 1664–1670. [CrossRef]
29. ten Tusscher, K.H.W.J.; Noble, D.; Noble, P.J.; Panfilov, A.V. A Model for Human Ventricular Tissue. *Am. J. Physiol. -Heart Circ. Physiol.* **2004**, *286*, H1573–H1589. [CrossRef]
30. Hodgkin, A.L.; Huxley, A.F. A Quantitative Description of Membrane Current and Its Application to Conduction and Excitation in Nerve. *J. Physiol.* **1952**, *117*, 500. [CrossRef]
31. Krogh-Madsen, T.; Christini, D.J. Action Potential Duration Dispersion and Alternans in Simulated Heterogeneous Cardiac Tissue with a Structural Barrier. *Biophys. J.* **2007**, *92*, 1138–1149. [CrossRef] [PubMed]
32. Bonnemeier, H.; Wiegand, U.K.H.; Brandes, A.; Kluge, N.; Katus, H.A.; Richardt, G.; Potratz, J. Circadian Profile of Cardiac Autonomic Nervous Modulation in Healthy Subjects: *J. Cardiovasc. Electrophysiol.* **2003**, *14*, 791–799. [CrossRef] [PubMed]
33. Li, G.; Zhang, L. The Role of Mexiletine in the Management of Long QT Syndrome. *J. Electrocardiol.* **2018**, *51*, 1061–1065. [CrossRef] [PubMed]
34. Weiss, J.N.; Chen, P.S.; Qu, Z.; Karagueuzian, H.S.; Garfinkel, A. Ventricular Fibrillation. *Circ. Res.* **2000**, *87*, 1103–1107. [CrossRef] [PubMed]
35. Jalife, J. Ventricular Fibrillation: Mechanisms of Initiation and Maintenance. *Annu. Rev. Physiol.* **2003**, *62*, 25–50. [CrossRef]
36. Janusek, D.; Svehlikova, J.; Zelinka, J.; Weigl, W.; Zaczek, R.; Opolski, G.; Tysler, M.; Maniewski, R. The Roles of Mid-Myocardial and Epicardial Cells in T-Wave Alternans Development: A Simulation Study. *Biomed. Eng. Online* **2018**, *17*, 57. [CrossRef]
37. Selvaraj, R.J.; Picton, P.; Nanthakumar, K.; Mak, S.; Chauhan, V.S. Endocardial and Epicardial Repolarization Alternans in Human Cardiomyopathy: Evidence for Spatiotemporal Heterogeneity and Correlation With Body Surface T-Wave Alternans. *J. Am. Coll. Cardiol.* **2007**, *49*, 338–346. [CrossRef]
38. Aras, K.; Gams, A.; Faye, N.R.; Brennan, J.; Goldrick, K.; Li, J.; Zhong, Y.; Chiang, C.-H.; Smith, E.H.; Poston, M.D.; et al. Electrophysiology and Arrhythmogenesis in the Human Right Ventricular Outflow Tract. *Circ. Arrhythm. Electrophysiol.* **2022**, *15*, e010630. [CrossRef]
39. Itzhaki, I.; Rapoport, S.; Huber, I.; Mizrahi, I.; Zwi-Dantsis, L.; Arbel, G.; Schiller, J.; Gepstein, L. Calcium Handling in Human Induced Pluripotent Stem Cell Derived Cardiomyocytes. *PLoS ONE* **2011**, *6*, e18037. [CrossRef]
40. Law, M.L.; Cohen, H.; Martin, A.A.; Angulski, A.B.B.; Metzger, J.M. Dysregulation of Calcium Handling in Duchenne Muscular Dystrophy-Associated Dilated Cardiomyopathy: Mechanisms and Experimental Therapeutic Strategies. *J. Clin. Med.* **2020**, *9*, 520. [CrossRef]
41. Zhang, J.; Chou, O.H.-I.; Tse, Y.-L.; Ng, K.-M.; Tse, H.-F. Application of Patient-Specific IPSCs for Modelling and Treatment of X-Linked Cardiomyopathies. *Int. J. Mol. Sci.* **2021**, *22*, 8132. [CrossRef] [PubMed]
42. Wang, Y.; Lei, W.; Yang, J.; Ni, X.; Ye, L.; Shen, Z.; Hu, S. The Updated View on Induced Pluripotent Stem Cells for Cardiovascular Precision Medicine. *Pflugers Arch.* **2021**, *473*, 1137–1149. [CrossRef] [PubMed]

bioengineering

MDPI

Article

Nanostructured ZnO-Based Electrochemical Sensor with Anionic Surfactant for the Electroanalysis of Trimethoprim

Vinoda B. Patil [1], Davalasab Ilager [2], Suresh M. Tuwar [1,*], Kunal Mondal [3,*] and Nagaraj P. Shetti [4,5,*]

1 Department of Chemistry, Karnatak Science College, Dharwad 580001, Karnataka, India
2 Department of Chemistry, K.L.E. Institute of Technology, Hubballi 580027, Karnataka, India
3 Idaho National Laboratory, Idaho Falls, ID 83415, USA
4 Department of Chemistry, School of Advanced Sciences, KLE Technological University, Vidyanagar, Hubballi 580031, Karnataka, India
5 University Center for Research & Development (UCRD), Chandigarh University, Gharuan, Mohali 140413, Panjab, India
* Correspondence: sm.tuwar@gmail.com (S.M.T.); Kunal.Mondal@inl.gov (K.M.); dr.npsheti@gmail.com (N.P.S.)

Abstract: In this research, detection of trimethoprim (TMP) was carried out using a nanostructured zinc oxide nanoparticle-modified carbon paste electrode (ZnO/CPE) with an anionic surfactant and sodium dodecyl sulphate (SDS) with the help of voltametric techniques. The electrochemical nature of TMP was studied in 0.2 M pH 3.0 phosphate-buffer solution (PBS). The developed electrode displayed the highest peak current compared to nascent CPE. Effects of variation in different parameters, such as pH, immersion time, scan rate, and concentration, were investigated. The electrode process of TMP was irreversible and diffusion controlled with two electrons transferred. The effective concentration range (8.0×10^{-7} M–1.0×10^{-5} M) of TMP was obtained by varying the concentration with a lower limit of detection obtained to be 2.58×10^{-8} M. In addition, this approach was effectively employed in the detection of TMP in pharmaceutical dosages and samples of urine with the excellent recovery data, suggesting the potency of the developed electrode in clinical and pharmaceutical sample analysis.

Keywords: trimethoprim; electrochemical sensor; zinc oxide nanoparticles; anionic surfactant; biomedical applications

Citation: Patil, V.B.; Ilager, D.; Tuwar, S.M.; Mondal, K.; Shetti, N.P. Nanostructured ZnO-Based Electrochemical Sensor with Anionic Surfactant for the Electroanalysis of Trimethoprim. *Bioengineering* **2022**, *9*, 521. https://doi.org/10.3390/bioengineering9100521

Academic Editors: Ali Zarrabi and Gou-Jen Wang

Received: 10 September 2022
Accepted: 28 September 2022
Published: 2 October 2022

1. Introduction

Trimethoprim (TMP), chemically referred to as 5-(3,4,5-trimethoxybenzyl) pyrimidine-2,4-diamine, is used as an antibacterial medication to prevent and treat infections of the urinary, intestinal, middle ear travelers' diarrhea, and respiratory tract [1]. In pharmaceutical manufacture, it is frequently used with sulfonamides (such as sulfamethoxazole) to induce synergistic effects [2]. TMP is also available in a single drug formulation. It is a synthetic antibiotic drug that prevents the formation of tetrahydro folic acid (a chemical required for bacteria and human cells to build proteins) by blocking the enzymes that convert dihydro folic acid into tetrahydrofolic acid. TMP hinders the bacterial enzyme higher than the human enzyme. Antibiotics, such as TMP, are given to animals and humans annually to cure and prevent illnesses and infections [3]. Many antibiotics are eliminated in conjugated forms that can be easily regenerated to their original condition and destroyed from dosed animals without metabolism. Antibiotics are referred to as "pseudo persistent" pollutants because they remain in the environment for a long time. As a result, antibiotics in the atmosphere have gotten a lot of attention. Additionally, this TMP leads to a very dangerously low level of thrombocytes by decreasing folic acid level and leads to the formation of bone marrow cells [4]. Hence, it is essential to propose a highly sensitive and selective technique to analyze TMP in biological and pharmaceutical samples.

In recent times, electrochemical sensors have piqued the interest of scientific researchers. Electrochemical sensors are used as sensitive detection techniques because of their fast response rate, better sensitivity, low cost, and easy fabrication [5,6]. In addition, electrochemical sensors are helpful in determining the biologically active molecule and gaining insight into its biochemical processes and a better knowledge of their interactions [7]. Carbon paste electrodes (CPEs) have gained a lot of consideration as a sensor material for a variety of reasons, including their chemical inertness, stability, robustness, minor ohmic resistance, lack of requirement for an internal solution, and are suitable in detection applications [8,9]. The CPEs are free from toxic, environmentally friendly electrodes. The passivation concerns are simply resolved throughout the scenario by a rapid and simple surface rejuvenation. However, the bare CPEs also exhibit some flaws in determination of electrochemical processes, including lesser sensitivity, slower electron transport, poor stability across a broad range of concentration solutions, and demand for a larger over-potential for the electro-catalytic technique. However, the electrodes' modification can help avoid such problems [10–12]. The electron transport rate can be accelerated using chemically modified electrodes by lowering the over-potential. The nanoparticle (NP)-oriented chemical-modified electrodes are currently the focus of interest because of their intensified signal response, improved sensitivity, and higher stability.

ZnO nanoparticles comprise a wurtzite structure along with a wide surface area, making them an excellent biosensor material. At room temperature, the ZnO possesses a large band gap with a high excitation binding energy. ZnO nanoparticles' chemical as well as physical properties differ from bulk ZnO nanoparticles. Electro and photoluminescent substances [13], gas sensors [14], solar cells, etc., are some examples of ZnO nanoparticle applications. The ZnO in powdered form shows excellent optical transmission and electrical conductivity. Furthermore, the electronegativities of the Zn and O atoms in ZnO nanoparticles range significantly, making the bonding extremely ionic. Most importantly, ZnO nanoparticles are less hazardous and cost efficient.

An anionic surfactant, sodium dodecyl sulphate (SDS), has a unique structure consisting of a lengthy hydrophobic tail on one side and a polar head on the other [15]. Adsorption of SDS on the sensing surface causes the accumulation of many negative ions on the electrode surface. Many works have reported the use of SDS in the field of electrochemistry [16,17]. As per the literature, some works are published regarding determination and detection of TMP, such as spectrophotometry [18–24], liquid chromatography [25–28], capillary electrophoresis [29], potentiometry [30], HPLC-UV [31], and HILIC-HPLC [32]. However, these methods are time consuming, need complexity in sample preparation, and are costly compared to the electrochemical techniques.

The literature survey reveals that no research was explored to investigate TMP employing ZnO/SDS/CPE electrodes. Furthermore, the modified electrode was easy to prepare, less costly, and gave stable results. Further, ZnO/SDS/CPE was highly sensitive and more selective. The electrochemical investigations were performed using voltametric strategies, such as CV and DPV. The observations were made by studying the different parameters, such as pH, scan rate, and concentration study, which suggested that the inclusion of SDS with ZnO nanoparticles increased the surface area and current response, thereby increasing the sensitivity with a low detection limit. Hence, this sensor was employed to determine TMP in tablets as well as urine samples and to study excipient interference; the findings revealed the relevance of the method in detecting TMP.

2. Experimental Section

2.1. Chemicals and Reagents

The TMP (98%) and SDS (99%) were acquired from Sigma-Aldrich (St. Louis, MO, USA) and utilized as received. Himedia chemicals Pvt. Ltd. (Mumbai, India) provided H_3PO_4 (98%), KH_2PO_4 (99%), and Na_2HPO_4 (98%), which were utilized for the preparation of electrolytic solution (0.2 M phosphate-buffer solutions) for various pHs (3.0–6.0). The required 0.5 mM stock solution of TMP was prepared in ethanol. The use of double-distilled water was maintained while performing all the investigations.

2.2. Instrumentation

The study of the electrochemical behavior of TMP was carried out by employing electrochemical workstation (CHI 1110C, USA) consisting of an electrochemical cell congaing three-electrode electrode system, namely platinum electrode (counter), Ag/AgCl (reference), and ZnO/SDS/CPE (working). To perform pH measurements Elico pH meter (Elico Ltd., Mumbai, India) was used. Nano surf (Switzerland) was used to take AFM measurements. JEOL (JSM-IT 500LA) (Japan) instrument was used for SEM analysis and for XRD diffraction studies, XRD instrument MODEL RIGAKU (Japan) was employed.

2.3. Preparation of Pharmaceutical Sample

Trimethoprim tablets (each having 200 mg TMP) were acquired from a nearby pharmacy. Using agitate and a mortar, tablets were finely grounded. Small portion of tablet powder equal to the TMP was dissolved in ethanol and diluted up to the 100 mL volumetric flask using double-distilled water, sonicated for 15 min. Under ideal experimental conditions, the tablet's TMP (1.0×10^{-5} M) analysis was quantified using DPV and by employing calibration graph concentration of the drug.

2.4. Preparation of Urine Samples

The samples were acquired from healthy participants and stored in the refrigerator for later analysis. Samples were diluted with PBS of pH 3.0. A known amount of TMP was spiked to a sample and was subjected to DPV measurements. Then, the concentration of the TMP sample was estimated using a standard plot.

2.5. Synthesis of Nanostructure ZnO Nanoparticle

The precipitation processes synthesized ZnO nanoparticles using zinc nitrate and potassium hydroxide as precursors. Using double-distilled water, a 0.2 M concentration of zinc nitrate ($Zn(NO_3)_2 6H_2O$) and 0.4 M concentration of potassium hydroxide (KOH) solution were prepared. Later, potassium hydroxide was added slowly to the zinc nitrate solution with constant stirring. The white precipitate of ZnO was formed, centrifuged for 30 min in 5000 rpm, and cleaned with double-distilled water and ethanol to remove impurities; ultimately, the obtained product was calcined for 3 h at 500 °C [33].

2.6. Fabrication of Electrode

Using a mortar, the CPE was developed by combining graphite (70%) and paraffin oil (30%). Then, the mixed paste was filled into the Teflon tube of 3 mm diameter consisting of a copper rod for electrical conductance; the prepared electrode was polished on a smooth surface and cleaned with double-distilled water. After every determination, the paste was expelled and fresh paste was used for further analysis. ZnO/CPE was developed by adding 0.05 g of ZnO, 0.65 g of graphite powder and 0.3 mL of paraffin oil (05:65:30). Later, the different concentrations of SDS (0.05 mM–0.25 mM) were added to the test solution containing TMP and buffered solution. Maximum peak current was observed for the 0.1 mM SDS concentration (Figure S1). Hence, further studies were carried out in this optimal condition.

2.7. Experimental Procedure

In the activation of the developed electrode, pH 3.0 was employed for the electrochemical investigation of TMP. Stock solution of TMP was prepared with the help of ethanol. The electrochemical investigation and impact of pH, accumulation time, surfactant concentration, and scan rate variation were studied by CV technique under optimized conditions. The effects of concentration, pharmaceutical sample analysis, urine sample analysis, and interference study were examined utilizing differential pulse voltammetry in a potential range 0.8–1.4 V, 3.0 pH at an accumulation time of 10 s.

3. Results and Discussion

3.1. Surface Area of Electrode and ZnO Characterization

To calculate the active area of the developed sensor, 1.0 mM $K_3(Fe\ (CN)_6)$ was used as a probe and 0.1 M KCl as a supporting buffer. The CV measurements were taken for various scan rates displayed in Figure 1. Randle Sevcik Equation (1) was employed in calculating the area of the electrode [34] with the help of the value of slope obtained from Ip vs. $\sqrt{\upsilon}$ plot. We calculated the electrode's area, which was found to be 0.058 cm^2, 0.094 cm^2, and 0.099 cm^2 for bare CPE, ZnO/CPE, and ZnO/SDS/CPE, respectively.

$$Ip = 2.69 \times 10^5 \times n^{3/2} \times A \times {D_0}^{1/2} \times \upsilon^{1/2} \times C_o \quad (1)$$

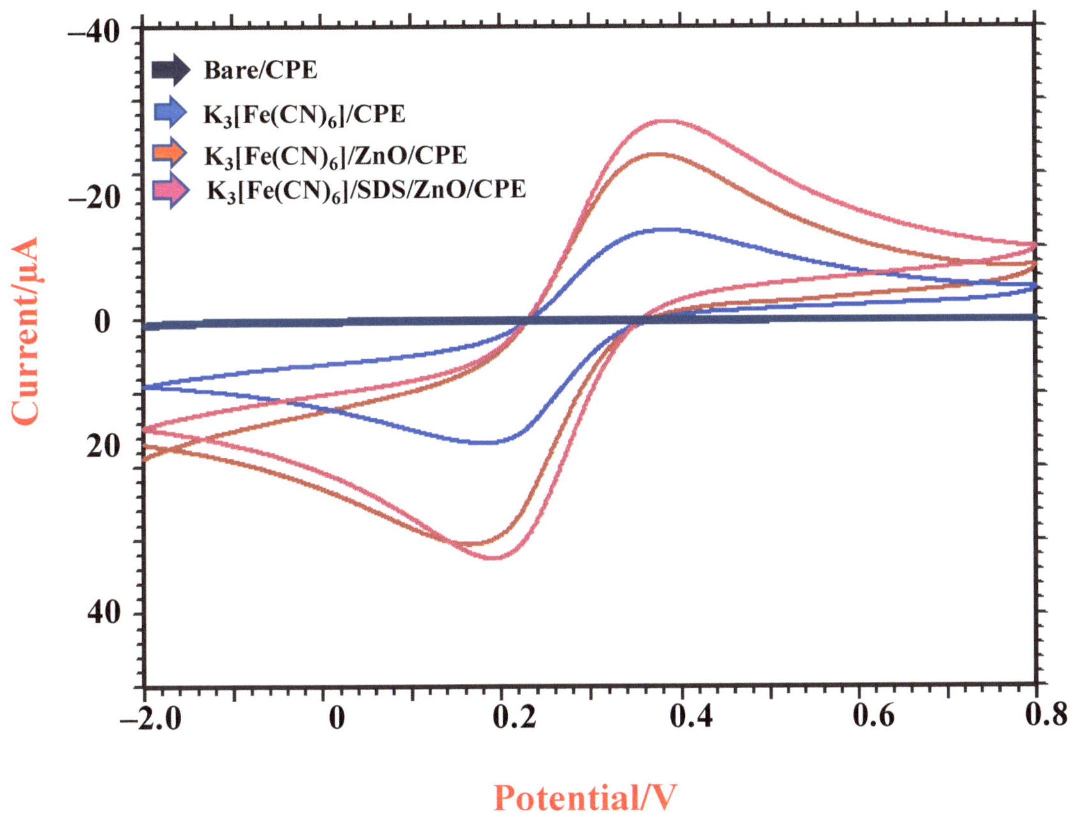

Figure 1. CV measurements of 1.0 mM K_3(Fe (CN)6) at bare CPE and ZnO/CPE and ZnO/SDS/CPE.

3.2. Characterisation of Modifier

Atomic force microscopy (AFM), scanning electron microscopy (SEM), and X-ray diffraction (XRD) studies were carried out to understand the morphological characteristics of the sensing surface. AFM images of bare and ZnO/CPE are displayed in Figure 2A,B. The total area roughness (Z) for bare and ZnO/CPE was found to be 0.4 and 1.12 nm. The XRD pattern of ZnO reveals firm diffraction peaks, indicating the sample's crystalline composition. The diffraction peaks were intense at 2θ° = 32.8, 34.4, and 36.2. However, the diffraction peaks further appeared at 47.5, 56.6, 62.8, 66.3, and 67.9, which were due to the various diffraction phases of ZnO exhibiting the hexagonal wurtzite structure. The absence of impurities in diffraction patterns revealed that the prepared ZnO is of high purity (Figure 2C). Further, SEM analysis showed that synthesized ZnO has an irregular shape and size with large surface area, which is depicted in the SEM image in Figure 2D.

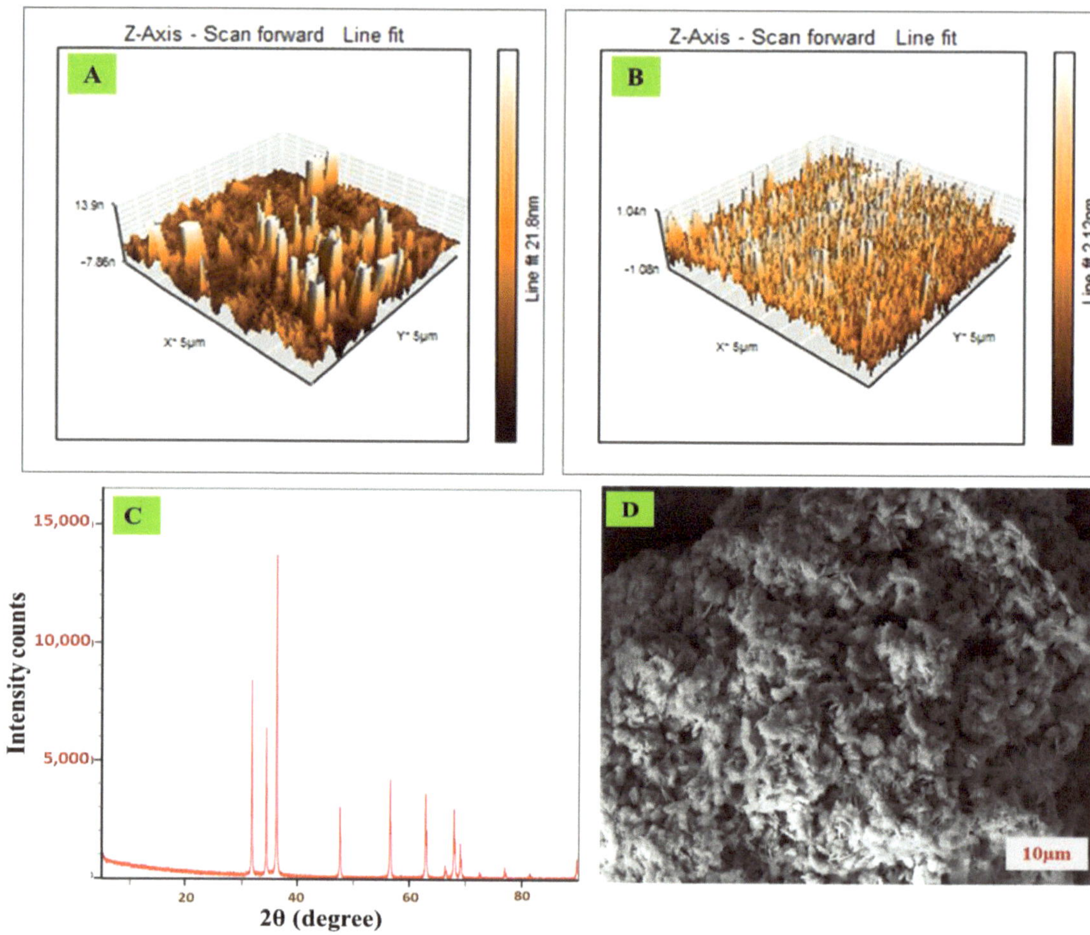

Figure 2. Characterization of ZnO nanoparticles (**A**) AFM of CPE; (**B**) AFM of ZnO/CPE; (**C**) XRD of ZnO; (**D**) SEM morphology of ZnO.

3.3. Electrochemical Detection of TMP

Electrochemical investigations of TMP were performed by employing the CV technique using 0.5 mM of TMP (in pH 3.0) at scan rate 0.05 V/s. Figure 3 shows the presence of an oxidation peak for all four electrodes (CPE, SDS/CPE, ZnO/CPE, ZnO/SDS-CPE). However, the double-fold-enhanced peak intensity was observed for ZnO/SDS-CPE at 153.9 μA, with a peak potential value of 1.280 V. In the presence of SDS, a good interaction occurred between the negatively charged monomers of SDS and they were strongly attracted towards the positively charged amine moiety on TMP (Scheme S1). On the reverse scan, the reduction peak was not detected and it indicates the process to be irreversible.

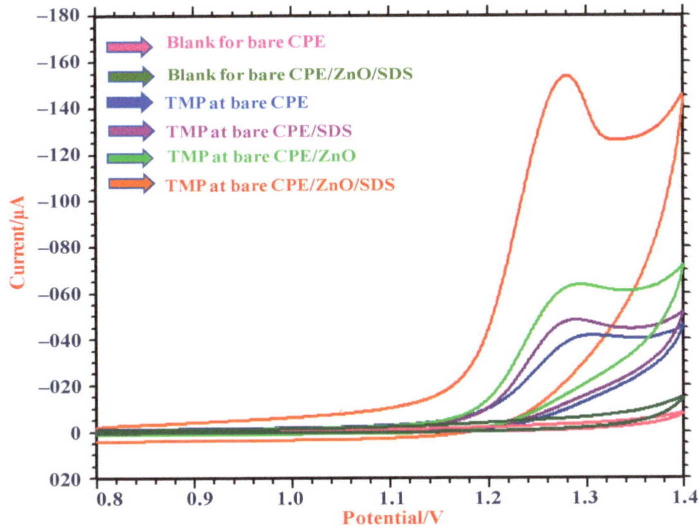

Figure 3. CV responses of 0.5 mM TMP in pH 3.0 (0.2 M PBS) at CPE, SDS/CPE, ZnO/CPE, and ZnO/SDS/CPE.

3.4. Immersion Time

The time required for an analyte molecule to accumulate at the electrode surface can influence the electrode's sensitivity. Hence, to study the effect of accumulation time, CV measurements were recorded for different time intervals from 0 to 70 s using 0.5 mM TMP at 0.05 V/s scan rate. The change in peak intensity was observed with respect to time and higher peak intensity was obtained at 10 s (Figure 4) and, hence, 10 s was preferred as an immersion time for further investigations.

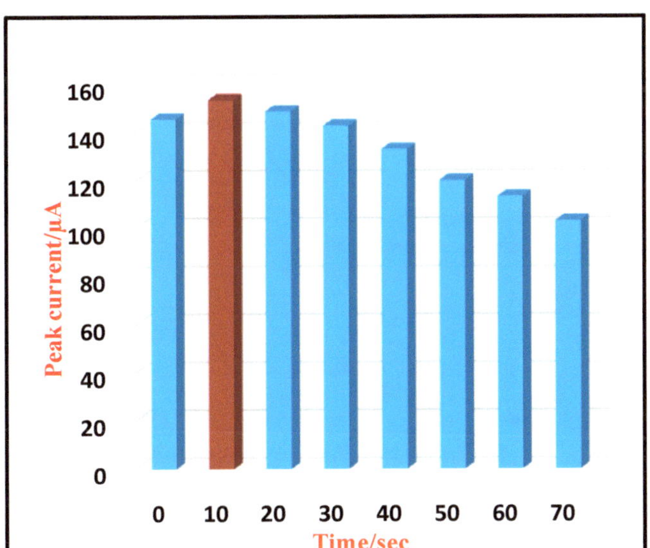

Figure 4. Plot of Ip and immersion time (t_{imm}) for 0.5 mM TMP.

3.5. pH Study

An electrochemical examination associates the control of peak potential by utilizing different supporting electrolytes. This was executed to enhance the conductivity of the solution. We initially examined the electrochemical behavior of TMP at ZnO/SDS/CPE in different electrolytic solutions, such as citrate buffer, Briton-Robinson buffer, sulphuric acid, and PBS at pH 3.0 (Figure S2). A well-defined oxidation peak was observed in the PBS-supporting electrolyte. Impact of PBS was studied for 0.5 mM TMP at ZnO/SDS/CPE for different pH ranges (from 3.0 to 6.0) using the CV technique at 0.8–1.4 V potential window. Voltammograms are displayed in Figure 5A. A well-resolved single oxidation peak was detected for all pH ranges and it can be noticed that with the rise in pH of supporting buffer concentration, a shift in peak potential (Ep) occurred, which is towards the less positive side, revealing the presence of H^+ in the oxidation of TMP [35]. After investigating the pH range above 6.0, we noticed the absence of a peak current, which concludes that the oxidation of TMP at ZnO/SDS/CPE can be acidic. The graphs for Ep and pH are plotted in Figure 5B, which gave a fitted equation Ep = −0.0219 pH + 1.340, $R^2 = 0.988$ and the slope value obtained was −0.0219 V/pH, which is nearest to the standard value of −0.021 V, which specifies that the electron and proton engaging in the reaction mechanism are uneven in number [36]. From Figure 5C, we can observe that at pH 3.0, the peak current (Ip) was intense; hence, the supporting electrolytic solution of pH 3.0 was chosen for further measurements. Further, we examined the impact of pH using the DPV technique and obtained results similar to the above CV results (Figure S3).

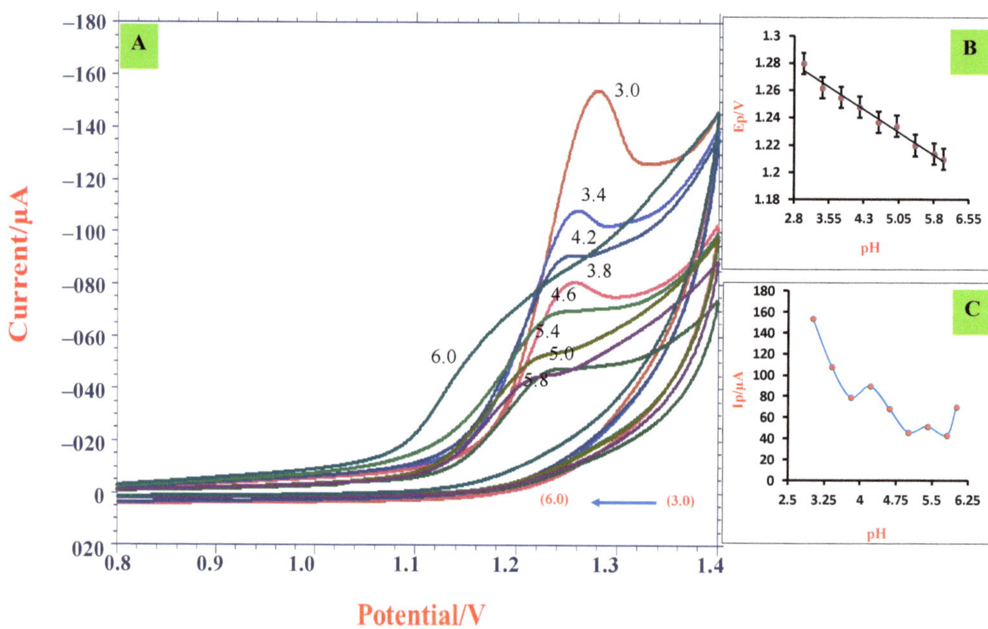

Figure 5. (**A**) CVs obtained for 0.5 mM TMP for various pH (pH 3.0–pH 6.0) employing ZnO/SDS/CPE at scan rate 0.05 V/s; (**B**) dependence of potential (Ep) on pH; (**C**) plot of peak current (Ip) vs. pH.

3.6. Scan Rate Study

The scan rate studies give insights into reaction mechanisms and the dependence of peak current and peak potential on scan rate. Therefore, the impact of scan rate on electro-oxidation of TMP was studied by employing CV technique at ZnO/SDS/CPE using 0.5 mM TMP by altering the scan rate from 0.01 to 0.41 V/s with its optimum pH at 3.0 (Figure 6A).

The voltammograms showed that with a peak current increase, peak potential slightly shifted towards the positive direction and peak current increased gradually, indicating the irreversible nature of reaction [37]. The increase in peak intensity with increasing scan frequency is due to the signal being anticipated due to improved charge transfer during system capacitance loading. In addition, a linear equation for the log of Ip vs. log ν (Figure 6C) acquired is shown below.

$$\log Ip = -0.384 \log \nu + 2.604;\ R^2 = 0.978$$

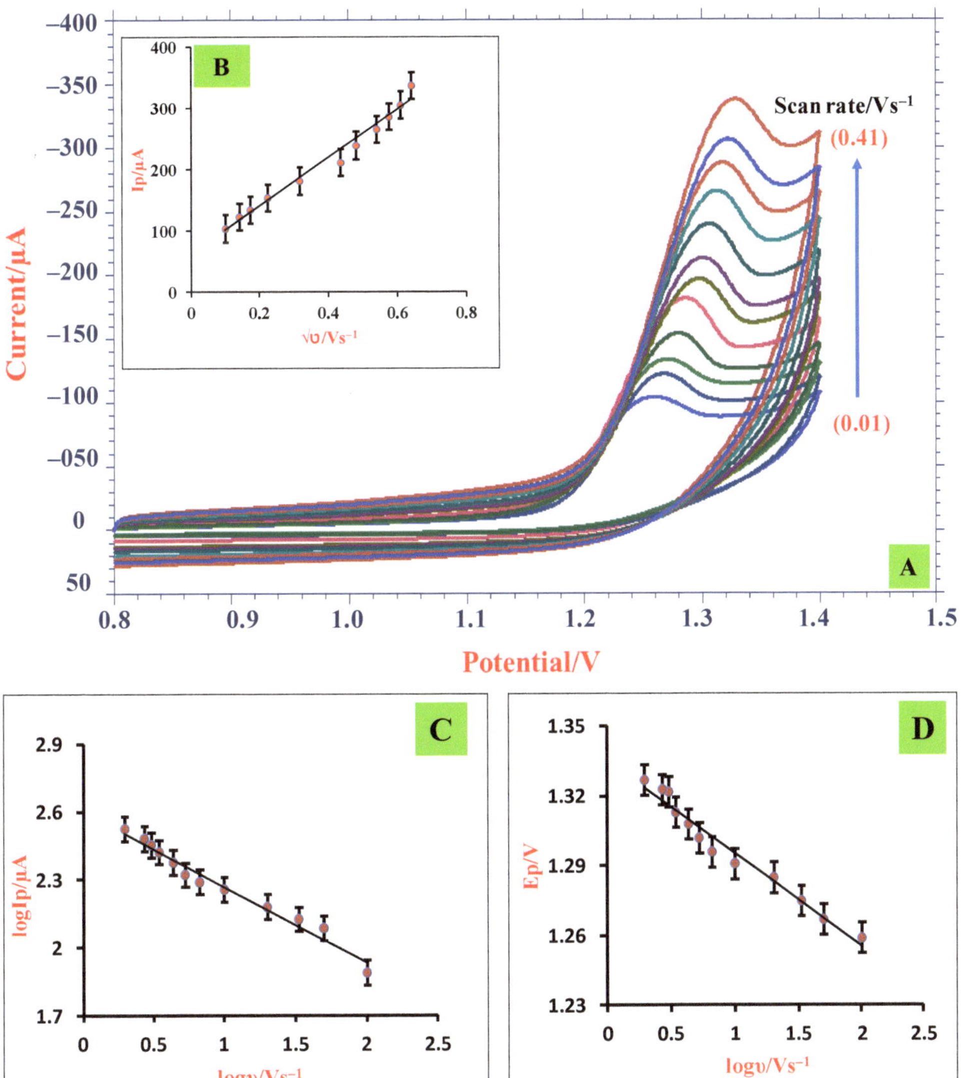

Figure 6. (**A**) CVs obtained for various scan rates (0.01–0.41 V/s); (**B**) dependence of Ip and $\sqrt{\upsilon}$; (**C**) dependence of log Ip and log υ; (**D**) plot of Ep vs. log υ.

Resultant slope (−0.384) was significantly closer to the predicted value 0.5; therefore, we can conclude that the reaction was governed by the process of diffusion [38]. From plot, Ip vs. $\sqrt{\nu}$ (Figure 6B) gave the linearity equation below.

$$Ip = 393.21\sqrt{\nu} + 61.575;\ R^2 = 0.980$$

$$Ep = -0.0402\ \log\nu + 1.356;\ R^2 = 0.969$$

The linear relationship was obtained from the plot peak potential (Ep) vs. log scan rate (log ν) (Figure 6D) and the intercept of this plot was substituted in the Laviron Equations (2) and (3), which were employed to measure the transfer coefficient (α) and heterogeneous rate constant (k°), respectively.

Further, the number of electrons was estimated to be two (2.42) [39].

$$Ep - Ep/2 = \left(\frac{47.7}{\alpha}\right)mV \tag{2}$$

$$E = E_o + \left(\frac{2.303RT}{(1-\alpha)nF}\right)\log\left(\frac{(1-\alpha)nF}{RTk^\circ}\right) + \left(\frac{2.303RT}{(1-\alpha)nF}\right)\log(\upsilon) \tag{3}$$

By calculation, the electrons (α) included in the oxidation of TMP were found to be 0.58 and k° value obtained was 1.28 s^{-1}.

Using the equations below, (4) and (5), the number of proton transfer and surface coverage concentration (Γ*) were calculated to be 1(0.89) and 7.58×10^{-6} mol cm^{-2}, respectively.

$$\frac{dEp}{dpH} = \frac{2.303RTm}{nF} \tag{4}$$

$$Ip = \frac{n^2F^2A\ n\ \Gamma * \ \nu}{4\ R\ T} \tag{5}$$

3.7. Possible Electrode Mechanism

In the electrochemical reaction of TMP, pH investigation confirmed the presence of protons, whereas per variation of scan rate studies, two electrons were transferred in the electrochemical oxidation reaction. Further, in this mechanism, electron losses occur initially from a non-bonding electron from nitrogen (non-bonding electrons are higher in energy). The electron-deficient nitrogen pulls the electron from the pyrimidine ring. As a result, positive charge was created and an unstable ion is formed. Hence, a rapid 1,2 hydride shift takes place, leading to the formation of secondary carbocation, which is attacked by the hydroxide ion. From the collected data, the plausible electrochemical mechanism can be predicted, as described in Scheme 1.

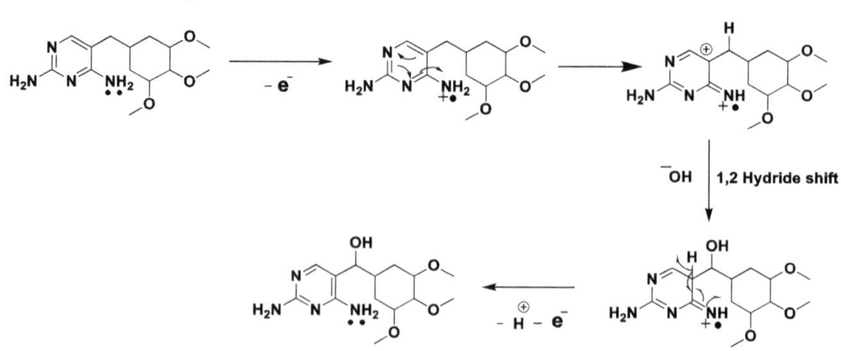

Scheme 1. Probable electrochemical mechanism of TMP at ZnO/SDS/CPE.

4. Analytical Applications

4.1. Variation in TMP Concentration

DPV approach was opted for concentration variation studies as the CV technique was not sensitive in detecting TMP in lower concentrations. Variation in concentration was observed using the developed electrode at its optimum pH 3.0. From Figure 7A, it can be noticed that with a decrease in TMP concentration, the peak current declined and the linearity range obtained was 8.0×10^{-7} M–1.0×10^{-5} M. By using the intercept and slope values acquired from plot of Ip and concentration (Figure 7B), the limit of detection and limit of quantification were measured employing the below equations [40]:

$$\mathrm{LOD} = 3\mathrm{s}/\mathrm{m} \tag{6}$$

$$\mathrm{LOQ} = 10\mathrm{s}/\mathrm{m} \tag{7}$$

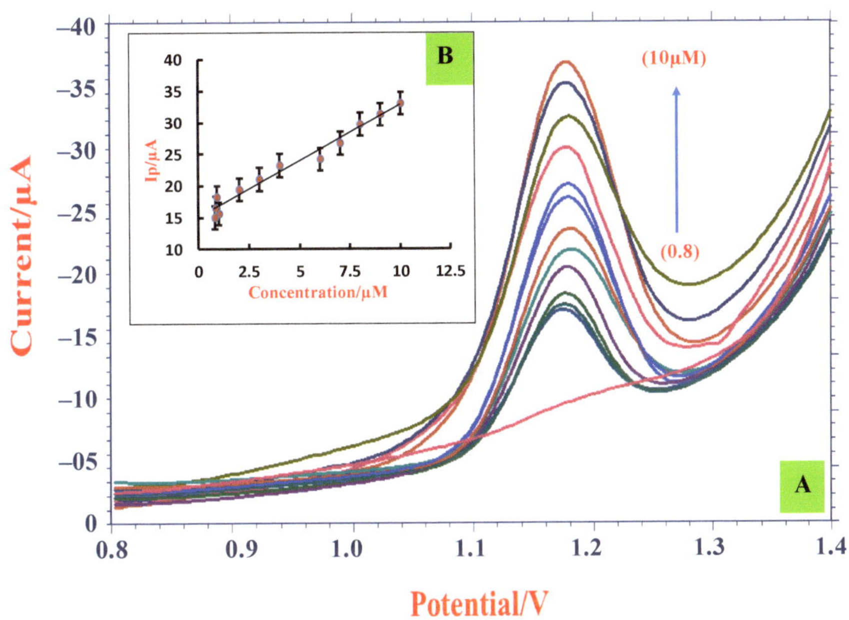

Figure 7. (**A**) DPVs of concentration variation (0.8 μM–10 μM) at ZnO/SDS/CPE; (**B**) graph of Ip and concentration.

The formula 's' is referred to as standard deviation of intercept value and 'm' corresponds to mean of the slope, which was calculated for three sets of measurements and the LOD and LOQ calculated data obtained were 2.58×10^{-8} M and 8.61×10^{-8} M, respectively. The obtained LOD value was compared with other reported methods (Table 1) and it was found that the present method was sensitive with a lower detection limit because of the novel properties of zinc oxide (semiconductor material with large excitation binding energy of 60 meV). Furthermore, ZnO is a favorable material for sensor applications due to its high catalytic ability, extraordinary electrical conductance, nontoxicity, and chemical stability. The resultant LOD and LOQ values showed that the developed electrode was sensitive, convenient, and efficient in TMP sensing and detection.

Table 1. Comparative study of detection limit of TMP.

Electrode Used	Technique Used	Linearity Range	LOD	References
Boron-doped diamond electrode	HPLC	85–1300 μM	5.5 μM	[24]
Methacrylic acid/molecularly imprinted polymer sensor	Potentiometry	0.03–26 μM	0.4 μM	[26]
molecularly imprinted polymer-graphene/Glassy carbon electrode	CV	1.0–100 μM	0.1 μM	[41]
Silver nanoparticle-graphene/Glassy carbon electrode	DPV	1.0–10 μM	0.4 μM	[42]
Graphene-Zinc oxide/Glassy carbon electrode	DPV	1–180 μM	0.3 μM	[43]
Cetyl trimethylammonium bromide/carbon paste electrode	CV	0.2–1.0 μM	0.1 μM	[44]
Printex L6 carbon black/copper (II) phthalocyanine film/Glassy carbon electrode	SWV	0.4–1.1 μM	0.6 μM	[45]
Multi-walled carbon nanotubes modified with antimony nanoparticles/CPE	DPV	0.1–0.7 μM	0.031 μM	[46]
ZnO/SDS/CPE	DPV	0.8–10 μM	0.025 μM	Present method

4.2. Pharmaceutical Sample Analysis

A pharmaceutical dosage of TMP was obtained from a local medical shop with the brand name Bacstol 200 mg. The tablet was crushed using a mortar and then the tablet powder was dissolved in ethanol and diluted up to the 100 mL volumetric flask, followed by sonication for a few minutes. Then, the solution was filtered to eliminate the undissolved solute. The filtrate obtained was subjected to DPV measurements, shown in Figure S4; the concentration was compared with the help of a calibration graph. The concentration and the recovery achieved are represented in Table 2.

Table 2. Tablet analysis of TMP.

TMP	Observations
Labelled claim (mg)	200
Obtained amount(mg)	198
RSD %	1.06
Added (mg)	1.0
Obtained (mg)	0.99
Recovery %	99.36

4.3. Analysis of Urine Sample

The proposed method was validated by analyzing the urine samples obtained from healthy people. The sample was diluted to using a supporting electrolyte of pH 3.0. The sample was then filtered and the supernatant solution was spiked with a known amount of TMP and analyzed using the DPV technique (Figure S5). The use of a calibration graph calculated the recovery range. As such, 97.5–99% recovery was obtained with RSD 1.62%. Table 3 provides details about analyzed data.

Table 3. Urine sample analysis of TMP.

TMP Sample	Added ($\times 10^{-5}$ M)	Obtained	Recovery (%)	RSD%
1	1.0	0.99	99	0.90
2	0.8	0.78	97.5	2.11
3	0.6	0.59	98.33	1.85

4.4. Effect of Interference Molecules

The excipient study was conducted to check the interference of some of the commonly used excipients, such as citric acid, lactose, dextrose, sucrose, glycine, glucose, urea, sodium chloride, potassium chloride, calcium chloride, and potassium sulphate. The concentration of 0.01 mM excipient solution was chosen and investigated. The changes in peak potential were observed to a smaller extent but were less than ±5% (Table 4), which confirmed that the excipients do not intervene much in TMP detection.

Table 4. Excipient study of TMP.

Excipients	Potential Observed (E_p)	Change in Signal (%)
Citric acid	1.190	1.01
Lactose	1.188	0.84
Dextrose	1.181	0.59
Sucrose	1.191	1.10
Glycine	1.184	0.50
Glucose	1.190	1.01
Urea	1.174	−0.33
	Salts	
Potassium chloride	1.171	−0.59
Sodium chloride	1.163	−1.27
Calcium chloride	1.183	0.50
Potassium sulphate	1.192	1.18

4.5. Repeatability and Reproducibility

To verify the functionality and viability of ZnO/SDS/CPE, inter-day and intra-day measurements were performed. Three replicates of measurements for 0.01 mM TMP were taken on the same day to study the repeatability nature of an electrode and the electrode retained its initial peak intensity, with 99% recovery, shown in Figure S6. To examine the reproducibility nature of an established electrode, it was kept in a dark place in an airtight jar and DPVs were recorded for 0.01 mM TMP. The results showed 97% recovery with an RSD value of 0.81%, which verified the efficiency of the sensor.

5. Conclusions

In this research, TMP was determined by electrochemical techniques, such as CV and DPV, at ZnO/CPE along with anionic surfactant SDS. The enhanced peak current for TMP was noticed at the proposed sensor in its optimum pH 3.0 of PBS. We performed scan rate investigations to gather information about reaction kinetics. The study suggested the process to be diffusion controlled and a total of two electrons was transferred during the reaction process, along with a single proton. Concentration variation studies observed the linear relationship between peak current and TMP concentration. Detection limit was calculated (2.58×10^{-8} M) and compared with other reported methods. Excipient investigations verified that they did not interfere with the excipients in signal response of TMP. Overall, the developed electrode was easy to fabricate, was cost effective, and could produce reproducible results. Furthermore, ZnO/SDS/CPE provides the possibility of a miniaturized system, offering high sensitivity and a quick response with a small amount of analyte. Hence, the developed method can be applied in the determination of clinical samples in trace quantities.

Supplementary Materials: The following supporting information can be downloaded at: https://www.mdpi.com/article/10.3390/bioengineering9100521/s1, Figure S1: Optimaiztion of SDS concentration; Figure S2: impact of different buffers on TMP; Figure S3: Effect of pH on behavior of TMP by DPV method; Figure S4: DPV for tablet analysis; Figure S5: DPV for urine samples; Figure S6: Reapatibility of SDS/ZnO/CPE sensor at 0.01 mM TMP; Scheme S1: Probable interaction of SDS and TMP.

Author Contributions: Conceptualization, V.B.P., D.I., S.M.T., K.M. and N.P.S.; methodology, V.B.P., D.I., S.M.T. and N.P.S.; formal analysis, V.B.P. and D.I.; investigation, V.B.P., D.I., S.M.T., K.M. and N.P.S.; resources, S.M.T. and N.P.S.; data curation, V.B.P. and D.I.; writing—original draft preparation, V.B.P., D.I., S.M.T., K.M. and N.P.S.; writing—review and editing, V.B.P., D.I., S.M.T., K.M. and N.P.S.; Supervision, S.M.T. and N.P.S. All authors have read and agreed to the published version of the manuscript.

Funding: This research received no external funding.

Institutional Review Board Statement: The institutional approval number is: INL/JOU-22-68791.

Informed Consent Statement: Not applicable.

Data Availability Statement: The original data were available from the corresponding author upon an appropriate request.

Acknowledgments: Vinoda B. Patil and Suresh M. Tuwar thank the Department of Chemistry, Karnatak Science College, Dharwad, Karnataka, India, for providing the lab facility and support. Davalasab Ilager and Nagaraj P. Shetti thank the Department of Chemistry, KLE Technological University, Hubballi, Karnataka, India, for providing the infrastructure and support. Kunal Mondal gratefully acknowledges the Department of Energy and Environment Science and Technology at the Idaho National Laboratory, USA, for their support.

Conflicts of Interest: The authors declare no conflict of interest.

References

1. Akinyele, O.F.; Adejayan, S.B.; Durosinmi, L.M.; Ayeni, A.O.; Ajayeoba, T.A. Interactions of Metal ions withTrimethoprim and Metformin. *Int. J. ChemTech Res.* **2020**, *13*, 38–46. [CrossRef]
2. Minato, Y.; Dawadi, S.; Kordus, S.; Sivanandam, A.; Aldrich, C.C.; Baughn, A.D. Mutual potentiation drives synergy between trimethoprim and sulfamethoxazole. *Nat. Commun.* **2018**, *9*, 1003. [CrossRef] [PubMed]
3. He, Q.; Ma, X.; Feng, S.; Miao, Y.; Du, M.; Xu, Y.; Wang, F. The performance and evolution of bacterial community of activated sludge exposed to trimethoprim in a sequencing batch reactor. *Bioresour. Technol.* **2017**, *244*, 872–879.
4. Rajith, L.; Kumar, K.G. Electroanalysis of trimethoprim on metalloporphyrin incorporated glassy carbon electrode. *Drug Test. Anal.* **2010**, *2*, 436–441. [CrossRef]
5. Fekry, A.; Abdel-Gawad, S.; Tammam, R.; Zayed, M. An electrochemical sensor for creatinine based on carbon nanotubes/folic acid/silver nanoparticles modified electrode. *Measurement* **2020**, *163*, 107958. [CrossRef]
6. Sarpong, K.A.; Zhang, K.; Luan, Y.; Cao, Y.; Xu, W. Development and application of a novel electrochemical sensor based on AuNPS and difunctional monomer-MIPs for the selective determination of Tetrabromobisphenol-S in water samples. *Microchem. J.* **2020**, *154*, 104526. [CrossRef]
7. Shanbhag, M.M.; Ilager, D.; Mahapatra, S.; Shetti, N.P.; Chandra, P. Amberlite XAD-4 based electrochemical sensor for diclofenac detection in urine and commercial tablets. *Mater. Chem. Phys.* **2021**, *273*, 125044. [CrossRef]
8. Yang, Z.; Zhang, X.; Guo, J. Functionalized Carbon-Based Electrochemical Sensors for Food and Alcoholic Beverage Safety. *Appl. Sci.* **2022**, *12*, 9082. [CrossRef]
9. Patil, V.B.; Malode, S.J.; Mangasuli, S.N.; Tuwar, S.M.; Mondal, K.; Shetti, N.P. An Electrochemical Electrode to Detect Theophylline Based on Copper Oxide Nanoparticles Composited with Graphene Oxide. *Micromachines* **2022**, *13*, 1166. [CrossRef]
10. Shetti, N.P.; Malode, S.J.; Nandibewoor, S.T. Electro-oxidation of captopril at a gold electrode and its determination in pharmaceuticals and human fluids. *Anal. Methods* **2015**, *7*, 8673–8682. [CrossRef]
11. Ilager, D.; Malode, S.J.; Shetti, N.P. Development of 2D graphene oxide sheets-based voltammetric sensor for electrochemical sensing of fungicide, carbendazim. *Chemosphere* **2022**, *303*, 134919. [CrossRef] [PubMed]
12. Shetti, N.P.; Sampangi, L.V.; Hegde, R.N.; Nandibewoor, S.T. Electrochemical oxidation of loop diuretic furosemide at gold electrode and its analytical applications. *Int. J. Electrochem. Sci.* **2009**, *4*, 104–121.
13. Hsueh, H.T.; Chang, S.J.; Hung, F.Y.; Weng, W.Y.; Hsu, C.L.; Hsueh, T.J.; Tsai, T.Y.; Dai, B.T. Fabrication of coaxial p-Cu2/n-ZnO nanowire photodiodes. *Super Lattices Microstruct.* **2011**, *49*, 572–580. [CrossRef]
14. Karimi, M.; Saydi, J.; Mahmoodi, M.; Seidi, J.; Ezzati, M.; Anari, S.S.; Ghasemian, M.B. A comparative study on ethanol gas sensing properties of ZnO and Zn0.94Cd0.06O nanoparticles. *J. Phys. Chem. Solids* **2013**, *74*, 1392–1398. [CrossRef]
15. Chandrashekar, B.N.; Swamy, B.E.K. Electrocatalysis of SDS surfactant Modified Carbon Paste Electrode for the Simultaneous Determination of Ascorbic Acid, Norepinephrine and Folic Acid. *Anal. Bioanal. Electrochem.* **2016**, *3*, 345–357.
16. Deepa, S.; Swamy, B.K.; Pai, K.V. A surfactant SDS modified carbon paste electrode as an enhanced and effective electrochemical sensor for the determination of doxorubicin and dacarbazine its applications: A voltammetric study. *J. Electroanal. Chem.* **2020**, *879*, 114748. [CrossRef]
17. Zhang, S.H.; Wu, K.B. Square Wave Voltammetric Determination of Indole-3-acetic Acid Based on the Enhancement Effect of Anionic Surfactant at the Carbon Paste Electrode. *Bull. Korean Chem. Soc.* **2004**, *25*, 1321–1325.

18. Adegoke, O.A.; Babalola, C.P.; Kotila, O.A.; Obuebhor, O. Simultaneous spectrophotometric determination of trimethoprim and sulphamethoxazole following charge-transfer complexation with chloranilic acid. *Arab. J. Chem.* **2017**, *10*, S3848–S3860. [CrossRef]
19. Hassib, S.T.; Farag, A.E.; Elkady, E.F. Bull, Liquid chromatographic and spectrophotometric methods for the determination of erythromycin sterate and trimethoprim in tablets. *Facual. Pharm.* **2011**, *49*, 81–89.
20. Hajjan, R.; Haghighi, R.; Shams, N. Combination of ratio derivative spectrophotometry with simultaneous standard additions method for determination of sulfamethoxazole and trimethoprim. *Asian J. Chem.* **2010**, *22*, 6569.
21. Stojković, G.; Dimitrieska-Stojković, E.; Soklevska, M.; Velev, R. Optimization, Validation and Application of UV-Vis Spectrophotometric-Colorimetric Methods for Determination of Trimethoprim in Different Medicinal Products. *Maced. Veter Rev.* **2016**, *39*, 65–76. [CrossRef]
22. Lecours, M.-A.; Eysseric, E.; Yargeau, V.; Lessard, J.; Brisard, G.M.; Segura, P.A. Electrochemistry-High Resolution Mass Spectrometry to Study Oxidation Products of Trimethoprim. *Environments* **2018**, *5*, 18. [CrossRef]
23. Raauf, A.M.R.; Ali, H.M.; Hameid, H. Spectrophotometric Determination of Trimethoprim in Pharmaceutical Formulation via Schiff base Reaction using Prepared Organic Reagents, Tikrit. *J. Pharm. Sci.* **2012**, *8*, 209–220.
24. Nagaraja, P.; Shrestha, A.; Shivakumar, A.; Gowda, A. Spectrophotometric determination of chloroquine, pyrimethamine and trimethoprim by ion pair extraction in pharmaceutical formulation and urine. *J. Food Drug Anal.* **2010**, *18*, 7. [CrossRef]
25. Yang, Y.-J.; Liu, X.-W.; Li, B.; Li, S.-H.; Kong, X.-J.; Qin, Z.; Li, J.-Y. Simultaneous determination of diaveridine, trimethoprim and ormetoprim in feed using high performance liquid chromatography tandem mass spectrometry. *Food Chem.* **2016**, *212*, 358–366. [CrossRef]
26. Berzas, J.J.; Penalvo, G.C.; Bernardo, F.J.G. Determination of sulfamethoxazole, sulfadiazine and associated compounds in pharmaceutical preparations by capillary zone electrophoresis. *J. Chromatogr. A* **2001**, *918*, 205. [CrossRef]
27. Beletskaya, I.; Tyurin, V.S.; Tsivadze, A.Y.; Guilard, R.; Stern, C. Supramolecular Chemistry of Metalloporphyrins. *Chem. Rev.* **2009**, *109*, 1659–1713. [CrossRef] [PubMed]
28. Andrade, L.S.; de Moraes, M.C.; Rocha-Filho, R.C.; Fatibello-Filho, O.; Cass, Q.B. A multidimensional high performance liquid chromatography method coupled with amperometric detection using a boron-doped diamond electrode for the simultaneous determination of sulfamethoxazole and trimethoprim in bovine milk. *Anal. Chim. Acta* **2009**, *654*, 127–132. [CrossRef]
29. Liu, L.; Wan, Q.; Xu, X.; Duan, S.; Yang, C. Combination of micelle collapse and field-amplified sample stacking in capillary electrophoresis for determination of trimethoprim and sulfamethoxazole in animal-originated foodstuffs. *Food Chem.* **2017**, *219*, 7–12. [CrossRef] [PubMed]
30. Rebelo, T.S.; Almeida, S.A.; Guerreiro, J.R.L.; Montenegro, M.C.B.; Sales, M.G.F. Trimethoprim-selective electrodes with molecularly imprinted polymers acting as ionophores and potentiometric transduction on graphite solid-contact. *Microchem. J.* **2011**, *98*, 21–28. [CrossRef]
31. Pedroso, R.C.R.; Peralba, M.D.R.; Santos, J.H.Z.d.; Pizzolato, T.M.; Froehlich, P.E. Desenvolvimento de metodos de analise Por CLAE-UV para os antimicrobianos tetraciclina, sulfamethoxazole e trimethoprim utilized materials a base de silica Como sistemas de pre-concentrical. *Quim. Nova.* **2011**, *34*, 206–212. [CrossRef]
32. Hussein, Z.M.; Rasheed, A.S. A Reliable Quantification Method for Trimethoprim in Pharmaceutical Samples by HILIC-HPLC. *Indian J. Forensic Med. Toxic.* **2021**, *15*, 2403.
33. Patil, V.B.; Malode, S.J.; Tuwar, S.M.; Shetti, N.P. Graphene sheet-based electrochemical sensor with cationic surfactant for sensitive detection of atorvastatin. *Sensors Int.* **2022**, *3*, 100198. [CrossRef]
34. Shanbhag, M.M.; Shetti, N.P.; Kalanur, S.S.; Pollet, B.G.; Upadhyaya, K.P.; Ayachit, N.H.; Aminabhavi, T.M. Hf-Doped Tungsten Oxide Nanorods as Electrode Materials for Electrochemical Detection of Paracetamol and Salbutamol. *ACS Appl. Nano Mater.* **2021**, *5*, 1263–1275. [CrossRef]
35. Huang, L.; Bu, L.; Zhao, F.; Zeng, B. Voltammetric behaviour of ethopropazine and the influence of sodium dodecyl sulphate on its accumulation on gold electrodes. *J. Solid State Electrochem.* **2004**, *8*, 976–981. [CrossRef]
36. Ilager, D.; Malode, S.J.; Kulkarni, R.M.; Shetti, N.P. Electrochemical sensor based on Ca- doped ZnO nanostructured carbon matrix for algicide dichlone. *J. Hazard. Mater. Adv.* **2022**, *7*, 100132. [CrossRef]
37. Erady, V.; Mascarenhas, R.J.; Satpati, A.K.; Bhakta, A.K.; Mekhalif, Z.; Delhalle, J.; Dhason, A. Carbon paste modified with Bi decorated multi-walled carbon nanotubes and CTAB as a sensitive voltammetric sensor for the detection of Caffeic acid. *Microchem. J.* **2019**, *146*, 73–82. [CrossRef]
38. Gosser, D.K. *Cyclic Voltammetry: Simulation and Analysis of Reaction Mechanisms*; VCH: New York, NY, USA, 1993.
39. Laviron, E. General expression of the linear potential sweep voltammogram in the case of diffusion less electrochemical systems. *J. Electroanal. Chem. Interfacial Electrochem.* **1979**, *101*, 19–28. [CrossRef]
40. Madhusudhana; Manasa, G.; Bhakta, A.K.; Mekhalif, Z.; Mascarenhas, R.J. Bismuth-nanoparticles decorated multi-wall-carbon-nanotubes cast-coated on carbon paste electrode; an electrochemical sensor for sensitive determination of Gallic Acid at neutral pH. *Mater. Sci. Energy Technol.* **2020**, *3*, 174–182. [CrossRef]
41. da Silva, H.; Pacheco, J.G.; Magalhães, J.M.; Viswanathan, S.; Delerue-Matos, C. MIP-graphene-modified glassy carbon electrode for the determination of trimethoprim. *Biosens. Bioelectron.* **2014**, *52*, 56–61. [CrossRef]
42. Golinelli, D.L.; Machado, S.A.; Cesarino, I. Synthesis of Silver Nanoparticle-Graphene Composites for Electroanalysis Applications using Chemical and Electrochemical Methods. *Electroanalysis* **2017**, *29*, 1014–1021. [CrossRef]

43. Yue, X.; Li, Z.; Zhao, S. A new electrochemical sensor for simultaneous detection of sulfamethoxazole and trimethoprim antibiotics based on graphene and ZnO nanorods modified glassy carbon electrode. *Microchem. J.* **2020**, *159*, 105440. [CrossRef]
44. Kumari, C.T.R.; Mamatha, G.P.; Santhosh, H.M. Voltammetric detection of trimethoprim at CTAB modified carbon paste electrode. *Chem. Sci. Trans.* **2016**, *5*, 619–626.
45. Guaraldo, T.T.; Goulart, L.A.; Moraes, F.C.; Lanza, M.R. Carbon black nanospheres modified with Cu (II)-phthalocyanine for electrochemical determination of Trimethoprim antibiotic. *Appl. Surf. Sci.* **2019**, *470*, 555–564. [CrossRef]
46. Cesarino, I.; Cesarino, V.; Lanza, M.R. Carbon nanotubes modified with antimony nanoparticles in a paraffin composite electrode: Simultaneous determination of sulfamethoxazole and trimethoprim. *Sensors Actuators B Chem.* **2013**, *188*, 1293–1299. [CrossRef]

bioengineering

Article

Rational Design of Porous Poly(ethylene glycol) Films as a Matrix for ssDNA Immobilization and Hybridization

Zhiyong Zhao, Saunak Das and Michael Zharnikov *

Angewandte Physikalische Chemie, Universität Heidelberg, 69120 Heidelberg, Germany
* Correspondence: michael.zharnikov@urz.uni-heidelberg.de; Tel.: +49-6221-54 4921

Abstract: Poly(ethylene glycol) (PEG) films, fabricated by thermally induced crosslinking of amine- and epoxy-terminated four-arm STAR-PEG precursors, were used as porous and bioinert matrix for single-stranded DNA (ssDNA) immobilization and hybridization. The immobilization relied on the reaction between the amine groups in the films and N-hydroxy succinimide (NHS) ester groups of the NHS-ester-decorated ssDNA. Whereas the amount of reactive amine groups in the films with the standard 1:1 composition of the precursors turned out to be too low for efficient immobilization, it could be increased noticeably using an excess (2:1) concentration of the amine-terminated precursor. The respective films retained the bioinertness of the 1:1 prototype and could be successfully decorated with probe ssDNA, resulting in porous, 3D PEG-ssDNA sensing assemblies. These assemblies exhibited high selectivity with respect to the target ssDNA strands, with a hybridization efficiency of 78–89% for the matching sequences and full inertness for non-complementary strands. The respective strategy can be applied to the fabrication of DNA microarrays and DNA sensors. As a suitable transduction technique, requiring no ssDNA labeling and showing high sensitivity in the PEG-ssDNA case, electrochemical impedance spectroscopy is suggested.

Keywords: poly(ethylene glycol); STAR-PEGs; ssDNA immobilization; ssDNA hybridization; electrochemical impedance spectroscopy; X-ray photoelectron spectroscopy

Citation: Zhao, Z.; Das, S.; Zharnikov, M. Rational Design of Porous Poly(ethylene glycol) Films as a Matrix for ssDNA Immobilization and Hybridization. *Bioengineering* **2022**, *9*, 414. https://doi.org/10.3390/bioengineering9090414

Academic Editor: Rossana Madrid

Received: 28 July 2022
Accepted: 23 August 2022
Published: 24 August 2022

1. Introduction

Immobilization of single-stranded DNA (ssDNA) on solid supports is an important issue in the physical chemistry of interfaces and biomedical research, widely applied in biological detection, microarray technology, and related fields [1–3]. Usually, ssDNA are immobilized on the target substrate by their decoration with a suitable anchoring group which has a strong affinity to the substrate. In particular, thiol-substituted ssDNA strands, capable of assembling on coinage-metal substrates (above all on gold), are frequently used [4–7]. Specifically modified ssDNA strands can also be attached to surfaces precoated with self-assembled monolayers (SAMs) [8,9] or polymers [10,11], relying on covalent bonding between specific docking groups of ssDNA and matching terminal groups of SAMs or polymers.

To achieve a high hybridization efficiency and to form individual sensing spots, it is frequently necessary to immobilize ssDNA into a biocompatible matrix resisting non-specific ssDNA-surface interactions [12–14]. In this way, one can control the density of the immobilized probe ssDNA and suppress nonspecific adsorption of target ssDNA beyond the predefined sensing spots, thus improving the specificity and efficiency of a particular assembly or a device. The most frequently used bioinert material in this context is poly(ethylene glycol) (PEG). PEG is a hydrophilic polymer with remarkable biocompatible properties, prohibiting nonspecific adsorption of proteins, oligonucleotides, bacteria, and other bioorganisms [15–19]. This material is used in particular for the decoration of SAMs, in the form of oligo(ethylene glycol) (OEG) tail groups, resulting in the fabrication of bioinert surfaces and interfaces [20–24]. Properly decorated ssDNA can then be immobilized onto the respective supports, either as a component of binary ssDNA-OEG SAMs or within a predefined spot, using a lithographic approach [25,26].

Along with OEG-decorated SAMs, all-PEG materials can also be used, such as PEG brushes and polymer films [27–30]. Alternatively, porous PEG and PEG-based films can be used, taking advantage of 3D immobilization of ssDNA in contrast to the standard 2D assemblies provided by SAM supports and usual OEG-based polymers [31–34]. In particular, such porous films can be efficiently formed by thermally activated crosslinking of multi-armed STAR-PEG precursors, decorated with amine (STAR-NH_2) or epoxy (STAR-EPX) groups, which build ethanol-amine bridges between individual arms of the precursors upon the crosslinking (Figure 1) [35]. The porosity of the resulting films is then predominantly determined by the length of the PEG arms which are linked together in the joint center of each of the precursors [36]. The thickness of these films can be flexibly adjusted in a range from several to hundreds of nanometers by varying the concentration of the precursors in the primary solutions [35,36]. Distinct bioinert, hydrogel, and elastic properties of these films, existing also in the form of free-standing membranes, make them useful for further modification and processing as well as for a variety of applications [35–42]. Due to their distinguished bioinertness, these films can also serve as a bioinert matrix which can be decorated with bioreceptors and subsequently used for specific biosensing. This ability has indeed been demonstrated in the case of proteins, relying on the well-known biotin-avidin key-lock affinity [41]. With this achievement in mind, it is interesting and promising to explore whether such a strategy is also suitable for the immobilization and hybridization of ssDNA, which is the subject of the present study. For this purpose, we used the STAR-NH_2 and STAR-EPX precursors with moderate molecular weights, adjusting their mixing ratio to optimize the immobilization efficiency of ssDNA, and applied X-ray photoelectron spectroscopy (XPS) and complementary electrochemical techniques to monitor the immobilization and hybridization processes.

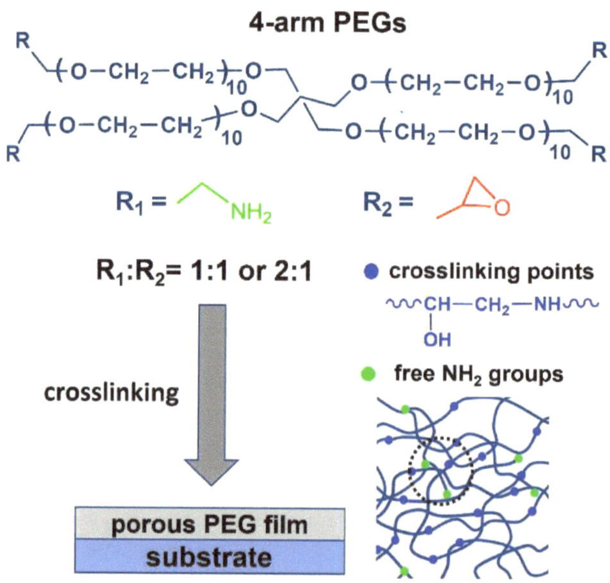

Figure 1. Structure of the STAR-NH_2 (R_1) and STAR-EPX (R_2) precursors and a schematic drawing of the PEG film fabrication procedure relying on the extensive crosslinking of the precursors, mediated by the reactions between their terminal amine (R_1) and epoxy (R_2) groups. The resulting crosslinking points (ethanol-amine bridges) are marked by blue circles in the porous PEG film scheme (bottom, right); free (non-reacted) NH_2 groups are depicted as green circles. The mixing ratio of the precursors was varied.

2. Materials and Methods

2.1. Chemicals

The 4-arm STAR-NH_2 and 4-arm STAR-EPX compounds (Figure 1) with a molecular weight of 2000 g/mol were purchased from Creative PEGWorks (Chapel Hill, NC, USA) and used as received. These compounds are characterized by low polydispersity and high purity, viz. 99% for STAR-NH_2 and 98% for STAR-EPX in terms of amine and epoxy substitution, respectively. According to the molecular weight, the PEG arms of these compounds contain 10-11 EG monomers, corresponding to an arm length of 3.5−3.9 nm. Desalted ssDNA sequences were purchased from Metabion International AG (Munich, Germany). The first group of these sequences included unmodified thymine (T) and adenine (A) homo-oligonucleotides, viz′−T5−3′ (T5), 5′−T10−3′ (T10), 5′−A5−3′ (A5), and 5′−A10−3′ (A10). The second group included substituted homo-oligonucleotides, viz. N-hydroxy succinimide ester-C10−T5−3′ (NHS-T5) and 5′−N-hydroxy succinimide ester-C10−T10−3′ (NHS-T10). Please note that the NHS esters are reactive groups formed by carbodiimide-activation of carboxylate molecules. NHS-ester-labeled compounds react with primary amines under physiologic to slightly alkaline conditions (pH 7.2 to 9) to yield stable amide bonds after the release of the NHS group [43]. Consequently, decoration of ssDNA strands with NHS ester should be a reasonable strategy to immobilize these strands into the PEG matrix over the NH_2 groups which did not participate in the crosslinking reaction and retained their reactivity. Other chemicals were purchased from Sigma-Aldrich.

2.2. Film Fabrication

The PEG films were prepared by the established protocol [35], schematically illustrated in Figure 1. Accordingly, the STAR-NH_2 and STAR-EPX precursors were separately dissolved in chloroform with specific concentrations, mixed together in a ratio of 1:1 (V/V), spin-coated onto SiO_2 passivated Si substrates (Siegert Wafer GmbH, Aachen, Germany), and crosslinked by thermal annealing (6 h, 80 °C). The resulting films were extensively rinsed with ethanol to remove possible weakly bound material. Two kinds of films were prepared. In the first case, the same concentration of the STAR-NH_2 and STAR-EPX precursors in the primary solutions was used (1:1 ratio), set to either 2 mg/mL or 25 mg/mL to obtain either thin (~15 nm) or thick (~100 nm) films. In the second case, the concentrations of 20 mg/mL for STAR-NH_2 and 10 mg/mL for STAR-EPX were used (2:1 ratio), to obtain PEG films (~80 nm thickness) with a noticeable amount of free NH_2 groups, suitable for the reaction with the NHS ester groups of the substituted homo-oligonucleotides. For the sake of brevity, we will refer to these systems further in the manuscript as the 1:1 and 2:1 films, respectively. The 2:1 PEG films were also fabricated on evaporated Au(111) substrates (30 nm Au on Si(100); Georg-Albert PVD-Beschichtungen, Silz, Germany)—specifically for electrochemical measurements (see below). These films had similar thickness and similar properties as the films on the Si/SiO_2 substrates. The stability and robustness of these films was in particular verified by the fabrication of large-area, free-standing PEG membranes which feature extreme elasticity [44].

2.3. ssDNA Immobilization and Hybridization

The procedures are schematically illustrated in Figure 2. For ssDNA immobilization, PEG films were immersed into 1M $CaCl_2$ −TE buffer (10 mM Tris-HCl and 1 mM EDTA, pH = 7.4) containing 10 μM of ssDNA for 40 h at 37 °C. Please note that the 1M $CaCl_2$-TE buffer was supplanted by PBS buffer (pH = 7.4) in the NHS-T5 and NHS-T10 cases to avoid reaction between the NHS ester group of ssDNA and the NH_2 group of Tris-HCl. After the incubation, the samples were rinsed with Milli-Q water for 1 min and-dried under N_2 flow.

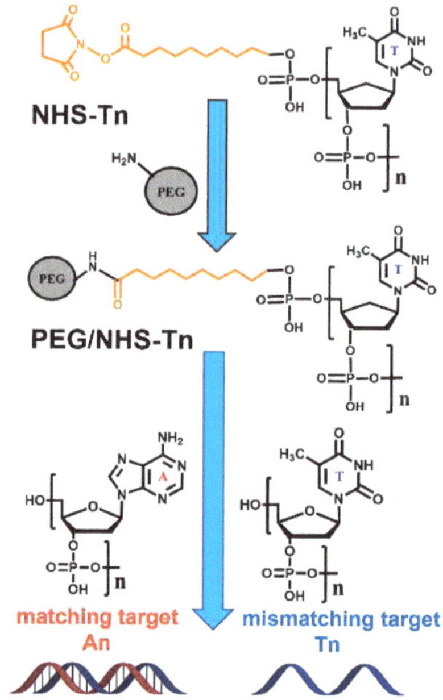

Figure 2. The structures of the NHS-Tn, An, and Tn compounds as well as a schematic illustration of the NHS-Tn immobilization in the PEG matrix (shown schematically as a gray circle) over the free (non-reacted) NH_2 groups and its subsequent reaction with the An and Tn targets.

For hybridization tests, the samples were immersed in a 1 M NaCl buffer containing 10 µM of the target sequences for 8 h at room temperature. After incubation, the samples were rinsed with 1 M NaCl buffer for 1 min, briefly dipped in a small amount of Milli-Q water (~0.5 mL) to remove excess salts, and finally dried with N_2.

2.4. X-ray Photoelectron Spectroscopy

Bioinert properties of the PEG films, ssDNA immobilization in these films, and the hybridization ability of the resulting hybrid films were monitored by XPS, which is a frequently used technique for this purpose [5,45–48]. The measurements were performed using a MAX 200 (Leybold –Heraeus, Köln, Germany) spectrometer equipped with a hemispherical analyzer (EA 200; Leybold–Heraeus, Köln, Germany) and a Mg Kα X-ray source (260 W; ca. 1.5 cm distance to the samples). The spectra were obtained in normal emission geometry with an energy resolution of ~0.9 eV. The binding energy (BE) scale of the spectra was referenced to the Au $4f_{7/2}$ peak at 84.0 eV [49].

XPS was also used to estimate the areal density of the immobilized probe ssDNA. As a reference, we used a custom-designed SAM of nitrile-substituted naphthalenethiolates on Au(111) with an areal density of $\sim 4.2 \times 10^{14}$ molecules/cm^2 [50], which in view of the molecular structure, is also the density of the terminal nitrogen atoms. Along with the intensities of the N 1s signals for the PEG-ssDNA and reference SAM, the difference in the number of the nitrogen atoms in the relevant molecules was taken into account.

2.5. Electrochemistry

Electrochemical measurements, which included cyclic voltammetry and electrochemical impedance spectroscopy (EIS), were performed using an IM6E potentiostat (Zahner-Elektrik GmbH & Co. KG, Kronach-Gundelsdorf, Germany) and a custom-made three-electrode electrochemical cell. Along with the working electrode (blank Au, Au/PEG, Au/PEG/ssDNA), an Ag/AgCl (non-aqueous) electrode and a platinum electrode (Osilla, Sheffield, UK) were used as the reference and counter electrodes, respectively. The blank Au substrates for the working electrode were purchased from Georg Albert, PVD-Beschichtungen (see also Section 2.1); the root mean square value of the surface roughness was estimated as ~0.5 nm as an average over the 0.5 × 0.5 μm^2 and 5 × 5 μm^2 scans (see Figure S1 in the Supporting Information). The cyclic voltammograms (CVs) and EIS data were recorded in a 10 mM $Fe(CN)_6{}^{3-/4-}$ electrolyte containing 0.1 M KCl. The exposed area of the electrodes was ~0.5 cm^2. For CV measurements, a scan rate of 300 mV/s in the range from −0.8 V to +0.7 V (vs. Ag/AgCl) was applied. The EIS measurements were conducted at an alternating voltage with an amplitude of 5 mV, in the frequency range from 10^{-1} to 10^5 Hz. Please note that an analogous approach was previously applied to monitor the performance of a ssDNA array immobilized on a glassy carbon electrode modified with a complex inorganic matrix on the basis of Sm_2O_3 nanoparticles and graphene oxide [51].

3. Results and Discussions

3.1. XPS

The monitoring of the relevant properties and processes within the given study relied on the well-known XP spectra of pristine PEG films and thymine and adenine homo-oligonucleotides. In accordance with the chemical composition, the PEG films are adequately represented by the C 1s, O 1s, and N 1s spectra [35,36,42], exhibiting the characteristic singular peaks at BEs of 286.8 eV (C 1s), 532.8 eV (O 1s), and 399.6 eV (N 1s), as shown in Figures S2 and S3 in the Supporting Information for the standard 1:1 case. The first two peaks are related to the PEG arms of the network and the third peak is representative of the nitrogen atoms in the ethanol-amine bridges. The homo-oligonucleotides can be best traced by the N 1s and P 2p spectra, representative of the nucleobases and phosphate groups in the ssDNA backbone, respectively. The P 2p spectra, usually showing a merged P $2p_{3/2,1/2}$ doublet at a BE of 133.5–133.7 eV [5,45,52–54], are not nucleobase-specific, but are a suitable fingerprint for the presence of ssDNA in the PEG matrix, which originally contains no phosphorus. The N 1s spectra are nucleobase-specific, which not only allows monitoring the presence of ssDNA but also allows to distinguish between thymine and adenine homo-oligonucleotides. In the case of thymine, the spectrum exhibits a single peak at a BE of ~400.5 eV, sometimes accompanied by a weak shoulder at a BE of ~398.5 eV associated with thymine moieties which are in direct contact with the substrate [5,52,55,56]. In the case of adenine, the spectrum consists of two peaks at BEs of ~398.7 and ~400.5 eV with the characteristic intensity ratio of 2:1 [5,52,56]. Significantly, the positions of these characteristic features do not change noticeably upon the T-A hybridization [56–58].

The PEG films are expected to be inert to ssDNA strands, similar to their behavior with respect to proteins [35,36]. To verify this assumption, 1:1 films were exposed to the unmodified ssDNA strands, viz. Tn and An (n = 5 and 10), and characterized by XPS. The respective C 1s, O 1s and N 1s XP spectra were found to be identical (within the experimental accuracy) to those of the original films (see Figure S2 in the Supporting Information), which demonstrates that the PEG films with the optimal mixing ratio of the precursors are indeed bioinert.

These films contain, however, only a small amount of free amine groups (~3% according to our estimate, based on the infrared spectroscopy data from ref [35]), which can be insufficient for an effective immobilization of the NHS-ssDNA. Indeed, after the exposure of the 1:1 PEG films to NHS-T5 and NHS-T10, no noticeable changes could be observed in the XP spectra of the samples (see Figure S3 in the Supporting Information), indicating a very small (if at all) immobilization efficiency. As an additional proof, these samples were subsequently exposed to the target ssDNA strands complementary to T5 and T10, viz. A5 and A10, and characterized by XPS. Again, the spectra remained unchanged (see Figure S3 in the Supporting Information), which fully exclude that the probe T5 and T10 strands, capable of hybridizing with A5 and A10, were present in the PEG matrix.

The above results suggest that the 1:1 PEG films are not suitable for the immobilization of NHS-ester-modified ssDNA. A promising solution could then be a deviation from the 1:1 mixing ratio, resulting in a non-negligible amount of free amine groups, capable of reacting with the NHS ester moieties of the NHS-ssDNA. To verify this hypothesis, the STAR-NH_2/STAR-EPX mixing ratio was set to 2:1 (see Section 2 for the experimental details) and 2:1 PEG films were prepared. Please note that the crosslinking reaction still works efficiently at even such a non-optimal mixing ratio, resulting in the formation of stable and robust PEG films, with the swelling and mechanical properties differing only slightly from those of the 1:1 films [41,44]. Additionally, the XP spectra of the 2:1 PEG films were found to be nearly identical to those of the 1:1 prototypes, with the characteristic C 1s, O 1s, and N 1s peaks at BEs of 286.8 eV, 532.8 eV, and 399.6 eV, respectively (Figure 3). The presence of only one N 1s peak in these spectra means that both crosslinked and free amine groups have nearly the same XPS binding energy. This circumstance simplifies the analysis of the spectra but make it difficult to provide an estimate for the amount of free amine groups.

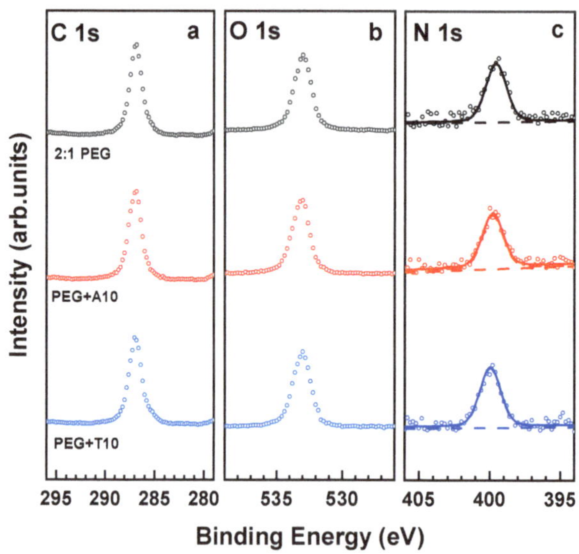

Figure 3. C 1s (**a**), O 1s (**b**) and N 1s (**c**) XP spectra of the 2:1 PEG films before (top curves) and after their incubation into the A10 and T10 solutions. The N 1s spectra are tentatively fitted by a single peak (solid lines) and a background (dashed lines).

For the next step, bioinert properties of the 2:1 PEG films were tested. For this purpose, these films were exposed to A10 and T10 and characterized afterwards by XPS, relying on the characteristic C 1s, O 1s, and N 1s spectra. The respective data are presented in Figure 3. The spectra of the films exposed to A10 and T10 turned out to be identical (within the experimental accuracy) to those of the original films. This observation means that a moderate deviation from the optimal mixing ratio does not result in a deterioration of bioinert properties. Thus, the 2:1 films can readily serve as a bioinert matrix for immobilization of probe ssDNA strands and subsequent hybridization with the target ssDNA strands, as far as immobilization and hybridization can be performed.

Both these processes turned out to be indeed possible. The immobilization of the probe ssDNA strands (T5 and T10) was carried out with the help of NHS-T5 and NHS-T10, relying on the reaction between the NHS ester group of the latter moieties and the free amine groups in the PEG films. The process was monitored by XPS, relying on the C 1s, N 1s and P 2p spectra. The respective data are shown in Figures 4 and 5 for the NHS-T5 and NHS-T10 case, respectively. Let us first discuss the data for NHS-T5 and later - for NHS-T10.

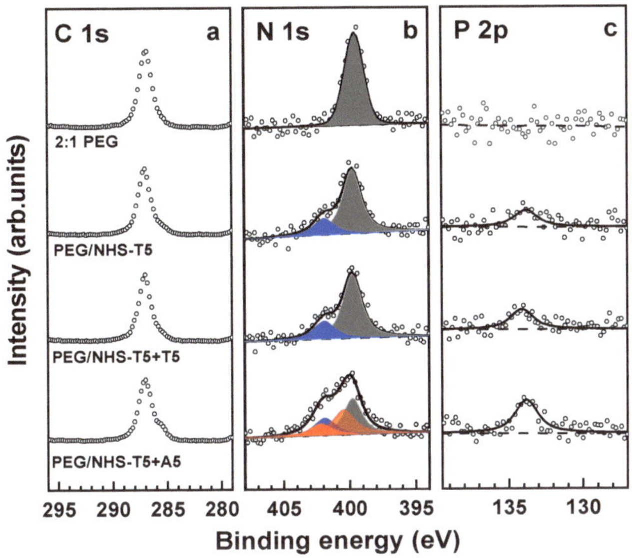

Figure 4. C 1s (**a**), N 1s (**b**), and P 2p (**c**) XP spectra of the original 2:1 PEG film, PEG film exposed to NHS-T5 (PEG/NHS-T5), and PEG/NHS-T5 probe film exposed to mismatching (T5) and matching (A5) target ssDNA. The N 1s spectra are decomposed into individual contributions related to the amine groups in the PEG matrix (dark gray), thymine (blue), and adenine (red). The P 2p spectra are tentatively fitted by a single peak (solid lines) and a linear background (dashed lines).

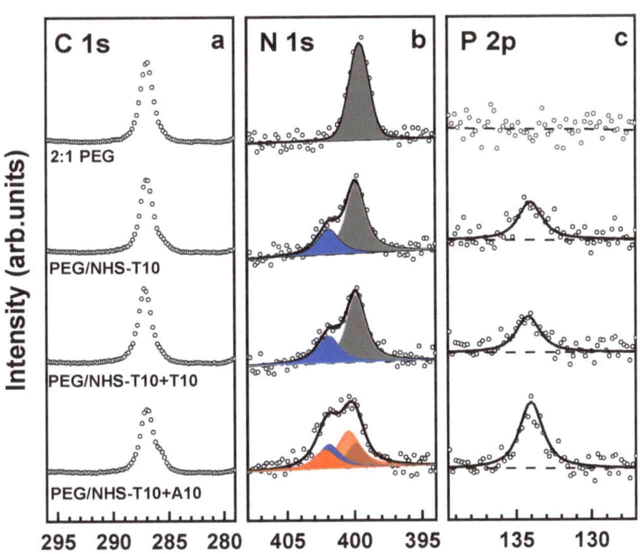

Figure 5. C 1s (**a**), N 1s (**b**), and P 2p (**c**) XP spectra of the original 2:1 PEG film, PEG film exposed to NHS-T10 (PEG/NHS-T10), and PEG/NHS-T10 probe film exposed to mismatching (T10) and matching (A10) target ssDNA. The N 1s spectra are decomposed into individual contributions related to the amine groups in the PEG matrix (dark gray), thymine (blue), and adenine (red). The P 2p spectra are tentatively fitted by a single peak (solid lines) and a linear background (dashed lines).

After the exposure of the 2:1 PEG films to NHS-T5, the C 1s spectrum of the resulting films (PEG/NHS-T5) looks similar to that of the original PEG films, whereas the N 1s and P 2p spectra change noticeably. In the N 1s spectrum, the peak at ~399.8 eV, associated with the PEG matrix, decreases in intensity, and becomes accompanied by the characteristic peak of thymine at a BE of ~401.9 eV (see refs [5,52,55,56]). In the P 2p spectrum, a characteristic signature of the phosphate groups in the ssDNA skeleton at a BE of ~133.7 eV is observed [5,45,52–54]. This joint evidence indicates that the probe T5 strands were successively immobilized into the PEG matrix.

Next, the ability of the T5-functionalized PEG films to probe a complimentary target ssDNA (A5) was tested by their incubation into A5 solution and subsequent characterization by XPS. Once again, the C 1s spectrum, representing predominantly the PEG matrix, did not change noticeably. In contrast, the N 1s and P 2p XP spectra, representing the ssDNA species, showed pronounced changes. In the P 2p spectrum, an increase in the intensity of the characteristic phosphate feature by a factor of ~1.78 is observed, corresponding to a high hybridization efficiency (~78%). In the N 1s spectrum, the shoulder at ~401.9 eV increases in intensity and becomes comparable to the main peak. Assuming that this increase stems from the T5-A5 hybridization, the spectrum was decomposed in three components associated with the NH_2 and NH groups in the PEG matrix, thymine, and adenine. Within the respective fit, the PEG matrix was represented by a single peak at a BE of ~399.8 eV, thymine—by a single peak at 401.9 eV, and adenine—by two peaks at ~400.4 eV and ~402.2 eV with an intensity ratio of 2:1. As shown in Figure 4, the N 1s spectrum could be fully reproduced by such a combination. The relative weights of the thymine and adenine components, corrected for the different contents of the nitrogen atoms in these bases (2 for thymine and 5 for adenine; see Figure 2), give then a hybridization efficiency of ~80%, in excellent agreement with the P 2p data.

To verify the selectivity of the T5-decorated PEG films to specific target, this film was exposed to a mismatching ssDNA sequence (T5) and examined by XPS. As shown in Figure 4, the C 1s, N 1s and P 2p XP spectra of the film taken before and after such an exposure are identical (within the experimental error), which indicates that the hybridization is indeed highly selective.

The data for the immobilization of NHS-T10 into the 2:1 PEG films using NHS-T10 and the related hybridization tests with the matching (A10) and mismatching (T10) ssDNA sequence are presented in Figure 5. The same behavior as in the case of NHS-T5, A5, and T5 is observed (Figure 4). However, changes in the XP spectra upon the immobilization of the probe strands and their hybridization with the matching target are even more pronounced, which is understandable in view of the longer ssDNA chain and, subsequently, a larger spectral weight of the respective fingerprint features. Based on the decomposition of the N 1s spectra, the hybridization efficiency was estimated as ~89%, which is even somewhat higher than that for the shorter T5/A5 strands, driven, most likely, by a larger energy gain. The ssDNA-backbone-representative P 2p spectra, which show an intensity increase by a factor of ~1.88 upon the specific hybridization (T10-A10), give nearly the same value of the hybridization efficiency, supporting the reliability of the derived value. In contrast, similar to the T5/A5 case, no changes in the XP spectra were observed after the exposure of the T10-decorated PEG films to a mismatching sequence (T10).

The XPS data for the T5/A5 and T10/A10 series can also be compared to each other. In particular, both for the T10-decorated PEG films and the films subjected to the specific hybridization, the intensity of the P 2p signal is approximately double with respect to that in the T5/A5 case. This relation suggests a similar amount of the immobilized ssDNA species in the T10/A10 and T5/A5 cases. Such a behavior indicates that the ssDNA immobilization ability of the 2:1 PEG film does not depend strongly on the length of ssDNA strands but is predominantly determined by the amount of free amine groups. A tentative evaluation of the areal densities of the immobilized T5 and T10 probe strands, performed on the basis of the N 1s XP spectra and the nitrile-terminated SAM as a reference (see Section 2.4 for the technical details), gives the areal densities of 3.6×10^{12} strands/cm^2 and

2.7×10^{12} strands/cm^2 for the PEG/NHS-T5 and PEG/NHS-T10 assemblies, respectively. Note, however, that both these values represent coarse estimates only and are most likely somewhat higher in reality since the N 1s photoemission signal from the quasi-bulk PEG-ssDNA samples is diminished by self-attenuation, in contrast to the signal from the terminal nitrogen atoms of the reference SAM, which is not affected by the attenuation at all.

The somewhat higher areal density for the PEG/NHS-T5 assembly compared to the PEG/NHS-T10 case is most likely related to a better permeability of the shorted NHS-T5 moieties in the PEG matrix. Nevertheless, the permeability is obviously still good enough for the NHS-T10 species, but can probably become a problem for noticeably longer ssDNA strands. Based on the length of the precursor arms (3.5–4 nm), a 3D PEG mesh with a characteristic pore size of 7–8 nm can be expected. This size is of course larger than the cross-sectional dimeter of ssDNA (~2 nm) but is, even for a short strand, much smaller than the ssDNA length, determined by the effective persistence length (~2 nm [59]) and the number of bases.

A related aspect is the behavior of the C 1s XP spectra. As was mentioned above and seen in Figures 4 and 5, these spectra do not exhibit noticeable changes on the immobilization of the probe T5 and T10 strands into the PEG film, except probably a small decrease in intensity. This means that the signal of the PEG matrix, represented by a single peak at a BE of ~286.6 eV (see above), dominates over the signal of the ssDNA strands, overlapping partly with the PEG feature and represented by several peaks with specific intensity ratios and dominant spectral weight at a BE of 284.6–285.5 eV [45,60]. Consequently, and most likely, the immobilization of ssDNA does not involve the entire PEG film but, predominantly, the topmost part of it, occurring in a gradient fashion. Only after the specific hybridization, a small ssDNA-stemming shoulder at the low BE side of the PEG-related C 1s peak is observed.

The permeability of ssDNA in the PEG matrix was additionally studied by exposure of comparably thin (15 nm) 1:1 PEG films to unmodified homo-oligonucleotides, T10 and A10. As demonstrated above, both 1:1 and 2:1 PEG films are generally inert to these biomolecules, so that any traces of T10 and A10 found in the spectra will most likely represent the strands penetrated through the film and adsorbed at the film-substrate interface, driven by their affinity to the non-bioinert Si/SiO_2 substrate. Indeed, such traces could be found in the XP spectra of both PEG/T10 and PEG/A10 (Figure S4 in the Supporting Information), suggesting that the permeability depth of these strand into the PEG film is at least 15 nm. The affinity of the Si/SiO_2 substrates to the ssDNA was additionally verified by their exposure to T10 and A10. The resulting XP spectra in Figure S5 (Supporting Information) show a noticeable increase in the intensity of the C 1s signal and appearance of the N 1s signal, which both indicate the adsorption of T10 and A10 onto the substrate. The C 1s spectra of both adsorbed ssDNA strands represent a single peak at a BE of 285.7–285.8 eV, accompanied by a weak high energy shoulder. Such spectra should indeed overlap significantly with the C 1s spectrum of the original PEG film, so that detection of ssDNA immobilization and hybridization on the basis of the C 1s XP spectra is hardly possible.

3.2. Electrochemical Studies

The immobilization of the ssDNA into the PEG matrix and hybridization ability of the resulting assemblies were also monitored by electrochemical measurements, which were carried out for the 2:1 films only. These films were specifically fabricated on Au substrates serving as the working electrode in the electrochemical cell (see Section 2 for the technical details). Then, the recorded cyclic voltammograms (CVs) provided a measure of the electrochemical passivating ability of the PEG films (Au/PEG) and PEG/ssDNA assemblies (Au/PEG/ssDNA) toward the $Fe(CN)_6^{3-/4-}$ redox couples in the electrolyte solution. In contrast, the EIS analysis provided information on the charge transfer resistance (R_{ct}) of the electrochemical cell.

As the first step, electrochemical passivating ability and bioinertness of the PEG films were tested. The respective data are shown in Figure 6 and the numerical results of the electrochemical measurements are summarized in Table 1. According to the CVs (Figure 6a) and Table 1, the presence of a ~80 nm PEG film on the Au electrode results in just a moderate suppression of the redox current and in just 33% decrease of the electrochemical capacitance, which is proportional to the area encircled by the respective CV [61]. Such a moderate reduction is most likely related to the porous structure of this film, which (structure) follows directly from the structure of the film precursors and the architecture of the PEG films (see Figure 1) as well as from the swelling and permeability properties of these films [35,36]. The porous structure is favorable for the efficient diffusion of the $Fe(CN)_6^{3-/4-}$ species toward the Au electrode. Additionally, the R_{ct} value did not change much after the introduction of the PEG film, increasing from 25 Ω to 40 Ω (Table 1), as follows from the Nyquist plots for the Au and Au/PEG samples in Figure 6b. The diameters of the semicircles in the high frequency region of these plots correspond to the R_{ct} values of the samples.

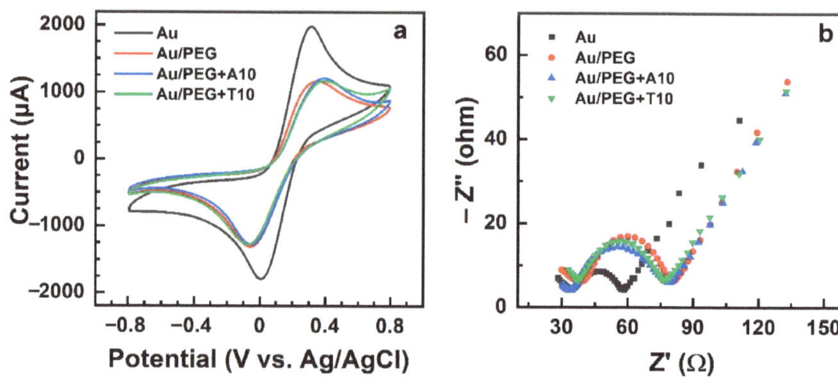

Figure 6. CVs (**a**) and Nyquist plots (**b**) for the blank Au electrode and Au/PEG electrode before and after its exposure to the unmodified ssDNA strands, A10 and T10.

Table 1. Capacitance with Respect to the Cell with the Blank Au Working Electrode and the Charge Transfer Resistance Associated with the Specific Samples.

Sample	Relative Capacitance	Charge Transfer Resistance
Au	100%	25 Ω
Au/PEG	67%	41 Ω
Au/PEG+A10	65%	40 Ω
Au/PEG+T10	65%	43 Ω
Au/PEG/NHS-T5	50%	95 Ω
Au/PEG/NHS-T5+T5	49%	97 Ω
Au/PEG/NHS-T5+A5	32%	195 Ω
Au/PEG/NHS-T10	34%	158 Ω
Au/PEG/NHS-T10+T10	33%	155 Ω
Au/PEG/NHS-T10+A10	19%	330 Ω

Exposure of the PEG films to the non-substituted ssDNA (A10 and T10) resulted in no obvious changes in their CVs (Figure 6a) and Nyquist plots (Figure 6b), with the nearly identical values of the relative capacitance and R_{ct} before and after exposure (Table 1). This behavior indicates the bioinert character of the 2:1 PEG matrix, in full agreement with the XPS data (see Section 3.1).

Subsequently, immobilization of NHS-T5 and NHS-T10 into the PEG matrix and the exposure of the resulting PEG-ssDNA films to the matching and non-matching target sequences was performed and the results were monitored by CV and EIS. The respective data are presented in Figures 7 and 8; the derived values of the relative capacitance and R_{ct} are compiled in Table 1. Let us first discuss the data for NHS-T5 and later for NHS-T10. After the exposure of the Au/PEG to NHS-T5, the redox currents in the electrochemical cell decreased (Figure 7a), indicating a higher resistance of the working electrode. This effect is even more obvious in the Nyquist plots (Figure 7b), which show a noticeable increase in the diameter of the semicircle corresponding to an increase in R_{ct} from 41 Ω to 95 Ω (Table 1). This increase manifests the immobilization of the probe T5 strands into the PEG matrix and is explained by the effect of the negatively charged phosphate groups of the ssDNA, which hinder $Fe(CN)_6{}^{3-/4-}$ from diffusing to the electrode surface [51,62].

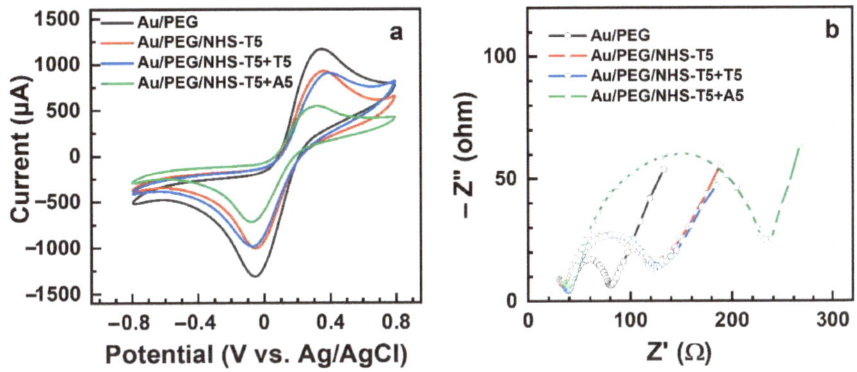

Figure 7. CVs (**a**) and Nyquist plots (**b**) for the Au/PEG working electrode and the Au/PEG/NHS-T5 electrode before and after its exposure to mismatching (T5) and matching (A5) target ssDNA.

The exposure of the Au/PEG/NHS-T5 to the matching sequence (A5) resulted in further reduction of the redox current (Figure 7a) and relative capacitance (Table 1) as well as in a noticeable increase in the diameter of the semicircle in the Nyquist plots (Figure 7b), corresponding to a significant increase in R_{ct} from 95 Ω to 195 Ω (Table 1). This behavior manifests the efficient hybridization of T5 and A5, in full agreement with the XPS data (see Section 3.1). In contrast, the exposure of Au/PEG/NHS-T5 to the non-matching sequence (T5) resulted in only minor change in the CV (Figure 7a) and nearly no change in the Nyquist plot (Figure 7b) and R_{ct} value (97 Ω, Table 1). This behavior indicates that the hybridization did not occur for the mismatched sequence, again - in full agreement with the XPS data.

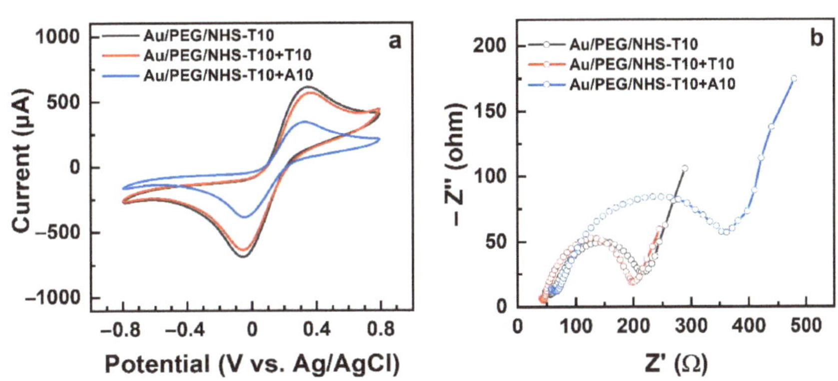

Figure 8. CVs (**a**) and Nyquist plots (**b**) for the Au/PEG/NHS-T10 working electrode before and after its exposure to mismatching (T10) and matching (A10) target ssDNA.

The electrochemical data for the immobilization of NHS-T10 into the PEG matrix and the subsequent exposure of the PEG/NHS-T10 probe to the mismatching (T10) and matching (A10) ssDNA sequences are shown in Figure 8. Both the CVs (Figure 8a) and the Nyquist plots (Figure 8b) exhibit the same behavior as the analogous data for the NHS-T5 case (Figure 7), which is also reflected by the relative capacitance and R_{ct} values in Table 1. In particular, the relative capacitance of Au/PEG decreased from 67% to 34% after the NHS-T10 exposure, while the R_{ct} value increased from 41 Ω to 158 Ω, manifesting the NHS-T10 immobilization in the PEG matrix. The exposure of the PEG/NHS-T10 probe to the matching sequence (A10) resulted in a further decrease of the relative capacitance from 34% to 19% and an increase of R_{ct} from 158 Ω to 330 Ω, manifesting a high hybridization efficiency. In contrast, no noticeable changes both in the experimental curves (Figure 8) and the derived fingerprint values (Table 1) were observed after the exposure of the PEG/NHS-T10 probe to the mismatching sequence (T10), manifesting thus a high selectivity of this probe.

Comparing the values for the NHS-T5 case with those for the NHS-T10 case in Table 1, viz. PEG/NHS-T5 vs PEG/NHS-T10 and PEG/NHS-T5+A5 vs PEG/NHS-T10+A10, we find that both the relative capacitance and R_{ct} do not reproduce exactly the factor of 2 describing the base number difference between T5/A5 and T10/A10. The observed relations can, on the one hand, be affected by the contributions from the PEG matrix and, on the other hand, reflect the somewhat different areal densities of the immobilized T5 and T10 moieties in the matrix.

In contrast, both in the NHS-T5 and NHS-T10 case, the values of R_{ct} increase by a factor close to 2 after the hybridization with the matching A5 and A10 sequences, which means that R_{ct} can be used as a tentative measure of hybridization efficiency. Generally, looking at the data in Figures 6–8, one can see that the Nyquist plots represent a much clearer and more distinct way to monitor the immobilization and hybridization processes in the PEG matrix than the CVs. Thus, EIS can be efficiently used as a transduction technique for these processes.

Finally, the sensitivity of this technique in the case of PEG/NHS-T10+A10 was tested. The concentration of A10 was varied from 10 μM (the standard value in this study; see Section 2.3) to 0.1 μM. The respective Nyquist plots are presented in Figure 9a and the derived values of R_{ct} are shown in Figure 9b. Accordingly and as expected, the R_{ct} value decreases progressively with the decreasing A10 concentration. This value is still noticeably higher than the reference value for PEG/NHS-T10 at 0.2 μM and nearly equal to the reference value at 0.1 μM. Consequently, the sensitivity of PEG/NHS-T10 to A10 is down to 0.1–0.2 μM, which can be probably improved even further by increasing the porosity of the PEG matrix and the areal density of the primary T10 probes.

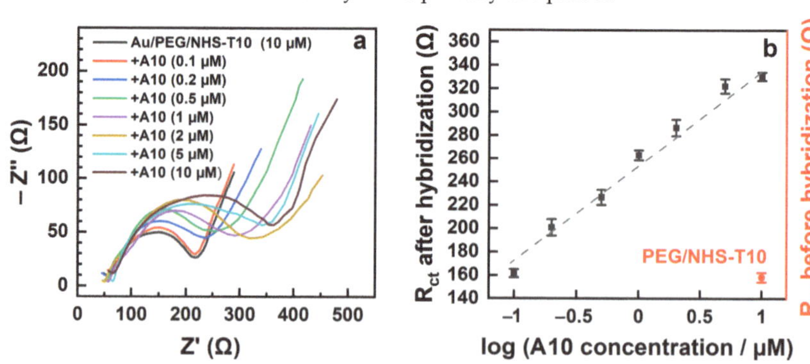

Figure 9. Nyquist plots (**a**) and the derived R_{ct} values (**b**) for the Au/PEG/NHS-T10 working electrode before (red symbol in **b**) and after (black symbols in b) its exposure to matching A10 target ssDNA. The concentration of A10 was varied. The R_{ct} behavior in (**b**) is tentatively traced by straight dashed line.

4. Conclusions

In the present work, we fabricated a series of bioinert and porous PEG films, comprised of the cross-linked, amine-/epoxy-terminated STAR-PEGs, to explore their application as a platform for immobilization of probe ssDNA strands, capable of hybridization with complementary target sequences. As test ssDNA compounds, non-substituted and NHS-ester-substituted thymine and adenine homo-oligonucleotides were used, aiming for the immobilization of NHS-ester-modified strands by the reaction of the NHS ester groups with the non-reacted amine moieties in the PEG films. The immobilization and hybridization processes were monitored by XPS, relying on the ssDNA-specific P 2p and base-specific N 1s signals, as well as on electrochemical techniques, viz. cyclic voltammetry and electrochemical impedance spectroscopy. The results of the XPS and electrochemistry experiments agree completely with each other, underlying the reliability of the results.

It is demonstrated that the standard PEG films with a crosslinking-optimal precursor mixing ratio (STAR-NH_2/STAR-EPX = 1:1) are not suitable for the NHS-ester-driven ssDNA immobilization, because of a very low concentration of the free (non-reacted) amine groups, used as coupling moieties for the NHS-ester-substituted ssDNA. To increase the amount of these groups, PEG films with the excess of the STAR-NH_2 precursors were prepared (STAR-NH_2/STAR-EPX = 2:1). These films showed the same complete inertness toward non-substituted ssDNA as the 1:1 layers but featured a much higher reactivity with respect to NHS-substituted ssDNA, allowing immobilizing probe ssDNA with a reasonable density. The films decorated with the probe ssDNA exhibited a high hybridization efficiency with respect to the matching target strands (78–89%), staying, at the same time, fully inert with respect to the non-matching ones. The efficiency was found to be somewhat higher for the longer strands compared to the shorter, presumably, because of a larger energy gain upon the hybridization. The most likely reason for the high hybridization efficiency is the 3D character of the probe ssDNA immobilization and sufficient separation of individual ssDNA probes in the PEG matrix, allowing their efficient accessibility by the target strands.

Both the immobilization and hybridization processes occurred predominately in the topmost part of the PEG films, which had a thickness of ca. 80 nm in most of the experiments. A significant reduction of this thickness should be avoided, since it results in the penetration of non-specific ssDNA strands to the substrate and their adhesion onto it, diminishing the bioinertness and specificity of the system.

Whereas the monitoring of ssDNA hybridization by XPS requires a cost-intensive and complex equipment, electrochemical impedance spectroscopy, relying on a comparably simple and non-expensive setup, can be readily used as a label-free transduction technique in this context. This technique is also favorable compared to the standard experimental tools used in the field, as fluorescence spectroscopy (FS) and surface plasmon resonance (SPR), which either require DNA labeling (FS) or quite expensive equipment (SPR). In our case, the probe ssDNA-decorated PEG films can both be directly prepared on suitable electrodes and transferred onto them using the established film separation and transfer procedures [36,38,39,44]. The porous character of these films is of advantage for efficient diffusion of redox species in electrochemical cell, enabling reliable and to a certain extent even quantitative monitoring of hybridization. The ultimate sensitivity of the approach is reasonable and can be further improved at its practical implementation and optimization. Additionally, theoretical simulations can be helpful, since they provide a deeper insight in the mechanisms of surface reactions, also in relation to DNA [63–66].

Supplementary Materials: The following supporting information can be downloaded at: https://www.mdpi.com/article/10.3390/bioengineering9090414/s1, Figure S1: Atomic force microscopy images of the working electrode, Figure S2: XP spectra of 1:1 PEG films at their incubation into the T5, A5, T10 and A10 solutions, Figure S3: XP spectra of 1:1 PEG films at their incubation into the T5 and T10 solutions and further incubation into matching target ssDNA solutions, Figure S4: XP spectra of the ultrathin 1:1 PEG film (15 nm) at their incubation into the A10 and T10 solutions, Figure S5: XP spectra of the Si/SiO_2 substrate at their incubation into the A10 and T10 solutions.

Author Contributions: Conceptualization and methodology, all authors; sample preparation, Z.Z.; XPS experiments and data evaluation, Z.Z.; electrochemistry experiments and data evaluation, Z.Z. and S.D.; data analysis, all authors; writing, M.Z. with an assistance of Z.Z. and S.D.; project administration, M.Z. All authors have read and agreed to the published version of the manuscript.

Funding: This research was funded by the own funds of the group. Z.Z. thanks the China Scholarship Council (CSC) for financial support.

Institutional Review Board Statement: Not applicable.

Informed Consent Statement: Not applicable.

Data Availability Statement: Data presented is contained within this paper. Additional data can be provided by the authors following a request.

Conflicts of Interest: The authors declare no conflict of interest.

References

1. Rao, A.N.; Grainger, D.W. Biophysical properties of nucleic acids at surfaces relevant to microarray performance. *Biomater. Sci.* **2014**, *2*, 436–447. [CrossRef] [PubMed]
2. Zhang, X.; Hu, H. DNA Molecules site-specific immobilization and their applications. *Cent. Eur. J. Chem.* **2014**, *12*, 977–993. [CrossRef]
3. Du, Y.; Dong, S. Nucleic acid biosensors: Recent advances and perspectives. *Anal. Chem.* **2017**, *89*, 189–215. [CrossRef] [PubMed]
4. Herne, T.M.; Tarlov, M.J. Characterization of DNA probes immobilized on gold surfaces. *J. Am. Chem. Soc.* **1997**, *119*, 8916–8920. [CrossRef]
5. Ballav, N.; Koelsch, P.; Zharnikov, M. Orientation and ordering in monomolecular films of sulfur-modified homo-oligonucleotides on gold. *J. Phys. Chem. C* **2009**, *113*, 18312–18320. [CrossRef]
6. Schreiner, S.M.; Hatch, A.L.; Shudy, D.F.; Howard, D.R.; Howell, C.; Zhao, J.; Koelsch, P.; Zharnikov, M.; Petrovykh, D.Y.; Opdahl, A. Impact of DNA-surface interactions on the stability of DNA hybrids. *Anal. Chem.* **2011**, *83*, 4288–4295. [CrossRef]
7. Leung, K.K.; Martens, I.; Yu, H.Z.; Bizzotto, D. Measuring and controlling the local environment of surface-bound DNA in self-assembled monolayers on gold when prepared using potential-assisted deposition. *Langmuir* **2020**, *36*, 6837–6847. [CrossRef]
8. Umek, R.M.; Lin, S.W.; Vielmetter, J.; Terbrueggen, R.H.; Irvine, B.; Yu, C.J.; Kayyem, J.F.; Yowanto, H.; Blackburn, G.F.; Farkas, D.H.; et al. Electronic detection of nucleic acids. *J. Mol. Diagn.* **2001**, *3*, 74–84. [CrossRef]
9. Raymundo-Pereira, P.A.; Oliveira Pedro, R.; Carr, O.; Melendez, M.E.; Gobbi, A.L.; Oliveira Piazzetta, M.H.; Oliveira, O.N., Jr. Influence of the molecular orientation and ionization of self-assembled monolayers in biosensors: Application to genosensors of prostate cancer antigen 3. *J. Phys. Chem. C* **2020**, *125*, 498–506. [CrossRef]
10. Broderick, A.H.; Carter, M.C.D.; Lockett, M.R.; Smith, L.M.; Lynn, D.M. Fabrication of oligonucleotide and protein arrays on rigid and flexible substrates coated with reactive polymer multilayers. *ACS Appl. Mater. Interfaces* **2013**, *5*, 351–359. [CrossRef]
11. Kargl, R.; Vorraber, V.; Ribitsch, V.; Köstler, S.; Stana-Kleinschek, K.; Mohan, T. Selective immobilization and detection of DNA on biopolymer supports for the design of microarrays. *Biosens. Bioelectron.* **2015**, *68*, 437–441. [CrossRef] [PubMed]
12. Cattani-Scholz, A.; Pedone, D.; Blobner, F.; Abstreiter, G.; Schwartz, J.; Tornow, M.; Andruzzi, L. PNA-PEG modified silicon platforms as functional bio-interfaces for applications in DNA microarrays and biosensors. *Biomacromolecules* **2009**, *10*, 489–496. [CrossRef] [PubMed]
13. Singh, A.; Sinsinbar, G.; Choudhary, M.; Kumar, V.; Pasricha, R.; Verma, H.N.; Arora, K. Graphene oxide-chitosan nanocomposite based electrochemical DNA biosensor for detection of typhoid. *Sens. Actuators B* **2013**, *185*, 675–684. [CrossRef]
14. Park, S.R.; Hauver, J.; Zhang, Y.; Revyakin, A.; Coleman, R.A.; Tjian, R.; Chu, S.; Pertsinidis, A. A Single-molecule surface-based platform to detect the assembly and function of the human RNA polymerase II transcription machinery. *Structure* **2020**, *28*, 1337–1343. [CrossRef]
15. Jeon, S.I.; Lee, J.H.; Andrade, J.D.; De Gennes, P.G. Protein-surface interactions in the presence of polyethylene oxide: I. Simplified theory. *J. Colloid Interface Sci.* **1991**, *142*, 149–158. [CrossRef]
16. Polyethylene Glycol Chemistry: Biotechnical and Biomedical Applications. Harris, J.M. (Ed.) Plenum Press: New York, NY, USA, 1992; pp. 1–406.
17. Holmberg, K.; Bergström, K.; Brink, C.; Österberg, E.; Tiberg, F.; Harris, J.M. Effects on protein adsorption, bacterial adhesion and contact angle of grafting PEG chains to polystyrene. *J. Adhes. Sci. Technol.* **1993**, *7*, 503–517. [CrossRef]
18. Lowe, S.; O'Brien-Simpson, N.M.; Connal, L.A. Antibiofouling polymer interfaces: Poly(ethylene glycol) and other promising candidates. *Polym. Chem.* **2015**, *6*, 198–212. [CrossRef]
19. Wang, F.; Zhang, H.; Yu, B.; Wang, S.; Shen, Y.; Cong, H. Review of the research on anti-protein fouling coatings materials. *Prog. Org. Coat.* **2020**, *147*, 105860. [CrossRef]
20. Prime, K.L.; Whitesides, G.M. Self-assembled organic monolayers: Model systems for studying adsorption of proteins at surfaces. *Science* **1991**, *252*, 1164–1167. [CrossRef]

21. Harder, P.; Grunze, M.; Dahint, R.; Whitesides, G.M.; Laibinis, P.E. Molecular conformation in oligo(ethylene glycol)-terminated self-assembled monolayers on gold and silver surfaces determines their ability to resist protein adsorption. *J. Phys. Chem. B* **1998**, *102*, 426–436. [CrossRef]
22. Herrwerth, S.; Eck, W.; Reinhardt, S.; Grunze, M. Factors that determine the protein resistance of oligoether self-assembled monolayers —internal hydrophilicity, terminal hydrophilicity, and lateral packing density. *J. Am. Chem. Soc.* **2003**, *125*, 9359–9366. [CrossRef]
23. Schilp, S.; Rosenhahn, A.; Pettitt, M.E.; Bowen, J.; Callow, M.E.; Callow, J.A.; Grunze, M. Physicochemical properties of (ethylene glycol)-containing self-assembled monolayers relevant for protein and algal cell resistance. *Langmuir* **2009**, *25*, 10077–10082. [CrossRef]
24. Rosenhahn, A.; Schilp, S.; Kreuzer, H.J.; Grunze, M. The Role of "inert" surface chemistry in marine biofouling prevention. *Phys. Chem. Chem. Phys.* **2010**, *12*, 4275–4286. [CrossRef]
25. Khan, M.N.; Tjong, V.; Chilkoti, A.; Zharnikov, M. Fabrication of ssDNA/oligo(ethylene glycol) monolayers and complex nanostructures by an irradiation-promoted exchange reaction. *Angew. Chem. Int. Ed.* **2012**, *51*, 10303–10306. [CrossRef]
26. Hu, W.-P.; Huang, L.-Y.; Kuo, T.-C.; Hu, W.-W.; Chang, Y.; Chen, C.-S.; Chen, H.-C.; Chen, W.-Y. Optimization of DNA-directed immobilization on mixed oligo(ethylene glycol) monolayers for immunodetection. *Anal. Biochem.* **2012**, *423*, 26–35. [CrossRef]
27. Nagasaki, Y. Construction of a densely poly(ethylene glycol)-chain-tethered surface and its performance. *Polymer J.* **2011**, *43*, 949–958. [CrossRef]
28. Jia, F.; Lu, X.; Tan, X.; Wang, D.; Cao, X.; Zhang, K. Effect of PEG architecture on the hybridization thermodynamics and protein accessibility of PEGylated oligonucleotides. *Angew. Chem. Int. Ed.* **2017**, *56*, 1239–1243. [CrossRef]
29. Lu, X.; Zhang, K. PEGylation of therapeutic oligonucletides: From linear to highly branched PEG architectures. *Nano Res.* **2018**, *11*, 5519–5534. [CrossRef]
30. Wang, Y.; Wang, D.; Jia, F.; Miller, A.; Tan, X.; Chen, P.; Zhang, L.; Lu, H.; Fang, Y.; Kang, X.; et al. Self-assembled DNA–PEG bottlebrushes enhance antisense activity and pharmacokinetics of oligonucleotides. *ACS Appl. Mater. Interfaces* **2020**, *12*, 45830–45837. [CrossRef]
31. Kivlehan, F.; Paolucci, M.; Brennan, D.; Ragoussis, I.; Galvin, P. Three-dimensional hydrogel structures as optical sensor arrays, for the detection of specific DNA sequences. *Anal. Biochem.* **2012**, *421*, 1–8. [CrossRef]
32. Wang, K.W.; Betancourt, T.; Hall, C.K. Computational study of DNA-cross-linked hydrogel formation for drug delivery applications. *Macromolecules* **2018**, *51*, 9758–9768. [CrossRef]
33. Xu, Y.; Wang, H.; Luan, C.; Fu, F.; Chen, B.; Liu, H.; Zhao, Y. Porous hydrogel encapsulated photonic barcodes for multiplex microRNA quantification. *Adv. Funct. Mater.* **2018**, *28*, 1704458. [CrossRef]
34. Qi, Y.; Li, K.; Zhao, C.; Ma, Y.; Yang, W. Preparation of a poly (PEGDA-co-GMA) thin hydrogel matrix for oligonucleotide microarray applications. *J. Chem. Technol. Biotechnol.* **2021**, *96*, 1902–1908. [CrossRef]
35. Meyerbröker, N.; Kriesche, T.; Zharnikov, M. Novel ultrathin polyethylene glycol films as flexible platform for biological applications and plasmonics. *ACS Appl. Mater. Interfaces* **2013**, *5*, 2641–2649. [CrossRef]
36. Zhao, Z.; Das, S.; Zharnikov, M. Tuning the properties of polyethylene glycol films and membranes by molecular weight of the precursors. *ACS Appl. Polymer Mater.* **2021**, *4*, 645–653. [CrossRef]
37. Meyerbröker, N.; Zharnikov, M. Modification and patterning of nanometer-thin poly(ethylene glycol) films by electron irradiation. *ACS Appl. Mater. Interfaces* **2013**, *5*, 5129–5138. [CrossRef] [PubMed]
38. Meyerbröker, N.; Zharnikov, M. Ultraflexible, Freestanding nanomembranes based on poly(ethylene glycol). *Adv. Mater.* **2014**, *26*, 3328–3332. [CrossRef]
39. Meyerbröker, N.; Zharnikov, M. Hydrogel nanomembranes as templates for patterned deposition of nanoparticles on arbitrary substrates. *ACS Appl. Mater. Interfaces* **2014**, *6*, 14729–14735. [CrossRef]
40. Khan, M.; Schuster, S.; Zharnikov, M. Giant electrical response of pristine and nanoparticle-loaded hydrogel nanomembranes to humidity variation. *J. Phys. Chem. C* **2015**, *119*, 14427–14433. [CrossRef]
41. Khan, M.; Schuster, S.; Zharnikov, M. Chemical derivatization and biofunctionalization of hydrogel nanomembranes for potential biomedical and biosensor applications. *Phys. Chem. Chem. Phys.* **2016**, *18*, 12035–12042. [CrossRef]
42. Zhao, Z.; Yan, R.; Zharnikov, M. The effect of ultraviolet light on biorepulsive hydrogel poly(ethylene glycol) films. *ACS Appl. Polymer Mater.* **2021**, *3*, 3446–3454. [CrossRef]
43. Mädler, S.; Bich, C.; Touboul, D.; Zenobi, R. Chemical cross-linking with NHS esters: A systematic study on amino acid reactivities. *J. Mass Spectrom.* **2009**, *44*, 694–706. [CrossRef] [PubMed]
44. Zhao, Z.; Zharnikov, M. Elastic properties of poly(ethylene glycol) nanomembranes and respective implications. *Membranes* **2022**, *12*, 509. [CrossRef]
45. Lee, C.-Y.; Gong, P.; Harbers, G.M.; Grainger, D.W.; Castner, D.G.; Gamble, L. Surface coverage and structure of mixed DNA/alkylthiol monolayers on gold: Characterization by XPS, NEXAFS, and fluorescence intensity measurements. *J. Anal. Chem.* **2006**, *78*, 3316–3325. [CrossRef]
46. Lee, C.-Y.; Nguyen, P.-C.T.; Grainger, D.W.; Gamble, L.J.; Castner, D.G. Structure and DNA hybridization properties of mixed nucleic acid/maleimide–ethylene glycol monolayers. *Anal. Chem.* **2007**, *79*, 4390–4400. [CrossRef]

47. Schreiner, S.M.; Shudy, D.F.; Hatch, A.L.; Opdahl, A.; Whitman, L.J.; Petrovykh, D.Y. Controlled and efficient hybridization achieved with DNA probes immobilized solely through preferential DNA-substrate interactions. *Anal. Chem.* **2010**, *82*, 2803–2810. [CrossRef]
48. Howell, C.; Zhao, J.; Koelsch, P.; Zharnikov, M. Hybridization in ssDNA films—a multi-technique spectroscopy study. *Phys. Chem. Chem. Phys.* **2011**, *13*, 15512–15522. [CrossRef]
49. Moulder, J.F.; Stickle, W.E.; Sobol, P.E.; Bomben, K.D. Handbook of X-ray Photoelectron Spectroscopy. Chastian, J., Ed.; Perkin-Elmer-Corporation: Eden Prairie, MN, USA, 1992; pp. 184–185.
50. Ossowski, J.; Wächter, T.; Silies, L.; Kind, M.; Noworolska, A.; Blobner, F.; Gnatek, D.; Rysz, J.; Bolte, M.; Feulner, P.; et al. Thiolate versus selenolate: Structure, stability and charge transfer properties. *ACS Nano* **2015**, *9*, 4508–4526. [CrossRef]
51. Mohammadian, N.; Farnoush, F. ALS Genosensing using DNA-hybridization electrochemical biosensor based on label-free immobilization of ssDNA on Sm_2O_3 NPs-rGO/PANI composite. *Sensor. Actuat. B Chem.* **2018**, *275*, 432–438. [CrossRef]
52. Saprigin, A.; Thomas, C.; Dulcey, C.; Patterson, C.; Spector, M. Spectroscopic quantification of covalently immobilized oligonucleotides. *Surf. Interface Anal.* **2004**, *36*, 24–32. [CrossRef]
53. Petrovykh, D.Y.; Kimura-Suda, H.; Tarlov, M.J.; Whitman, L.J. Quantitative characterization of DNA films by X-ray photoelectron spectroscopy. *Langmuir* **2004**, *20*, 429–440. [CrossRef]
54. Rei Vilar, M.; Botelho do Rego, A.M.; Ferraria, A.M.; Jugnet, Y.; Nogues, C.; Peled, D.; Naaman, R.J. Interaction of self-assembled monolayers of DNA with electrons: HREELS and XPS studies. *J. Phys. Chem. B* **2008**, *112*, 6957–6964. [CrossRef]
55. Petrovykh, D.Y.; Perez-Dieste, V.; Opdahl, A.; Kimura-Suda, H.; Sullivan, J.M.; Tarlov, M.J.; Himpsel, F.J.; Whitman, L. Nucleobase orientation and ordering in films of single-stranded DNA on gold. *J. Am. Chem. Soc.* **2006**, *128*, 2–3. [CrossRef]
56. Opdahl, A.; Petrovykh, D.Y.; Kimura-Suda, H.; Tarlov, M.J.; Whitman, L.J. Independent control of grafting density and conformation of single-stranded DNA brushes. *Proc. Natl. Acad. Sci. USA.* **2007**, *104*, 9–14. [CrossRef]
57. Howell, C.; Jeyachandran, Y.L.; Koelsch, P.; Zharnikov, M. Orientation and Ordering in Sequence- and Length-Mismatched Surface-Bound DNA Hybrids. *J. Phys. Chem. C* **2012**, *116*, 11133–11140. [CrossRef]
58. Howell, C.; Hamoudi, H.; Zharnikov, M. Thymine/adenine diblock-oligonucleotide monolayers and hybrid brushes on gold: A spectroscopic study. *Biointerphases* **2013**, *8*, 6. [CrossRef]
59. Roth, E.; Glick Azaria, A.; Girshevitz, O.; Bitler, A.; Garini, Y. Measuring the conformation and persistence length of single-stranded DNA using a DNA origami structure. *Nano Lett.* **2018**, *18*, 6703–6709. [CrossRef]
60. Petrovykh, D.Y.; Kimura-Suda, H.; Whitman, L.J.; Tarlov, M.J. Quantitative analysis and characterization of DNA immobilized on gold. *J. Am. Chem. Soc.* **2003**, *125*, 5219–5226. [CrossRef]
61. Zhang, S.; Pan, N. Supercapacitors performance evaluation. *Adv. En. Mater.* **2015**, *5*, 1401401. [CrossRef]
62. Liu, R.; Pan, S.; Liu, M.; Huang, W.; Lv, Z.; He, A. A Label-free electrochemical biosensor with magnetically induced self-assembly for the detection of CYP2C9* 3 gene. *Appl. Surf. Sci.* **2021**, *537*, 147868. [CrossRef]
63. Irving, D.; Gong, P.; Levicky, R. DNA surface hybridization: Comparison of theory and experiment. *J. Phys. Chem. B* **2010**, *114*, 7631–7640. [CrossRef] [PubMed]
64. Wong, I.Y.; Melosh, N.A. An electrostatic model for DNA surface hybridization. *Biophys. J.* **2010**, *98*, 2954–2963. [CrossRef] [PubMed]
65. Schmitt, T.J.; Knotts IV, T.A. Thermodynamics of DNA hybridization on surfaces. *J. Chem. Phys.* **2011**, *134*, 205105. [CrossRef] [PubMed]
66. Freitas, R.R.Q.; Rivelino, R.; de Mota, F.B.; Gueorguiev, G.K.; de Castilho, C.M.C. Energy barrier reduction for the double proton-transfer reaction in guanine-cytosine DNA base pair on a gold surface. *J. Phys. Chem. C* **2015**, *119*, 15735–15741. [CrossRef]

bioengineering

MDPI

Article

Development and Characterization of Zein/Ag-Sr Doped Mesoporous Bioactive Glass Nanoparticles Coatings for Biomedical Applications

Syeda Ammara Batool [1], Ushna Liaquat [1], Iftikhar Ahmad Channa [2], Sadaf Jamal Gilani [3], Muhammad Atif Makhdoom [4,*], Muhammad Yasir [1], Jaweria Ashfaq [2], May Nasser bin Jumah [5,6,7] and Muhammad Atiq ur Rehman [1,*]

1 Department of Materials Science and Engineering, Institute of Space Technology Islamabad, Islamabad 44000, Pakistan
2 Thin Film Laboratory, Department of Metallurgical Engineering, NED University of Engineering and Technology, Off University Road, Karachi 75270, Pakistan
3 Department of Basic Health Sciences, Preparatory Year, Princess Nourah Bint Abdulrahman University, Riyadh 11671, Saudi Arabia
4 Institute of Metallurgy and Materials Engineering, University of the Punjab, Lahore 54590, Pakistan
5 Biology Department, College of Science, Princess Nourah Bint Abdulrahman University, Riyadh 11671, Saudi Arabia
6 Environment and Biomaterial Unit, Health Sciences Research Center, Princess Nourah Bint Abdulrahman University, Riyadh 11671, Saudi Arabia
7 Saudi Society for Applied Science, Princess Nourah Bint Abdulrahman University, Riyadh 11671, Saudi Arabia
* Correspondence: atif.imme@pu.edu.pk (M.A.M.); atique1.1@hotmail.com (M.A.u.R.)

Citation: Batool, S.A.; Liaquat, U.; Channa, I.A.; Gilani, S.J.; Makhdoom, M.A.; Yasir, M.; Ashfaq, J.; Jumah, M.N.b.; Rehman, M.A.u. Development and Characterization of Zein/Ag-Sr Doped Mesoporous Bioactive Glass Nanoparticles Coatings for Biomedical Applications. *Bioengineering* **2022**, *9*, 367. https://doi.org/10.3390/bioengineering9080367

Academic Editor: Teen-Hang Meen

Received: 30 June 2022
Accepted: 2 August 2022
Published: 4 August 2022

Abstract: Implants are used to replace damaged biological structures in human body. Although stainless steel (SS) is a well-known implant material, corrosion of SS implants leads to the release of toxic metallic ions, which produce harmful effects in human body. To prevent material degradation and its harmful repercussions, these implanted materials are subjected to biocompatible coatings. Polymeric coatings play a vital role in enhancing the mechanical and biological integrity of the implanted devices. Zein is a natural protein extracted from corn and is known to have good biocompatibility and biodegradability. In this study, zein/Ag-Sr doped mesoporous bioactive glass nanoparticles (Ag-Sr MBGNs) were deposited on SS substrates via electrophoretic deposition (EPD) at different parameters. Ag and Sr ions were added to impart antibacterial and osteogenic properties to the coatings, respectively. In order to examine the surface morphology of coatings, optical microscopy and scanning electron microscopy (SEM) were performed. To analyze mechanical strength, a pencil scratch test, bend test, and corrosion and wear tests were conducted on zein/Ag-Sr doped MBGN coatings. The results show good adhesion strength, wettability, corrosion, and wear resistance for zein/Ag-Sr doped MBGN coatings as compared to bare SS substrate. Thus, good mechanical and biological properties were observed for zein/Ag-Sr doped MBGN coatings. Results suggested these zein/Ag-Sr MBGNs coatings have great potential in bone regeneration applications.

Keywords: antibacterial; biomedical implants; electrophoretic deposition; osteogenesis

1. Introduction

Biomaterials are materials that can reside in a biological system to perform a certain function without inducing any toxic effect. They have a wide range of applications in bioengineering, pharmaceutics, biomedical implants, etc. The global biomedical implants market was reported to have a worth of ~USD 86 billion in 2019 and was anticipated to reach ~USD 147 billion by 2027, with 7.2% growth in the period 2020–2027. Growth in demographic aging and dire medical conditions influence recent technological progression in the bioimplants market [1]. The consistent growth of the implants market brought the interest of many researchers to this field. Metals, ceramics, polymers, and composites are potentially being used in the fabrication of different types of implants, e.g., cardiovascular, orthopedic, cochlear, dental, etc. Among all the suggested materials, metallic implants are broadly used in the fabrication of implant devices because, apart from being compatible in a biological system, they possess superior mechanical strength as compared to other materials [2]. Foreign bodies trigger the immune system to respond, which may result in fibrous encapsulation of a foreign body to isolate its interaction with the biological system [3]. For successful implantation, the implanted material should be familiar to the immune system so that the latter will respond positively.

Among all metallic implants, stainless steel (SS) implants have the highest proportion in the fabrication of orthopedic implants due to their cost effectiveness and accessibility. Poor corrosion resistance and relatively higher yield strength of SS implants in a physiological environment are the major drawbacks [4,5]. These implants, when interacting with a physiological environment, exhibit either corrosion or no response. Upon corrosion, SS implants release toxic metal ions (Ni and Cr ions) responsible for infections [6,7]. Corrosion impairs the function of implants by decreasing cell adhesion and mechanical strength, which ultimately leads to implant loosening [8]. The surface coating of SS implants with bio-composites can enhance corrosion resistance and bioactivity of the implants [4,9].

Biomaterial coatings on metallic implants improve the mechanical as well as biological properties of the implanted devices [10]. These coated materials can either be natural or synthetic as per the required properties of the final product. After implantation, the first component that interacts with the implant surface is water, followed by proteins. Thus, it is important to tune the surface of the implant to improve the initial protein attachment [11,12]. Zein is an alcohol soluble protein that largely contains prolamin. Zein was approved by the FDA in 1985 as a "generally recognized as safe" excipient. Zein is a natural polymer found in the endosperm cells of corn [13,14]. Zein contains hydrophobic, neutral amino acids (such as leucine, proline, and alanine), and some polar amino acid residues (glutamine). Zein outperforms other proteins because of the complete absence of lysine and tryptophan. The solubility of zein is restricted to acetone, acetic acid, aqueous alcohols, and aqueous alkaline solutions due to the amino acids [15]. Biopolymers are usually used with bioceramics due to their low mechanical strength, such as hydroxyapatite (HA) and bioactive glasses (BGs) [16]. BGs are third generation biomaterials widely used in bone tissue engineering applications. BGs can form strong bonds with the natural bone, thus they are considered as bioactive ceramics, which are usually used with other polymers [17–19]. Mesoporous bioactive glass nanoparticles (MBGNs) have small pores (2–7 nm) in them that increase the surface area of particles, which largely promotes its reactivity. MBGNs can be loaded with different drugs or metallic ions to facilitate bone tissue engineering, targeted drug delivery, and wound healing [20,21]. Tabia et al. [22] doped sol-gel derived MBGNs with magnesium and loaded them with amoxicillin. Results showed controlled drug release and high bioactivity due to the mesoporous structure and composition of BGs. Similarly, Ag and Sr ion doped MBGNs were prepared by [23] using a modified Stöber process. The presence of Ag ions imparts antibacterial properties, and Sr ions enhance the osteogenesis. An osteogenic material promotes bone regeneration and reduce bone resorption. Various coating techniques are being used to coat bioactive glasses for surface modification, such as thermal spraying [24], plasma spraying [25,26], radio frequency sputtering (RFS) [27,28], physical vapor deposition (PVD) [29], and electrophoretic deposition (EPD) [30,31]. EPD is well known for its applications in biomedical coatings [30]. EPD involves the movement and deposition of charged particles from the colloidal suspension in the presence of an electric field. EPD is a cost-effective process conducted at room temperature [32]. Numerous studies have been conducted on zein/bioactive glasses composite coatings deposited via EPD [32,33], but deposition of zein/MBGNs doped with metallic ions via EPD is a new approach to study the synergetic effect of dual ions. Evidently, we are the first to present a work on the development of zein/Ag-Sr doped MBGN coatings deposited on 316L SS via EPD.

In this study, we developed zein/Ag-Sr doped MBGN coatings on 316L SS substrates via EPD. After coatings were developed with designated parameters, coated substrates were subjected to material characterization. Optical microscopy and scanning electron microscopy (SEM) images verified the uniform deposition on 316L SS substrates. The coating showed antibacterial and bioactivity potential for biomedical applications.

2. Materials and Methods

2.1. Materials

Zein powder (CAS: 9010-66-6), absolute ethanol, and acetic acid were all purchased from Sigma-Aldrich® (Taufkirchen, Germany). AISI 316L stainless steel (addressed 316L SS in this article) foil of ~1 mm thickness was used to prepare substrates. The composition of 316L SS was Ni-14.15, Cr-17.75, Mo-2.72, Mn-1.87, Si-0.58, P-0.015, S-0.008, C-0.025, Fe-balance (wt.%). Ag-Sr doped MBGNs were synthesized via a modified Stöber process; the details are given in our previous work [23].

2.2. Suspension Preparation

Prior to the EPD process, a stable suspension of zein/Ag-Sr doped MBGNs was made by following the study conducted by Rivera et al. [33,34]. Zein powder (6 wt.%) was added in a 100 mL beaker, followed by the addition of distilled water (20 wt.%) and absolute ethanol (74 wt.%). The mixture was stirred for 30 min at 35–40 °C on a magnetic hot plate, followed by another 30 min round of stirring at room temperature. Acetic acid (~10 mL) was added dropwise to maintain the pH value (~3.0) of the solution. Subsequently, Ag-Sr doped MBGNs (3 g/L) were added to the zein solution, followed by 30 min of magnetic stirring. The suspension was finally ultrasonicated for approximately 60 min to produce a stable suspension of zein/Ag-Sr doped MBGNs required for the EPD process. The concentration of Ag-Sr doped MBGNs was chosen on the basis of initial studies, which showed that the concentrations higher than 3 g/L results in the non-uniform and thick coatings. If the concentration of Ag-Sr doped MBGNs is less than 3 g/L, it would not be enough to impart bioactivity and antibacterial activity.

The stability of the suspension was determined by measuring the zeta potential of the pure zein, Ag-Sr doped MBGNs, and zein/Ag-Sr doped MBGNs suspensions in ethanol. A zetasizer (Malvern Instruments, Malvern, UK) was utilized for this purpose. The measurement was taken in triplicate for each suspension, and the average value with standard deviation is reported here.

2.3. EPD Process

The 316L SS foil was cut into pieces of 30 × 25 mm^2 size. Subsequently, the pieces were cleaned in a mixture of acetone and absolute ethanol. It is important to note that no surface treatment was carried out on 316L SS foil. Substrates were rinsed with distilled water and dried at room temperature. The 316L SS was used as both working and counter electrodes at 10 mm inter-electrode spacing. Both electrodes were then submerged (half of total area) in 50 mL of initially prepared zein/Ag-Sr doped MBGN suspension. The coatings were developed by applying direct current (DC) on the substrates at different EPD parameters. EPD parameters were chosen on the basis of previous literature, which showed that lower deposition voltages (<10 V) led to the inhomogeneous coatings. In contrast to this, higher deposition voltages (>25 V) led to pronounced hydrolysis. The range 10–25 V of deposition voltages was investigated in the current study. The deposition time was kept fixed at 180 s on the basis of our previous studies [16].

2.4. Materials Characterization

The morphology and chemical composition of synthesized Ag-Sr doped MBGNs were examined via a scanning electron microscope (SEM-MIRA, TESCAN) equipped with energy dispersive spectroscopy (EDS).

Samples coated at different parameters were examined under an optical microscope (Novex) and SEM to study the morphology and thickness of the coatings. The samples were cut into $5 \times 5\ mm^2$ size prior to the examination.

A pencil scratch test (ASTM D3363-20) was conducted to measure the hardness of coatings. This test was performed manually by using graphite pencils of different hardness grade (8B to 2H). Scratches were made by the pencils held at an angle of 45° between the lead tip and the surface of the coated substrates. Starting from most hard (2H) to the softest grade (8B), pencils were pushed under uniform pressure to create a scratch throughout the sample width. The scratches were made until the pencil could not scratch the surface further. The lowest hardness grade of the pencil that is able to scratch the coating is considered the hardness of the coating. A bend test (ASTM-B571-97) was performed to check the adhesion/bending strength of the coatings. In order to perform the bend test, the coated surface was bent at 180° with the help of tweezers. After the bending test, the bending sites were examined to check for defects, i.e., cracks and delamination.

Contact angle measurements were done to determine surface wettability. A fixed volume (5 μL) of distilled water was dropped via microliter pipette on the surface of coated substrates. Digital images of the droplet were captured within 5 s of dropping, and the contact angles between the droplet and the surface of the coatings were measured by using Image J™. The test was performed at five different spots on the same coating, and the values of contact angles were averaged out. The mean values of contact angles were plotted with their standard deviations.

A tribometer (MT/60/NI, Spain) was used to conduct wear tests on the coated substrates. The ball-on-disk method was used, in which a steel ball indenter with a fixed load of 1 N was rotated for a distance of 50 m over the surface of the clamped substrates at 32 rpm. A graph between friction coefficient (COF; μ) and partial distance (m) was drawn to understand the wear behavior of the coating.

Corrosion tests were conducted using a three-electrode system Gamry Instrument (potentiostat reference 600, USA). A potentiodynamic polarization scan was recorded in simulated body fluid (SBF). The SBF was prepared according to the protocol reported in [35]. The composition of SBF is also noted in [35]. The coated sample was mounted as the working electrode with a graphite counter electrode. Ag-AgCl was used as the reference electrode. The test was conducted between the potential range of −0.5 V and +0.5 V at the scan rate of ~1 mV/s. The uncoated side of the substrate was covered with epoxy and dried prior to the test in order to make sure that corrosion results were obtained only from the selected surface area ($1\ cm^2$) of the coating. Before conducting this test, open circuit potential (OCP) was measured for an hour. A Tafel plot was constructed by extrapolating the cathodic and anodic branches. The corrosion potential (E_{corr}) and corrosion current density (I_{corr}) were measured for coated and uncoated samples.

2.5. Biological Characterization

The biological behavior of the deposited coatings was assessed via antibacterial and bioactivity analyses. Antibacterial tests were carried out against Gram-positive (*Staphylococcus aureus; S. aureus*) and Gram-negative (*Escherichia coli; E. coli*) bacterial strains. The coated samples were placed on agar plates containing 20 μL spread of each bacterial dilution. The samples were placed facing downwards to facilitate contact between coating and bacterial spread. The plates were then incubated at 37 °C. After 24 h, the inhibition zone on each plate was observed.

In vitro bioactivity test was carried out using SBF. Coated samples were placed in SBF (30 mL) and incubated in an orbital shaker at 37 °C for 1, 3, 5, and 7 days. The SBF was refreshed at alternate days to maintain the ionic concentration around the immersed samples. The coatings were removed at the designated time from the SBF, washed with de-ionized water, and air dried. The presence of an HA layer on the coated surface was confirmed via SEM/EDS analysis.

3. Results and Discussion

3.1. EPD Kinetics and Suspension Stability

Zein polymer is insoluble in water and organic solvents. It can be dissolved after inducing charge on its functional groups with the addition of acidic or basic medium. The polar and nonpolar groups present in zein are protonated in highly pure ethanol (<99%). The next step is to make a stable suspension of zein in ethanol. The stability of the suspension is analyzed using zeta potential measurements. The zeta potential of pure zein, Ag-Sr doped MBGNs, and zein/Ag-Sr doped MBGN suspensions were measured as +25 $\pm$ 10 mV, +16 $\pm$ 4 mV, and +22 $\pm$ 6 mV, respectively. According to the literature, polymer suspension should be sufficiently charged (positively or negatively) for effective electrophoretic mobility [36]. Particles with very low zeta potential tend to agglomerate and settle down due to strong attractive forces of interaction, whereas the particles with high values of zeta potential exhibit strong repulsive forces that inhibit the mobility of particles. In case of zein/Ag-Sr doped MBGN suspension, overall zein and Ag-Sr doped MBGNs are positively charged. Polymeric chains of zein contain some carboxylic side groups that are negatively charged, hence, the zein polymer wraps around the Ag-Sr doped MBGNs due to the attractive forces between oppositely charged side groups [37]. However, resultant zeta potential is positive, and zein molecules along with Ag-Sr doped MBGNs move towards the cathode and get deposited there following a mechanism termed charge stabilization [38], explained later in the kinetics of EPD.

Due to the addition of highly concentrated ethanol (>99%), the zein molecules get protonated and move towards the negatively charged electrode, hence cathodic deposition occurs. Hydrophilic groups of zein make bonds with Ag-Sr doped MBGNs and are deposited as a single entity. When the electric field is applied, water decomposes into H^+ and OH^- ions in the presence of ethanol. The H^+ ions towards the cathode due to their opposite charge and hydrogen gas evolves at the cathode, whereas the OH^- gets attached to the positively charged zein/Ag-Sr doped MBGNs in the suspension. The pH here increases due to the presence of OH^- ions. Overall, positively charged zein/Ag-Sr doped MBGNs get attracted towards the cathode, and, along with the OH^- ions, become unstable and get deposited on the cathode [39].

3.2. Morphology of Synthesized Ag-Sr Doped MBGNs and Coatings

Figure 1A shows the morphology of the as-synthesized Ag-Sr doped MBGNs. The image confirms that the particles were spherical in shape, and the average diameter was approximately 92 nm. The Ag and Sr were detected in the EDS of the particles, as shown in Figure 1B.

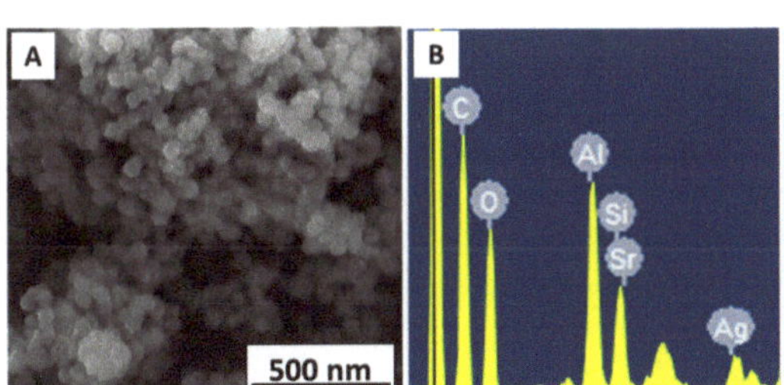

Element	Weight%	Atomic%
C K	51.25	61.27
O K	36.50	32.76
Al K	7.08	3.77
Si K	3.97	2.03
Sr L	0.41	0.07
Ag L	0.79	0.10
Totals	100.00	

Figure 1. (**A**) SEM image showing spherical morphology of Ag-Sr doped MBGNs, (**B**) EDS spectrum confirms the doping of Ag and Sr in the bioactive glass network.

Figure 2 shows the digital (a–g), optical (h–n), and SEM images (o–u) of zein/Ag-Sr doped MBGNs coatings developed on SS at different voltages. During the coating process, a slight change in the color of the coatings was observed at 18, 20, and 25 V (Figure 2e–g). Hydrolysis of suspension seems to be the reason for change in the color of coatings at higher voltages. Hydrolysis causes the chemical breakdown of materials by water. At higher voltages, hydrolysis increases, which causes the temperature of the suspension to rise. This increase in temperature may lead to the slight change in the color of coatings developed on SS substrates [40].

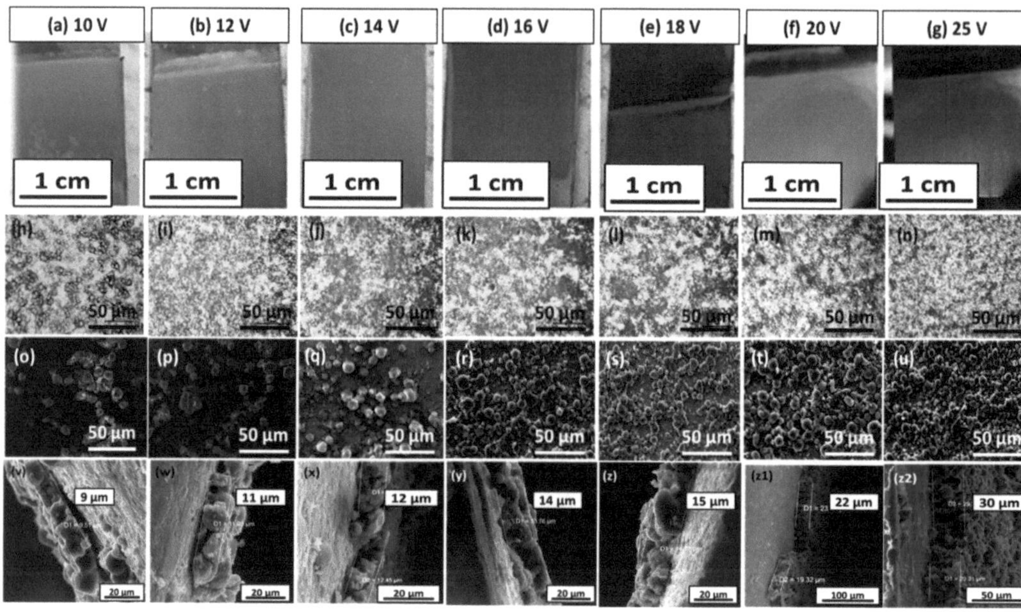

Figure 2. Digital images of zein/Ag-Sr doped MBGN coatings deposited on SS at (**a**) 10 V, (**b**) 12 V, (**c**) 14 V, (**d**) 16 V, (**e**) 18 V, (**f**) 20 V, and (**g**) 25 V. Optical and SEM images of coatings deposited at 10 V (**h**,**o**), 12 V (**i**,**p**), 14 V (**j**,**q**) and 16 V (**k**,**r**) showed non-uniform and less dense coating on SS, respectively. The coatings deposited at 18 V (**l**,**s**), 20 V (**m**,**t**), and 25 V (**n**,**u**) were more uniform and densely packed. The cross-sectional images show gradual increase in the thickness of coatings along with the voltage (**v**–**z2**).

Further morphological analysis of zein/Ag-Sr doped MBGN coatings was done via optical microscope and SEM. Figure 2h–n show the images of zein/Ag-Sr doped MBGN composite coatings deposited at different parameters taken from an optical microscope. Optical micrographs showed uniform dispersion of composite coatings on 316L SS at all the deposition voltages. However, the densest coating was obtained at 25 V/180 s (Figure 2n) due to high deposition rate at a higher value of applied electric field. A similar trend was observed in [16], where the increase in applied electric field led to the increase in deposition yield and density of the coatings. The results of the current study are in agreement with Hamaker's law [41].

Figure 2o–u show the SEM micrographs of zein/Ag-Sr doped MBGN coatings at different voltages. Figure 2o,p show that Ag-Sr doped MBGNs deposited at 10 V and 12 V were dispersed non-homogenously. A slight increase in voltage up to 14 V and 16 V (Figure 2q,r) shows large number of zein/Ag-Sr doped MBGNs deposited on SS with enhanced uniformity. At higher voltages of 18, 20, and 25 V, micrographs (Figure 2s–u) show the highly uniform and homogeneous dispersion of Ag-Sr doped MBGNs within the zein matrix, and the coatings were more densely packed. Large spherical agglomerates were formed during the electrophoretic deposition process [42]. The thickness of all coatings was measured from the cross section. It was observed that the thickness gradually increased along with the voltage. Figure 2v–z2 show the thickness of coatings deposited at various voltages. The coatings of 9, 11, 12, 14, 15, 22, and 30 μm thickness (average of three values) were obtained at 10, 12, 14, 16, 18, 20, and 25 V, respectively. As the applied voltage increased, more charged particles were forced to move towards the working electrode and deposit there.

3.3. Deposition Yield

Deposition yield of coatings was calculated by using following formula:

$$\text{Deposition yield} = (\Delta W)/A\ (mg/cm^2)$$

where ΔW is the change in the weight of substrates before and after coating, and 'A' represents the area of coated surfaces. Deposition yield was calculated for each parameter in triplicate, and the mean deposition yield (%) was plotted against applied voltages.

Figure 3 shows a consistent increase in the values of deposition yield (%) with the increase in voltage, which means that deposition yield is responsive towards the change in voltages. Thus, higher values of deposition yield were obtained at higher voltages (18, 20, and 25 V). The results of this study are in agreement with Hamaker's law [41].

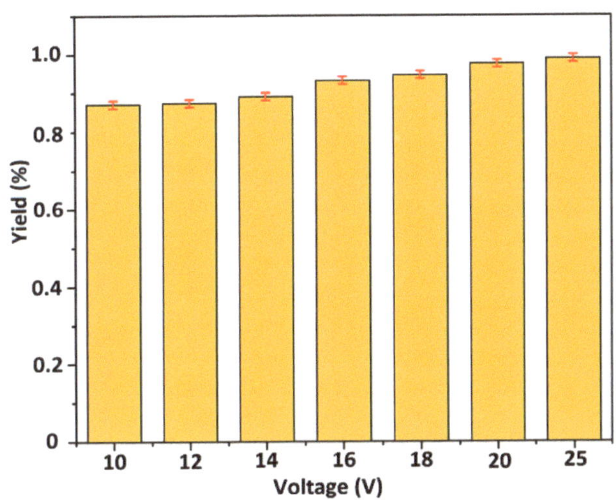

Figure 3. Deposition yield (%) graph with respect to the applied voltages. The highest deposition yield was achieved at higher voltage values, i.e., 18, 20, and 25 V, with slight standard deviations.

3.4. Adhesion Strength

The adhesion strength of zein/Ag-Sr doped MBGN coatings deposited on SS was measured via pencil scratch test and bend test. Scratches were made on the surface of coated substrates by graphite pencils of different hardness grades (8B–2H), starting from the hardest grade, as shown in Figure 4a–g. Results of pencil scratch hardness tests were then recorded, as shown in Table 1. It was observed that coatings developed at higher voltages had higher hardness compared to those developed at lower voltages. The coating deposited at 25 V was graded 'F' according to the ASTM standard, which showed adequate adhesion strength between coating and substrate [43].

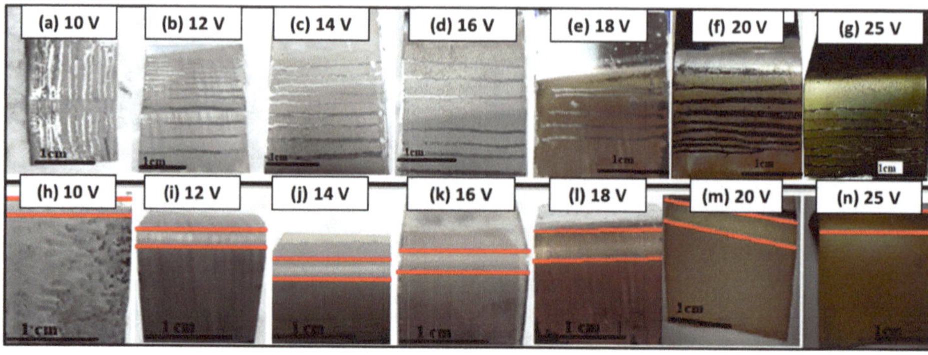

Figure 4. Pencil scratch test (**a–g**) and bend test results (**h–n**). Coatings deposited at 10 V and 12 V (**a**,**b**), detached even with softest pencil, i.e., grade 8B, coatings deposited at 14 V (**c**) detached at 5B, and coatings deposited at 16 V, 18 V (**d**,**e**), and 20 V (**f**) detached at 3B and 1B, respectively. The highest grade was achieved for the coating deposited at 25 V (**g**). The coatings deposited at 10 V (**h**), 12 V (**i**), and 14 V (**j**) showed delamination and micro-cracks around the bending site. Coatings deposited at 16 V (**k**) and 18 V (**l**) showed delamination around the edges of the bending site. The coatings deposited at 20 V (**m**) and 25 V (**n**) had no delamination or micro-cracks.

Table 1. Results of pencil scratch test.

Voltage (V)	Time (s)	Hardness Grade
10	180	8B
12	180	8B
14	180	5B
16	180	3B
18	180	3B
20	180	1B
25	180	F

Adhesion strength between coatings and substrates was further examined via bend tests. The tests were performed manually by bending coated substrates at 180° with tweezers. Figure 4h–n show the images of the coated substrates taken after bend test. Upon bending, coatings deposited at 10, 12, 14, 16, and 18 V (Figure 4h–l) showed minor delamination around the edges at the bend site. No prominent crack or delamination was observed for coatings deposited at 20 and 25 V (Figure 4m,n), showing good adhesion strength between zein/Ag-Sr doped MBGN coatings and SS substrates [37,43]. These coatings were graded '4B' according to the ASTM standard.

Morphological analysis of the coatings revealed that embedded Ag-Sr doped MBGNs in the zein matrix gain in homogeneity with the rise in deposition voltage, as shown in Figure 2o–u. Furthermore, the amount of Ag-Sr doped MBGNs within the coatings appear to rise proportionally with the increase in voltage, which could improve the mechanical strength of the coatings. Due to the increased amount of Ag-Sr doped MBGNs in the coatings, the adhesion strength of the coatings produced at higher voltages was higher than that of the coatings produced at lower voltages [44]. Moreover, the polymeric matrix (i.e., zein) may act as binder, holding the Ag-Sr doped MBGNs on the surface, giving a boost to the mechanical stability of the coatings.

3.5. Wettability Studies

Contact angle measurements were carried out to determine wettability of the coated substrates. Surface wettability of implanted materials is vital for successful implantation. When an implanted material enters into a physiological environment, it first interacts with body fluids, which further allow protein attachment. It is believed that cell adhesion and proliferation depend on the initial protein attachment [45]. If a surface allows initial protein attachment, it will subsequently allow cell adhesion and proliferation. Studies showed the surfaces with a contact angle in the range of 35–80° demonstrated good initial protein adsorption and osteoblast cell attachment [46,47].

Contact angle tests were performed on bare SS and zein/Ag-Sr doped MBGN coated substrates. The results are plotted between contact angle (○) and applied voltage (V) along with the standard deviation, as shown in Figure 5. The contact angle was measured five times for each type of coating. The bare SS showed an average contact angle of 58 ± 2°, which means that it is highly hydrophilic in nature. All the coated samples showed higher contact angle values as compared to the bare SS sample. Initially, a slight decrease in the values of contact angle was observed up until 16 V and, after that, values started to increase. Contact angle at 25 V was measured to be 72 ± 2°, which was the greatest among other coatings but perfectly within the suggested range for initial protein and osteoblast cell attachment.

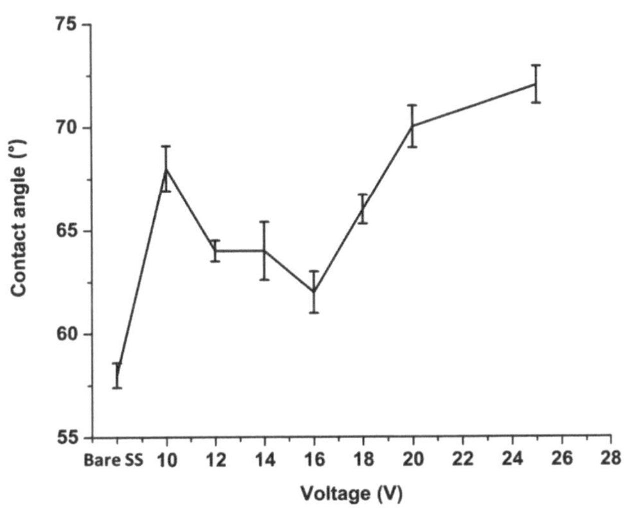

Figure 5. Contact angle graph for bare SS and coatings deposited at different voltages. Highest contact value was obtained for coating deposited at 25 V.

The coatings exhibited lower hydrophilicity as compared to the bare SS. This effect is attributed to the presence of zein molecules. Zein is mostly composed of hydrophobic proteins [15], hence, the wetting behavior of coating shift towards hydrophobicity. However, the contact angles are still in the hydrophilic domain due to the enhanced deposition of Ag-Sr doped MBGNs at higher voltages. The MBGNs are majorly composed of a silica network. Silanol (Si-OH) groups present on the surface of silica make it intrinsically hydrophilic [48]. As the amount of zein deposition increases in the coating along with the voltage, it retains the hydrophilic character due to the presence of MBGNs [49].

3.6. Wear Studies

As discussed previously, the highest deposition yield and best adhesion results were found for the coating deposited at 25 V for 180 s. Thus, wear studies were conducted only for the coatings deposited at 25 V by using the ball-on-disk method on a tribometer. A steel ball indenter was rotated over the surface of a coating under a constant load of 1 N for a partial distance of 50 m. A graph was plotted between COF (μ) and partial distance (m), as shown in Figure 6. The graph shows that COF remained almost constant initially, however, there was an abrupt increase in COF around 15 m of partial distance. This behavior may be attributed to the removal of the transfer layer between contacting surfaces [50]. A transfer layer is formed due to the accumulation of the wear debris around the wear track, leading to abrasive wear. Later, COF lowered and became constant again throughout the partial distance, which means that the coating deposited at higher voltage displayed sufficient resistance to wear [51]. The average wear rate of zein/Ag-Sr doped MBGN coatings was calculated to be 0.179 mm^3/Nm, which shows adequate wear resistance for a biocompatible implant [52].

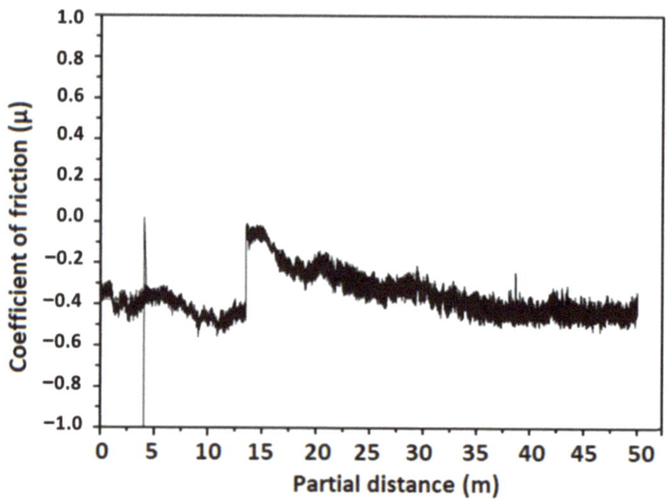

Figure 6. Graph between coefficient of friction (μ) and partial distance (m).

The negative value of the friction co-efficient is due to the adhesive interaction between two contacting surfaces, i.e., the steel ball indenter and the zein/Ag-Sr doped MBGN coating. When the force is applied, the friction force initially increases. This may occur due to the strong adhesion of coating molecules (zein and Ag-Sr MBGNs) among themselves as compared to the adhesion between coating and substrate. The debris of coating starts to accumulate until the sliding distance of 15 m and then, upon persisting load, the layer is detached from the surface as a whole and not in the form of small patches. Hence, the presence of detached coating in the path of the wear track can increase the frictional force, which could lead to the negative COF value. The same phenomenon is noted by Thormann et al. [53] and Dedinaite et al. [54]. We also observed this behavior of negative COF in our previously published paper [55].

3.7. Corrosion Studies

To evaluate the corrosion performance of materials, corrosion current (I_{Corr}) is measured in a relevant electrolyte; I_{Corr} is the amount of current flow while corrosion is taking place in an electrochemical cell. Metallic implants with low values of I_{Corr} and high corrosion potential are considered suitable for implantation [56]. Corrosion behavior of zein/Ag-Sr doped MBGN coatings deposited at 25 V was studied in SBF and compared with the corrosion behavior of bare SS with the help of potentiodynamic polarization curves, as shown in Figure 7. A potentiodynamic curve consists of anodic (upper) and cathodic (lower) curves. To better understand the corrosion behavior, the anodic curve is interpreted here. The graph shows the corrosion behavior of both bare SS and coated substrate. The anodic curve of bare SS shows an abrupt increase in the corrosion potential after a certain potential value. This abrupt increase indicates breakage of a passive layer that prevents corrosion, also reported in our previous study [57]. The breaking of this barrier layer results in the accelerated corrosion rate of the substrate material. However, the anodic curve of coated substrate shows no blunt increase in the values of corrosion potential.

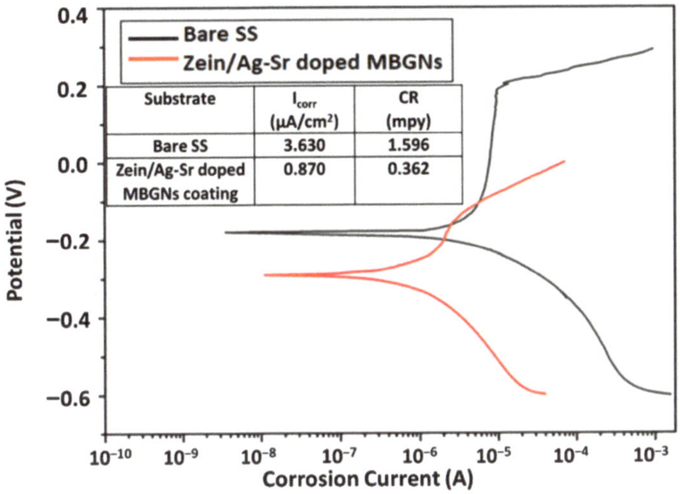

Figure 7. Potentiodynamic curves for bare and coated SS substrates.

The Tafel plot was fitted on these potentiodynamic curves with the help of Echem™ software. The I_{Corr} and corrosion rate (CR) for both bare and coated SS were calculated by the software. The values of I_{Corr} and CR for zein/Ag-Sr doped MBGN coatings deposited at 25 V were quite low as compared to those of bare SS substrates. Similar trends were observed in another study carried out by Ahmed et al. [16]. Zein/hydroxy apatite coating was deposited over 316L SS. The I_{Corr} was significantly lowered due to the presence of zein as compared to the bare substrate. It was concluded that zein coatings effectively increase the resistance against corrosion by fully covering the surface of the substrate. The Ag-Sr doped MBGNs were embedded inside the zein matrix, and both were strongly adhered to the substrate. Thus, in the present study, the corrosion resistance of the coated substrate was inferred to be higher in the physiological environment as compared to the bare SS substrates due to the presence of both zein and Ag-Sr doped MBGNs.

3.8. Biological Characterization

Zein/Ag-Sr doped MBGN coatings were tested against *S. aureus* and *E. coli* to determine the growth potential of bacteria. Figure 8A,B present the bacterial inhibition zones of tested samples. It was seen that a clear zone of inhibition (≈15 mm, measured from the widest side) formed in *S. aureus* (Figure 8A), whereas only a narrow inhibition area was observed around the two corners of the sample placed in *E. coli*, as marked in Figure 8B. This could have occurred due to the different cell membranes around the *S. aureus* and *E. coli*. It is reported that the outer membrane plays a significant role in protecting the bacteria from toxic materials [58]. The cell membrane around *E. coli* protects it from the antibacterial effect of the coating, whereas the absence of cell membranes around *S. aureus* renders it prone to coating effectiveness against it. It is expected that increasing the Ag amount in the MBGNs may increase the antibacterial efficiency of coating against *E. coli*; however, the exposure of Ag in the body above a safe level may cause cytotoxicity [23]. Therefore, optimizing the Ag quantity in MBGNs for efficiency against *E. coli* without exceeding its cytotoxic level could result in a very intriguing study topic.

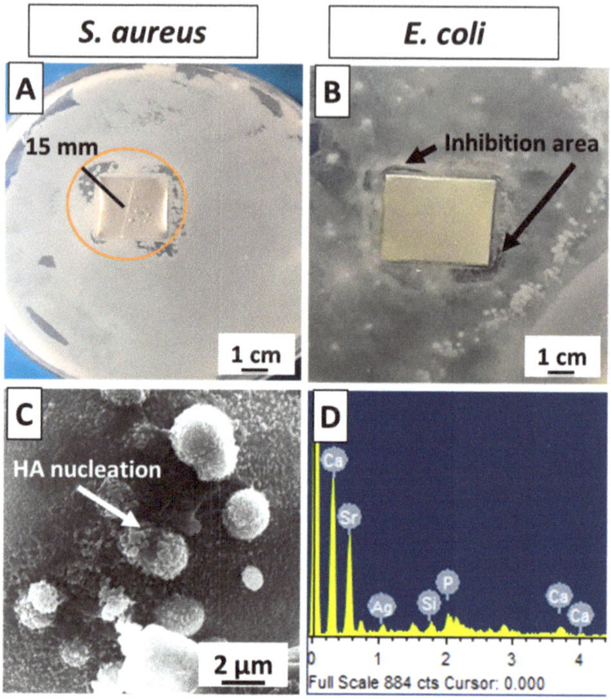

Figure 8. Antibacterial effect of zein/Ag-Sr doped MBGN coatings against (**A**) *S. aureus*, (**B**) *E. coli*, (**C**) HA nucleation starts in SBF at day 7; (**D**) EDS confirms presence of Ca and P on the surface of sample.

Bioactivity is an important criterion in the selection of materials for tissue regeneration. In vitro bioactivity test in SBF was performed to detect the formation of an HA layer on the surface of the coating. A calcium phosphate-based HA layer is similar to natural bone mineral and facilitates the biological bonding between implant surface and surrounding tissues [55]. SEM analysis revealed the morphology of the coated sample after immersion in SBF for 7 days, as shown in Figure 8C. It was observed that HA crystals started to nucleate, and EDS (Figure 8D) also showed the presence of Ca and P in the spectra, which indicated the presence of calcium phosphate on the surface of coating. Hence, the zein/Ag-Sr doped MBGN coatings exhibited appreciable bioactivity.

4. Conclusions

In this work, zein/Ag-Sr doped MBGN coatings were developed on 316L SS by EPD at designated parameters of voltage and deposition time. The following conclusions were obtained at the end of the study.

- High deposition yield of the coatings was obtained at higher voltages, i.e., 25 V.
- Optical microscopic images showed uniform deposition of coatings on the surface of SS substrates (at optimum deposition parameters). SEM images illustrated the homogenous distribution of Ag-Sr doped MBGNs throughout the zein matrix along with the presence of spherical agglomerates, indicating good mechanical integration of zein/Ag-Sr doped MBGN coatings.
- Pencil scratch test results showed increased hardness of zein/Ag-Sr doped MBGN coatings deposited at 25 V, from which it was inferred that coatings developed at higher voltage showed improved adhesion strength. Furthermore, zein/Ag-Sr doped MBGN coatings exhibited good adhesion strength during bend tests.

- Zein/Ag-Sr doped MBGN coatings deposited at 25 V demonstrated good wettability properties (contact angle of 72 ± 2°), suitable for initial protein and subsequent osteoblast cell attachment.
- Moreover, zein/Ag-Sr doped MBGN coatings showed good wear and corrosion resistance as compared to that of bare SS substrates.
- Coatings exhibited good antibacterial and bioactive potential.

The above conclusions imply that zein/Ag-Sr doped MBGN coatings developed in this study via EPD at 25 V exhibited commendable mechanical and surface properties for biomedical applications.

Author Contributions: Conceptualization, S.A.B., U.L. and M.A.u.R.; data curation, I.A.C.; formal analysis, U.L., I.A.C., M.A.M., M.Y., J.A. and M.N.b.J.; investigation, S.A.B.; methodology, S.A.B., S.J.G., M.Y. and M.A.u.R.; project administration, M.A.u.R.; resources, U.L. and J.A.; writing—original draft, S.A.B. and M.A.u.R.; writing—review and editing, S.J.G., M.A.M. and M.N.b.J. All authors have read and agreed to the published version of the manuscript.

Funding: This research was funded by Princess Nourah Bint Abdulrahman University Researchers Supporting Project number (PNURSP2022R108), Princess Nourah Bint Abdulrahman University, Riyadh, Saudi Arabia.

Institutional Review Board Statement: Not applicable.

Informed Consent Statement: Not applicable.

Data Availability Statement: Data available upon request from the corresponding authors.

Acknowledgments: This research project is supported by Princess Nourah bint Abdulrahman University Researchers Supporting Project number (PNURSP2022R108), Princess Nourah Bint Abdulrahman University, Riyadh, Saudi Arabia.

Conflicts of Interest: The authors declare no conflict of interest.

References

1. Shinde, A.S.S.; Sumant, O. *Medical Implant Market by Product Type (Orthopedic Implants, Cardiovascular Implants, Spinal Implant, Neurostimulators, Ophthalmic Implants, Dental Implants, Facial Implants, and Breast Implants) and Biomaterial Type (Metallic Biomaterials, Ceramic Biomaterials, Polymers Biomaterials, and Natural Biomaterials): Global Opportunity Analysis and Industry Forecast, 2020–2027*; Allied market Research: Portland, OR, USA, 2020.
2. Kuhn, L.T. 6—Biomaterials. In *Introduction to Biomedical Engineering*, 2nd ed.; Enderle, J.D., Blanchard, S.M., Bronzino, J.D., Eds.; Academic Press: Boston, MA, USA, 2005; pp. 255–312.
3. Kastellorizios, M.; Tipnis, N.; Burgess, D.J. *Foreign Body Reaction to Subcutaneous Implants*; Springer: Cham, Switzerland, 2015; pp. 93–108.
4. Godbole, N.; Yadav, S.; Manickam, R.; Belemkar, S. A Review on Surface Treatment of Stainless Steel Orthopedic Implants. *Int. J. Pharm. Sci. Rev. Res.* **2016**, *36*, 190–194.
5. Balestriere, M.A.; Schuhladen, K.; Seitz, K.H.; Boccaccini, A.R.; Cere, S.M.; Ballarre, J. Sol-gel coatings incorporating borosilicate bioactive glass enhance anti corrosive and surface performance of stainless steel implants. *J. Electroanal. Chem.* **2020**, *876*, 114735. [CrossRef]
6. Narmada, I.B.; Baya, R.A.; Hamid, T.J.J.o.I.D.; Research, M. Nickel and Chromium Ions Release from Stainless Steel Bracket Immersed in Fluoridated Mouthwash. *J. Int. Dent. Med. Res.* **2018**, *11*, 294–298.
7. Patnaik, L.; Maity, S.R.; Kumar, S. Status of nickel free stainless steel in biomedical field: A review of last 10 years and what else can be done. *Mater. Today Proc.* **2020**, *26*, 638–643. [CrossRef]
8. Hebert, C.K.; Williams, R.E.; Levy, R.S.; Barrack, R.L. Cost of Treating an Infected Total Knee Replacement. *Clin. Orthop. Relat. Res.* **1996**, *331*, 140–145. [CrossRef]
9. Manivasagam, G.; Dhinasekaran, D.; Rajamanickam, A. Biomedical Implants: Corrosion and its Prevention—A Review. *Recent Patents Corros. Sci.* **2010**, *2*, 40–54. [CrossRef]
10. Nouri, A.; Wen, C. 1—Introduction to surface coating and modification for metallic biomaterials. In *Surface Coating and Modification of Metallic Biomaterials*; Wen, C., Ed.; Woodhead Publishing: Sawston, UK, 2015; pp. 3–60.
11. Felgueiras, H.P.; Antunes, J.C.; Martins, M.C.L.; Barbosa, M.A. 1—Fundamentals of protein and cell interactions in biomaterials. In *Peptides and Proteins as Biomaterials for Tissue Regeneration and Repair*; Barbosa, M.A., Martins, M.C.L., Eds.; Woodhead Publishing: Sawston, UK, 2018; pp. 1–27.
12. Gagner, J.E.; Kim, W.; Chaikof, E.L. Designing protein-based biomaterials for medical applications. *Acta Biomater.* **2013**, *10*, 1542–1557. [CrossRef]

13. Paliwal, R.; Palakurthi, S. Zein in controlled drug delivery and tissue engineering. *J. Control. Release* **2014**, *189*, 108–122. [CrossRef]
14. Dong, J.; Sun, Q.; Wang, J.-Y.J.B. Basic study of corn protein, zein, as a biomaterial in tissue engineering, surface morphology and biocompatibility. *Biomaterials* **2004**, *25*, 4691–4697. [CrossRef]
15. Tortorella, S.; Maturi, M.; Vetri Buratti, V.; Vozzolo, G.; Locatelli, E.; Sambri, L.; Franchini, M.C. Zein as a versatile biopolymer: Different shapes for different biomedical applications. *RSC Adv.* **2021**, *11*, 39004–39026. [CrossRef]
16. Ahmed, Y.; Rehman, M.A.U. Improvement in the surface properties of stainless steel via zein/hydroxyapatite composite coatings for biomedical applications. *Surf. Interfaces* **2020**, *20*, 100589. [CrossRef]
17. El-Ghannam, A.; Ducheyne, P. 1.9 Bioactive Ceramics. In *Comprehensive Biomaterials II*; Ducheyne, P., Ed.; Elsevier: Oxford, UK, 2017; pp. 204–234.
18. Van Vugt, T.A.; Geurts, J.A.P.; Arts, J.J.; Lindfors, N.C. 3—Biomaterials in treatment of orthopedic infections. In *Management of Periprosthetic Joint Infections (PJIs)*; Arts, J.J.C., Geurts, J., Eds.; Woodhead Publishing: Sawston, UK, 2017; pp. 41–68.
19. Zahid, S.; Shah, A.T.; Jamal, A.; Chaudhry, A.A.; Khan, A.S.; Khan, A.F.; Muhammad, N.; Rehman, I.U. Biological behavior of bioactive glasses and their composites. *RSC Adv.* **2016**, *6*, 70197–70214. [CrossRef]
20. Huang, W.; Yang, J.; Feng, Q.; Shu, Y.; Liu, C.; Zeng, S.; Guan, H.; Ge, L.; Pathak, J.L.; Zeng, S. Mesoporous Bioactive Glass Nanoparticles Promote Odontogenesis and Neutralize Pathophysiological Acidic pH. *Front. Mater.* **2020**, *7*, 241. [CrossRef]
21. Neščáková, Z.; Zheng, K.; Liverani, L.; Nawaz, Q.; Galusková, D.; Kaňková, H.; Michálek, M.; Galusek, D.; Boccaccini, A.R. Multifunctional zinc ion doped sol–gel derived mesoporous bioactive glass nanoparticles for biomedical applications. *Bioact. Mater.* **2019**, *4*, 312–321. [CrossRef]
22. Tabia, Z.; El Mabrouk, K.; Bricha, M.; Nouneh, K. Mesoporous bioactive glass nanoparticles doped with magnesium: Drug delivery and acellular in vitro bioactivity. *RSC Adv.* **2019**, *9*, 12232–12246. [CrossRef]
23. Bano, S.; Akhtar, M.; Yasir, M.; Maqbool, M.S.; Niaz, A.; Wadood, A.; Rehman, M.A.U. Synthesis and Characterization of Silver–Strontium (Ag-Sr)-Doped Mesoporous Bioactive Glass Nanoparticles. *Gels* **2021**, *7*, 34. [CrossRef] [PubMed]
24. Heimann, R.B.J.S.; Technology, C. Thermal spraying of biomaterials. *Surf. Coat. Technol.* **2006**, *201*, 2012–2019. [CrossRef]
25. Heimann, R.B. *Plasma-Spray Coating: Principles and Applications*; John Wiley & Sons: Hoboken, NJ, USA, 2008.
26. Bourne, R.B.; Rorabeck, C.H.; Burkart, B.C.; Kirk, P.G. Ingrowth surfaces. Plasma spray coating to titanium alloy hip replacements. *Clin. Orthop. Relat. Res.* **1994**, *298*, 37–46. [CrossRef]
27. Yang, Y.; Kim, K.-H.; Ong, J.L. A review on calcium phosphate coatings produced using a sputtering process—An alternative to plasma spraying. *Biomaterials* **2005**, *26*, 327–337. [CrossRef]
28. Ong, J.; Lucas, L.; Lacefield, W.; Rigney, E. Structure, solubility and bond strength of thin calcium phosphate coatings produced by ion beam sputter deposition. *Biomaterials* **1992**, *13*, 249–254. [CrossRef]
29. Sergi, R.; Bellucci, D.; Cannillo, V. A Comprehensive Review of Bioactive Glass Coatings: State of the Art, Challenges and Future Perspectives. *Coatings* **2020**, *10*, 757. [CrossRef]
30. Besra, L.; Liu, M. A review on fundamentals and applications of electrophoretic deposition (EPD). *Prog. Mater. Sci.* **2007**, *52*, 1–61. [CrossRef]
31. Boccaccini, A.R.; Keim, S.; Ma, R.; Li, Y.; Zhitomirsky, I. Electrophoretic deposition of biomaterials. *J. R. Soc. Interface* **2010**, *7*, S581–S613. [CrossRef]
32. Ahmed, Y.; Nawaz, A.; Virk, R.S.; Wadood, A.; Rehman, M.A.U. Fabrication and characterization of zein/bioactive glass deposited on pre-treated magnesium via electrophoretic deposition. *Int. J. Ceram. Eng. Sci.* **2020**, *2*, 254–263. [CrossRef]
33. Ramos-Rivera, L.; Dippel, J.; Boccaccini, A.R. Formation of Zein/Bioactive Glass Layers Using Electrophoretic Deposition Technique. *ECS Trans.* **2018**, *82*, 73–80. [CrossRef]
34. Demir, M.; Ramos-Rivera, L.; Silva, R.; Nazhat, S.N.; Boccaccini, A.R. Zein-based composites in biomedical applications. *J. Biomed. Mater. Res. Part A* **2017**, *105*, 1656–1665. [CrossRef]
35. Kokubo, T.; Takadama, H. How useful is SBF in predicting in vivo bone bioactivity? *Biomaterials* **2006**, *27*, 2907–2915. [CrossRef]
36. Cheng, C.J.; Jones, O.G. Stabilizing zein nanoparticle dispersions with ι-carrageenan. *Food Hydrocoll.* **2017**, *69*, 28–35. [CrossRef]
37. Kaya, S.; Boccaccini, A.R. Electrophoretic deposition of zein coatings. *J. Coat. Technol. Res.* **2017**, *14*, 683–689. [CrossRef]
38. Pishbin, F.; Simchi, A.; Ryan, M.; Boccaccini, A. Electrophoretic deposition of chitosan/45S5 Bioglass®composite coatings for orthopaedic applications. *Surf. Coatings Technol.* **2011**, *205*, 5260–5268. [CrossRef]
39. Rehman, M.A.U. Zein/Bioactive Glass Coatings with Controlled Degradation of Magnesium under Physiological Conditions: Designed for Orthopedic Implants. *Prosthesis* **2020**, *2*, 211–224. [CrossRef]
40. Speight, J.G. (Ed.) Industrial Organic Chemistry. In *Environmental Organic Chemistry for Engineers*; Butterworth-Heinemann: Oxford, UK, 2017; Chapter 3; pp. 87–151.
41. Hamaker, H.C. Formation of a deposit by electrophoresis. *Trans. Faraday Soc.* **1940**, *35*, 279–287. [CrossRef]
42. Nawaz, Q.; Fastner, S.; Rehman, M.A.U.; Ferraris, S.; Perero, S.; di Confiengo, G.G.; Yavuz, E.; Ferraris, M.; Boccaccini, A.R. Multifunctional stratified composite coatings by electrophoretic deposition and RF co-sputtering for orthopaedic implants. *J. Mater. Sci.* **2021**, *56*, 7920–7935. [CrossRef]
43. Rehman, M.A.U.; Bastan, F.E.; Haider, B.; Boccaccini, A.R. Electrophoretic deposition of PEEK/bioactive glass composite coatings for orthopedic implants: A design of experiments (DoE) study. *Mater. Des.* **2017**, *130*, 223–230. [CrossRef]
44. Rehman, M.A.U.; Munawar, M.A.; Schubert, D.W.; Boccaccini, A.R. Electrophoretic deposition of chi-tosan/gelatin/bioactive glass composite coatings on 316L stainless steel: A design of experiment study. *Surf. Coat. Technol.* **2019**, *358*, 976–986. [CrossRef]

45. Aqib, R.; Kiani, S.; Bano, S.; Wadood, A.; Rehman, M.A.U. Ag–Sr doped mesoporous bioactive glass nanoparticles loaded chitosan/gelatin coating for orthopedic implants. *Int. J. Appl. Ceram. Technol.* **2021**, *18*, 544–562. [CrossRef]
46. Bumgardner, J.D.; Wiser, R.; Elder, S.H.; Jouett, R.; Yang, Y.; Ong, J.L. Contact angle, protein adsorption and osteoblast precursor cell attachment to chitosan coatings bonded to titanium. *J. Biomater. Sci. Polym. Ed.* **2003**, *14*, 1401–1409. [CrossRef]
47. Wei, J.; Igarashi, T.; Okumori, N.; Igarashi, T.; Maetani, T.; Liu, B.; Yoshinari, M. Influence of surface wettability on competitive protein adsorption and initial attachment of osteoblasts. *Biomed. Mater.* **2009**, *4*, 045002. [CrossRef]
48. Laskowski, J.; Kitchener, J.A. The hydrophilic—hydrophobic transition on silica. *J. Colloid Interface Sci.* **1969**, *29*, 670–679. [CrossRef]
49. Wu, J.; Xue, K.; Li, H.; Sun, J.; Liu, K. Improvement of PHBV Scaffolds with Bioglass for Cartilage Tissue Engineering. *PLoS ONE* **2013**, *8*, e71563. [CrossRef]
50. Bin Masripan, N.A.; Miyahira, Y.; Nishimura, H.; Tokoroyama, T.; Umehara, N.; Fuwa, Y. Effect of Transfer Layer on Ultra Low Friction of CNx Coating under Blowing Dry Ar. *Tribol. Online* **2013**, *8*, 219–226. [CrossRef]
51. Waqar, S.; Wadood, A.; Mateen, A.; Rehman, M.A.U. Effects of Ni and Cr addition on the wear performance of NiTi alloy. *Int. J. Adv. Manuf. Technol.* **2020**, *108*, 625–634. [CrossRef]
52. Virk, R.S.; Rehman, M.A.U.; Munawar, M.A.; Schubert, D.W.; Goldmann, W.H.; Dusza, J.; Boccaccini, A.R. Curcumin-Containing Orthopedic Implant Coatings Deposited on Poly-Ether-Ether-Ketone/Bioactive Glass/Hexagonal Boron Nitride Layers by Electrophoretic Deposition. *Coatings* **2019**, *9*, 572. [CrossRef]
53. Thormann, E. Negative friction coefficients. *Nat. Mater.* **2013**, *12*, 468. [CrossRef]
54. Dedinaite, A.; Thormann, E.; Olanya, G.; Claesson, P.M.; Nyström, B.; Kjøniksen, A.-L.; Zhu, K. Friction in aqueous media tuned by temperature-responsive polymer layers. *Soft Matter* **2010**, *6*, 2489–2498. [CrossRef]
55. Batool, S.A.; Ahmad, K.; Irfan, M.; Rehman, M.A.U. Zn–Mn-Doped Mesoporous Bioactive Glass Nanoparticle-Loaded Zein Coatings for Bioactive and Antibacterial Orthopedic Implants. *J. Funct. Biomater.* **2022**, *13*, 97. [CrossRef]
56. Eliaz, N. Corrosion of Metallic Biomaterials: A Review. *Materials* **2019**, *12*, 407. [CrossRef]
57. Rehman, M.A.U.; Bastan, F.E.; Nawaz, A.; Nawaz, Q.; Wadood, A. Electrophoretic deposition of PEEK/bioactive glass composite coatings on stainless steel for orthopedic applications: An optimization for in vitro bioactivity and adhesion strength. *Int. J. Adv. Manuf. Technol.* **2020**, *108*, 1849–1862. [CrossRef]
58. Silhavy, T.J.; Kahne, D.; Walker, S. The bacterial cell envelope. *Cold Spring Harb. Perspect. Biol.* **2010**, *2*, a000414. [CrossRef]

bioengineering

MDPI

Article

A Study on Lower Limb Asymmetries in Parkinson's Disease during Gait Assessed through Kinematic-Derived Parameters

Federico Arippa [1,2], Bruno Leban [1], Marco Monticone [2,3], Giovanni Cossu [4], Carlo Casula [5] and Massimiliano Pau [1,*]

1 Department of Mechanical, Chemical and Materials Engineering, University of Cagliari, 09123 Cagliari, Italy; federico.arippa@unica.it (F.A.); bruno.leban@unica.it (B.L.)
2 Neurorehabilitation Unit, Department of Neuroscience and Rehabilitation, ARNAS "G. Brotzu", 09134 Cagliari, Italy; marco.monticone@unica.it
3 Department of Medical Sciences and Public Health, University of Cagliari, 09042 Monserrato, Italy
4 Neurophysiology and Movement Disorders Unit, Department of Neurology, ARNAS "G. Brotzu", 09134 Cagliari, Italy; giovannicossu1@gmail.com
5 Physical Medicine and Rehabilitation Unit, ARNAS "G. Brotzu", 09134 Cagliari, Italy; carlocasula@aob.it
* Correspondence: massimiliano.pau@unica.it

Abstract: Unilaterality of motor symptoms is a distinctive feature of Parkinson's Disease (PD) and represents an important co-factor involved in motor deficits and limitations of functional abilities including postural instability and asymmetrical gait. In recent times, an increasing number of studies focused on the characterization of such alterations, which have been associated with increased metabolic cost and risk of falls and may severely compromise their quality of life. Although a large number of studies investigated the gait alterations in people with PD (pwPD), few focused on kinematic parameters and even less investigated interlimb asymmetry under a kinematic point of view. This retrospective study aimed to characterize such aspects in a cohort of 61 pwPD (aged 68.9 ± 9.3 years) and 47 unaffected individuals age- and sex-matched (66.0 ± 8.3 years), by means of computerized 3D gait analysis performed using an optical motion-capture system. The angular trends at hip, knee and ankle joints of pwPD during the gait cycle were extracted and compared with those of unaffected individuals on a point-by-point basis. Interlimb asymmetry was assessed using angle–angle diagrams (cyclograms); in particular, we analyzed area, orientation, trend symmetry and range offset. The results showed that pwPD are characterized by a modified gait pattern particularly at the terminal stance/early swing phase of the gait cycle. Significant alterations of interlimb coordination were detected at the ankle joint (cyclogram orientation and trend symmetry) and at the hip joint (range offset). Such findings might be useful in clinical routine to characterize asymmetry during gait and thus support physicians in the early diagnosis and in the evaluation of the disease progression.

Keywords: Parkinson's disease; gait; kinematics; symmetry

Citation: Arippa, F.; Leban, B.; Monticone, M.; Cossu, G.; Casula, C.; Pau, M. A Study on Lower Limb Asymmetries in Parkinson's Disease during Gait Assessed through Kinematic-Derived Parameters. *Bioengineering* **2022**, *9*, 120. https://doi.org/10.3390/bioengineering9030120

Academic Editor: Ali Zarrabi

Received: 21 February 2022
Accepted: 14 March 2022
Published: 16 March 2022

1. Introduction

Cardinal motor symptoms such as bradykinesia, rest tremor and rigidity represent some of the most distinctive features of Parkinson's Disease (PD) and originate from degeneration of nigral dopaminergic neurons [1]. Their presentation is typically asymmetric [2], as also confirmed by comparing data derived from imaging techniques of asymptomatic patients and those with mild-early symptoms [3–5], leading unilaterality to be considered as one of the main clinical features useful to discriminate PD from other Parkinsonisms [6,7]. It has been reported that unilaterality persists throughout the clinical course of the disease in many cases [8] as marked differences between motor functions of right and left sides remain evident for 30 years and up [6]. Such asymmetry also reflects on the Unified Parkinson's Disease Rating Scale (UPDRS) score [9–12] and, usually, does not significantly change during the progression of the disease. This is confirmed by several studies which reported that worsening in the UPDRS motor scores (UPDRS-III part) progresses similarly on both sides [8].

In people with PD (pwPD) unilaterality causes postural instability and asymmetrical gait [13], which are associated with increased metabolic cost and risk of falls, and thus negatively affect the quality of life [14]. Given the pivotal role played by the locomotor abilities in several activities of daily living (ADL) and, generally speaking, on the quality of life of pwPD, in the last decade, researchers and clinicians highlighted the need to have available objective tools for timely detection of gait alterations (even when subtle), to characterize the disease progression and to monitor the effectiveness of pharmacologic and rehabilitative treatments. In such context, some studies attempted to investigate and quantify gait asymmetries in pwPD with particular focus on spatiotemporal parameters. Unfortunately, their findings are quite mixed: in fact, although some of them detected larger gait asymmetries in step length and step time parameters in pwPD with respect to unaffected individuals [14] as well as the existence of correlations between asymmetry of gait and disease severity [15], others did not [16,17]. However, all these studies share an important limitation, namely the fact that they focus their attention on discrete values of spatiotemporal parameters. Although such approach has the advantage to provide clinicians with an easily interpretable summary of the entire gait performance, discrete values may not always be sufficiently reflective of the complex alterations of lower limb movement connected to pathological gait conditions [18]. Moreover, as pointed out in a recent review [19], walking-related information in pwPD needs to be improved. Thus, methods that focus on the kinematics of the lower limb during the whole gait cycle may be able to better gather the complexity of locomotor alterations in pwPD.

To these authors' knowledge, the existing study on lower limb kinematics of pwPD did not investigate on a point-by-point basis the difference in hip, knee and ankle joint angular trends with respect to unaffected individuals and, similarly, only few data exist in terms of interlimb symmetry. Since detection of asymmetry may support an early diagnosis of the disease, this additional information could be relevant for the clinician who first evaluates the pwPD, especially to support suitable recommendation of specific rehabilitation protocols, training programs, as well as healthier lifestyles. On the basis of such considerations, the main purpose of the present study was to extensively characterize lower limb kinematics in individuals diagnosed with idiopathic PD by providing summary indexes of gait quality, and symmetry parameters calculated from the angular trend associated with the entire gait cycle for each joint of interest.

2. Materials and Methods

2.1. Participants

Sixty-one pwPD admitted at the Neurologic Department of the ARNAS "G. Brotzu" General Hospital (Cagliari, Italy) underwent a 3D gait analysis at the Laboratory of Biomechanics and Industrial Ergonomics of the University of Cagliari (Cagliari, Italy). They were all diagnosed according to the UK Brain Bank criteria [20] by a trained expert neurologist (G.C.) and free from any other neurologic and orthopedic condition able to significantly influence gait or balance. Their motor functions were assessed using the motor section of UPDRS (UPDRS part III). The experimental trials were carried out in "ON" state (i.e., approx. 60 to 90 min after taking an appropriate oral dose of dispersible Levodopa). Forty-seven unaffected age- and sex-matched individuals recruited among the University and Hospital staff served as the control group (CG).

The study was conducted according to the principles expressed in the World Medical Association Declaration of Helsinki. At the time of the tests, all participants signed a written informed consent form which included detailed information about the aims of the study and the experimental methodology.

2.2. Spatiotemporal and Kinematic Data Collection and Processing

Spatiotemporal and kinematics parameters of gait were assessed by means of an optical motion-capture system composed of 8 infrared cameras (Smart-D, BTS Bioengineering, Milan, Italy) running at a 120 Hz frequency. Before starting the experimental tests, anthropometric data (i.e., height, weight, anterior superior iliac spine distance, pelvis thickness, knee and ankle width, leg length) were acquired, and then 22 spherical reflective passive markers were placed on subjects' skin in accordance with the protocol defined by Davis et al. [21]. All participants were instructed to walk at a self-selected speed as naturally as possible along a 10 m walkway, while the 3D marker's trajectories were acquired by the cameras. The test was considered valid if at least 6 trials were correctly recorded, in order to have available an adequate number of gait cycles for the subsequent processing. Suitable periods of rest between consecutive trials were allowed on request. At the end of the tests, raw data were processed with a dedicated software (Smart Analyzer, BTS Bioengineering, Milan, Italy) to calculate:

- Spatiotemporal gait parameters (i.e., gait speed, cadence, step length, step width, stance, swing and double support phase duration);
- Kinematic parameters (pelvic tilt, rotation and obliquity; hip flexion–extension, adduction–abduction and rotation; knee flexion–extension, ankle dorsi–plantarflexion, and foot progression). From these parameters, additional indexes on gait deviation from normality were obtained, namely the Gait Variable Scores (GVS) and Gait Profile Score (GPS) [22];
- Dynamic range of motion (ROM) for hip and knee flexion–extension and ankle dorsi–plantarflexion. Values were obtained as the difference between the maximum and minimum angle value recorded during the gait cycle;
- Sagittal kinematics of hip, knee and ankle (i.e., hip and knee flexion–extension and ankle dorsi–plantarflexion angles during the gait cycle) which were also employed to calculate the interlimb symmetry parameters as described later in detail.

Additionally, asymmetry between right and left limb in terms of spatiotemporal parameters was quantified on the basis of the Symmetry Index (SI) proposed by Robinson et al. [23]:

$$SI = ABS\left(2 \times \frac{V_R - V_L}{V_R + V_L} \times 100\right)$$

where V_R and V_L represent the values of the gait variable (in our case stance, swing, double support duration phases and step length) for the right and left limb. Such a method, originally proposed for the evaluation of symmetry in ground reaction force during gait, is one of the most used indexes in gait symmetry studies, and has been also modified so as to include spatiotemporal, kinematic parameters, as well as muscle activity data [18].

2.3. Inter-Limb Symmetry Quantification by Means of Waveform-Based Method

Bilateral cyclograms were calculated using a dedicated software developed under Matlab environment basing on the procedure proposed by Goswami [24] which requires right and left limb angles at hip, knee and ankle joints during the gait cycle to build left–right-angle diagrams from which the following symmetry parameters were calculated (Figure 1):

- Cyclogram area (degrees^2): area enclosed by the curve obtained from the left–right angle diagram [25]. A hypothetical symmetrical gait would lead left and right joints to assume the same angular position during the gait cycle. In this way, cyclogram points would lie on a 45° line in the diagram with a null area;
- Cyclogram orientation (degrees): this parameter is expressed as the absolute value of the angular difference φ between the perfect symmetry line (45° line) and the orientation of the principal axis of inertia [24,26], which is the direction of the eigenvector of the inertial matrix for the cyclogram points in the x–y (left vs. right joint angle) reference system. Low φ angles indicate higher interlimb symmetry;

- Trend Symmetry (dimensionless): Calculated to assess the similarity of two waveforms (i.e., right and left leg angular trend across the gait cycles for each joint) by means of an eigenvector analysis [27]. Trend Symmetry index is obtained by dividing the variability about the eigenvector to the variability along the eigenvector and is not affected by a shift or magnitude differences in two considered waveforms. Low or null values indicate higher symmetry, and interlimb asymmetry results in high Trend Symmetry values;
- Range offset, a measure of the differences in operating range of each limb, is calculated as the absolute value of the difference between the average of the right-side waveform from the average of the left-side waveform [27]. In particular, this parameter indicates if one side operates in a wider flexion range than the opposite side; zero values indicate that both sides work within the same ROM.

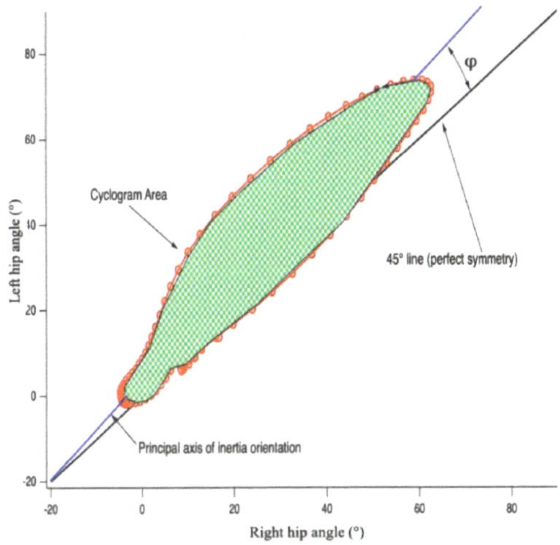

Figure 1. Graphic representation of a cyclogram and its main features considered for the present study.

2.4. Statistical Analysis

A statistical analysis was conducted to evaluate the effect of the disease on gait parameters of interest. In particular, all outcome measures were analyzed in order to investigate the existence of differences originated by the presence of PD. Separate one-way multivariate analysis of variance (MANOVA) were performed, considering group (PD/CG) as the independent variable while the spatiotemporal parameters, SI, GPS and GVSs, ROM were set as dependent variables. In the case of spatiotemporal parameters, they were separated into two groups:

1. Gait speed, cadence and step width, for which both limbs are involved;
2. Stance, swing, double support phases and step length, where only one limb is involved.

To investigate symmetry in joint kinematics, and to assess in which periods of the gait cycle significant differences associated with PD occurred, the angle-cycle curves for PD vs. CG were compared on a point-by-point basis using a one-way ANOVA, setting the group as independent variable. This analysis was performed for each of the 3 joints of interest [28].

Finally, the existence of significant differences in inter-limb symmetry due to PD was also investigated by means of a MANOVA, with group (PD or CG) as the independent variable and the 4 symmetry parameters (cyclogram area and orientation, trend symmetry and range offset) as dependent variables.

A preliminary analysis was performed to exclude the existence of significant differences in the investigated parameters between left and right limbs. Since no significant differences were found, the mean value of each parameter calculated across the two limbs was considered for each participant.

In all above cases, the level of significance was set at $p = 0.05$ and the effect sizes were assessed using the eta-squared (η^2) coefficient. Univariate analysis of variance (ANOVA) was carried out, when necessary, as a post hoc test by reducing the level of significance according to the Bonferroni correction. All analyses were performed using the SPSS version 26 software (IBM SPSS Statistics, Armonk, New York, NY, USA).

3. Results

Demographic, anthropometric and clinical characteristics of the participants are reported in Table 1, while the results of the comparison for spatial–temporal parameters, GPS, GVSs, dynamic ROM and symmetry indexes between pwPD and the CG are reported in Tables 2–5.

Table 1. Demographic, anthropometric, and clinical characteristics of participants. Values are expressed as mean ± SD.

	Control Group (19 F, 28 M)	PD Group (24 F, 37 M)
Age (years)	66.0 ± 8.3	68.9 ± 9.3
Body mass (kg)	66.9 ± 11.1	67.1 ± 10.9
Height (cm)	164.7 ± 6.9	164.5 ± 7.8
Disease Duration (years)	-	7.7 ± 5.6
UPDRS III score	-	19.9 ± 9.3

3.1. *Spatiotemporal Parameters of Gait*

Significant effect originated by the presence of PD on spatiotemporal parameters of gait was detected by MANOVA for both, single limb and double limb related parameters [F(4,106) = 4.286, $p = 0.003$, Wilks λ = 0. 857, $\eta^2 = 0.143$] and [F(3,106) = 4.378, $p = 0.006$, Wilks λ = 0.888, $\eta^2 = 0.112$], respectively. In particular, the follow-up ANOVA (Table 2) indicated that pwPD exhibit reduced speed, step length and swing phase duration and increased double support phase duration when compared to unaffected individuals.

Table 2. Spatiotemporal parameters of gait. Stance, swing and double support phases are expressed as percentage of the gait cycle duration. Values are expressed as mean ± SD.

	Control Group	PD Group
Speed (m/s)	1.18 ± 0.22	1.06 ± 0.26 **
Cadence (steps/min)	112.32 ± 10.24	111.49 ± 12.99
Step Length (m)	0.63 ± 0.08	0.55 ± 0.11 **
Step Width (m)	0.20 ± 0.02	0.19 ± 0.04
Stance Phase (% of the gait cycle)	59.96 ± 1.65	60.77 ± 2.62
Swing Phase (% of the gait cycle)	40.06 ± 1.65	38.67 ± 2.47 **
Double Support Phase (% of the gait cycle)	20.07 ± 3.29	22.60 ± 4.73 **

The symbol ** denotes a significant difference with respect to the Control Group (in all cases $p < 0.01$).

Significant effect of PD for spatiotemporal SI (Table 3) was detected by MANOVA analysis [F(4,106) = 5.574, $p = 0.000$, Wilks λ = 0. 822, $\eta^2 = 0.178$]. In particular, SI values were significantly higher in PD subjects for double support and step length parameters ($p = 0.017$ and $p = 0.001$, respectively).

Table 3. SI parameters of gait. Stance, swing and double support phases are expressed as percentage of the gait cycle duration. Values are expressed as mean ± SD.

Symmetry Index	Control Group	PD Group
Step Length	2.90 ± 1.92	4.90 ± 3.52 **
Stance Phase Duration	1.66 ± 1.20	2.39 ± 2.94
Swing Phase Duration	2.45 ± 1.79	3.62 ± 4.04
Double Support Phase Duration	7.90 ± 6.29	14.22 ± 17.03 *

The symbols * and ** denote a significant difference with respect to the Control Group (* $p < 0.05$, ** $p < 0.01$).

3.2. Gait Kinematics, GPS and GVS

The statistical analysis revealed a significant main effect of group on GPS and GVS indexes [F(10,104) = 2.622, $p = 0.007$, Wilks $\lambda = 0.784$, $\eta^2 = 0.216$]. The follow-up ANOVA showed that pwPD exhibit increased GPS in comparison to CG ($p < 0.01$) and increased GVS for pelvic obliquity and rotation ($p < 0.05$ and $p < 0.001$, respectively) and knee flex-extension ($p < 0.01$). Mean values along with standard deviations for each group are reported in Table 4.

Table 4. GPS and GVS indexes (in degrees). Values are expressed as mean ± SD.

	Control Group	PD Group
GPS	6.60 ± 1.35	7.37 ± 1.31 **
Pelvic Obliquity GVS	2.23 ± 0.88	2.60 ± 0.95 *
Pelvic Tilt GVS	5.45 ± 3.14	6.04 ± 3.55
Pelvic Rotation GVS	3.35 ± 1.02	4.18 ± 1.34 **
Hip Abduction–Adduction GVS	3.69 ± 1.29	3.94 ± 1.29
Hip Flexion–Extension GVS	7.96 ± 3.35	8.54 ± 4.14
Hip Rotation GVS	7.83 ± 3.23	8.86 ± 3.14
Knee Flexion–Extension GVS	7.61 ± 2.51	8.98 ± 2.69 **
Ankle Dorsi–plantarflexion GVS	5.84 ± 1.99	6.34 ± 2.21
Foot Progression GVS	7.94 ± 2.60	8.47 ± 3.64

The symbols * and ** denote a significant difference with respect to the Control Group (* $p < 0.05$, ** $p < 0.01$).

3.3. Dynamic ROM

The statistical analysis detected the existence of significant main effect of group [F(3,105) = 5.015, $p = 0.003$, Wilks $\lambda = 0.873$, $\eta^2 = 0.127$] on dynamic ROM during gait (Table 5). In particular, the follow-up ANOVA indicated that pwPD are characterized by significantly reduced ROM at hip and knee joints with respect to CG ($p < 0.001$ and $p < 0.01$, respectively).

Table 5. Dynamic range of motion during gait (in degrees). Values are expressed as mean ± SD.

Joint	Control Group	PD Group
Hip ROM	46.52 ± 6.11	42.02 ± 5.89 **
Knee ROM	58.80 ± 4.67	55.69 ± 5.53 **
Ankle ROM	26.47 ± 4.94	24.93 ± 4.98

The symbol ** denotes a significant difference with respect to the control group ($p < 0.01$).

3.4. Point-by-Point Analysis of Kinematic Curves

The analysis of hip, knee, and ankle kinematics in the sagittal plane (Figure 2) revealed the existence of:

- At the hip joint level, significant differences between pwPD and CG from 30 to 67% of the gait cycle;
- At the knee joint, between 1 and 3%, between 20 and 56% and between 87 and 100% of the gait cycle;
- At the ankle joint from 51 to 64% of the gait cycle.

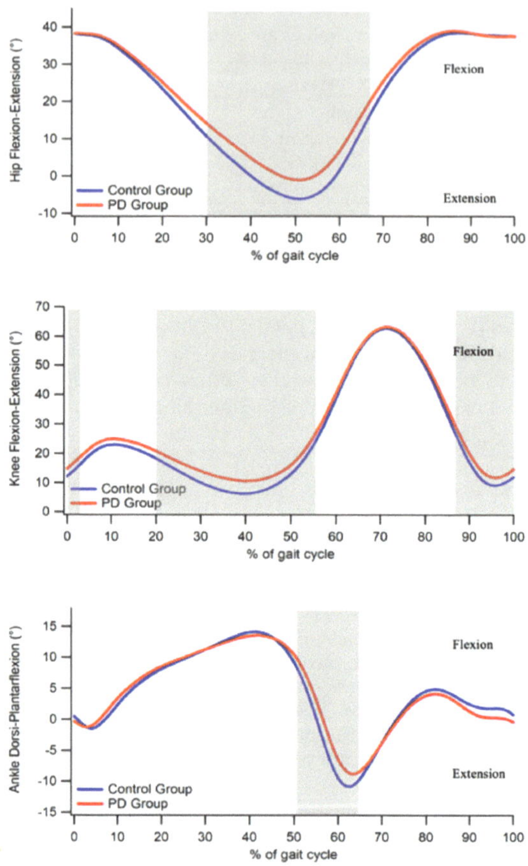

Figure 2. Gait kinematics in the sagittal plane. From top to bottom: hip flexion–extension, knee flexion–extension and ankle dorsi–plantar-flexion angles during gait cycle. Gray-shaded areas denote the periods of the gait cycle in which a significant difference between groups existed ($p < 0.05$).

3.5. Waveform-Based Symmetry Indexes

MANOVA detected a significant effect of group on symmetry indexes at hip and ankle joints (hip [$F(4,103) = 4.825$, $p = 0.001$, Wilks $\lambda = 0.838$, $\eta^2 = 0.162$]; ankle [$F(4,105) = 8.355$, $p < 0.001$, Wilks $\lambda = 0.753$, $\eta^2 = 0.247$]), while no main effect was found for the knee joint. In particular, the analysis showed that the range offset is significantly larger in pwPD at the hip joint level ($p < 0.05$), as well as for cyclogram orientation and trend symmetry at the ankle joint ($p < 0.001$ and $p < 0.01$, respectively) with respect to the unaffected individuals (Table 6).

Figure 3 shows an example of the different shapes and orientations of the cyclograms for PDs and unaffected individuals.

Table 6. Comparison between symmetry indexes of PD and CG subjects. Values are expressed as mean ± SD.

	Cyclogram Parameter	Control Group	PD Group
Hip	Area	116.57 ± 88.11	87.95 ± 72.18
	Orientation φ	2.26 ± 2.36	1.92 ± 1.76
	Trend Symmetry	0.24 ± 0.21	0.29 ± 0.26
	Range Offset	2.27 ± 2.02	3.22 ± 2.08 *
Knee	Area	268.76 ± 213.97	213.53 ± 156.65
	Orientation φ	1.47 ± 1.40	1.62 ± 1.35
	Trend Symmetry	0.49 ± 0.42	0.48 ± 0.32
	Range Offset	4.52 ± 3.97	5.50 ± 3.22
Ankle	Area	62.52 ± 51.59	84.58 ± 63.71
	Orientation φ	1.99 ± 1.44	3.92 ± 2.80 *
	Trend Symmetry	1.54 ± 1.21	2.27 ± 1.48 *
	Range Offset	2.83 ± 2.05	3.57 ± 2.63

The symbol * denotes a significant difference with respect to the control group.

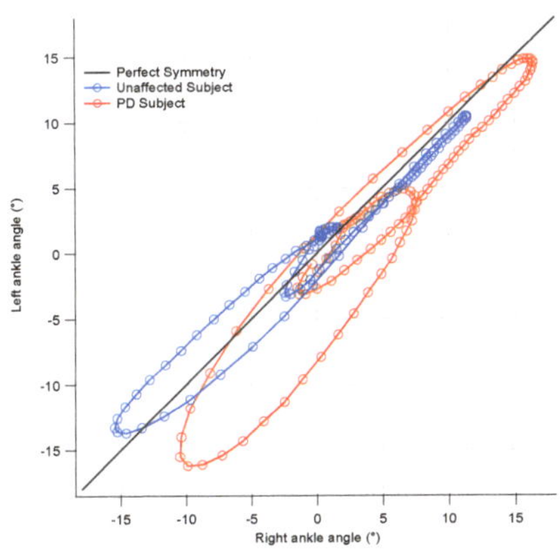

Figure 3. Comparison between cyclograms of an individual affected by PD and an unaffected individual. The diagram refers to the ankle joint.

4. Discussion

The aim of the present study was to characterize alterations in gait kinematics among pwPD, with particular focus on the interlimb symmetry of hip, knee and ankle joints with respect to unaffected individuals. At first, it should be noted that the gait patterns of pwPD are substantially consistent with most previous studies in terms of spatiotemporal parameters, as they exhibit increased double support phase duration [29–31], reduced swing phase duration, step length and gait speed [32–35].

We found marked differences between pwPD and CG with regards to SI step length and SI double support, a result partly consistent with previous studies which also reported significantly larger asymmetries in pwPD for swing duration and step time other than step length and double support duration [13,14]. It is possible that such differences are due to the different conditions in which the participants were tested. In fact, while the quoted studies investigated gait with pwPD in OFF levodopa state, in our case we evaluated them while in the ON phase. It is thus likely that effect of medication somehow attenuated the gait alterations associated with the disease.

From a kinematic point of view, the analysis of the GVS parameters highlighted differences at the level of the hip and knee joints. In particular, consistent with what was reported by previous studies [29], pwPD were characterized by higher values in pelvic obliquity and rotation, and knee flexion–extension. Overall, such alterations originated a significantly higher value of the GPS with respect to unaffected individuals. Higher GVS scores for hip and knee flexion–extension may be due to reduced strength of muscles acting on these joints [36].

The point-by-point analysis of the sagittal kinematics showed significant alterations in pwPD for all the three joints with respect to the unaffected individuals, though of different magnitude. As regards the ankle joint, the main differences involved the terminal stance and swing phases of the gait cycle, where an increased dorsiflexion in pwPD was detected. This is consistent with previous studies [35,37], which indicated that reduced plantarflexion in PD group during the toe-off phase is caused by a decreased ankle power generation [37] due to a reduced amplitude of gastrocnemius activity [38]. Similarly, reduced knee extension was observed in the terminal stance phase and between 87 to 100% of the gait cycle. The author of [39] attributed this behavior to the reduced gait velocity and to the reduced ROM at the hip level, which was reflective of an increased rigidity among the pathological group.

One of the main aims of this study was the characterization of interlimb joint kinematics symmetry by means of a waveform-based method, which was previously employed with encouraging results for the characterization of asymmetries among people with osteoarthritis and multiple sclerosis [40,41]. Even though a certain asymmetry of gait kinematics exists even in a healthy population [42], neurodegenerative diseases such as PD are characterized by unilaterality [43], which is likely to result in larger asymmetries. The waveform-derived parameters employed in the present study showed that pwPD are characterized by a marked asymmetry at the ankle joint level, while slight differences exist for the hip joint level. In particular, at ankle joint cyclogram orientation and trend symmetry were twice and one and a half time higher, respectively, in PDs, who also showed higher values for range offset at the hip level. On the other hand, no significant differences were found for the knee joint. Despite some differences which were evidenced for the hip joint, our results somehow differ from Goswami [24] who also reported significant increases for cyclogram orientation values at the hip.

The observed asymmetries may be influenced by several factors. At first, as previously mentioned, in pwPD reduction in the microstructural integrity of the transcallosal fibers connecting homologous regions of the pre-supplementary motor and supplementary motor areas were observed, which were previously recognized as responsible for step length asymmetry [14]. A relevant role might also be played by the existence of asymmetries in muscular strength, which originate from right–left hemispheric asymmetry of the functional organization of basal ganglia [44]. Moreover, as previously reported, EMG activity of the gastrocnemius is reduced while walking [38], thus amplifying asymmetry at the ankle joint level, which reflects on alterations of cyclograms parameters.

Some limitations of the study should be acknowledged. First, the relative limited size of the sample here tested implies that generalization of the obtained results should be performed cautiously. Secondly, we had no information on the first or more affected limb in the PD group, and for this reason we used averages of left and right limbs when assessing differences in spatiotemporal parameters and ROM with respect to a healthy population. Information on the affected side may result in differences for these parameters for more vs. less affected side. Lastly, even though none of the participants exhibited freezing of gait during the tests, this phenomenon cannot be adequately captured by means of the setup employed here. Thus, specific tests should be planned to consider freezing of gait episodes in the analysis.

5. Conclusions

In this study we investigated the gait patterns of pwPD, focusing on lower limb joint kinematics, by comparing their joint angle curves with those of unaffected individuals and by calculating symmetry parameters derived from a waveform-based approach. The obtained results show that pwPD exhibit modified gait patterns characterized by severe modifications of the physiologic kinematic trend at the hip, knee and ankle level, especially during the terminal stance-early swing phase and final part of the gait cycle. The symmetry analysis revealed that the effect of the disease on interlimb coordination is present at the ankle joint and is moderate in the hip, while the knee joint appears relatively exempt from specific negative effects from this point of view. Such findings could be useful in clinical routine, since the quantitative information on asymmetry may represent an additional tool that helps clinicians to diagnose PD earlier and/or evaluate its development.

Author Contributions: Conceptualization, M.P.; methodology, F.A. and B.L.; software, B.L.; formal analysis, F.A., M.P. and B.L.; clinical assessment, M.M., G.C. and C.C.; writing—original draft preparation, F.A. and B.L.; writing—review and editing, M.P. All authors have read and agreed to the published version of the manuscript.

Funding: This research received no external funding.

Institutional Review Board Statement: Data used for this retrospective study were collected during previous studies conducted according to the guidelines of the Declaration of Helsinki, and approved by the Local Ethics Committee (authorization number PG/2014/19654).

Informed Consent Statement: Informed consent was obtained from all subjects involved in the study.

Data Availability Statement: Data will be made available on request.

Acknowledgments: The authors are grateful to Federica Corona, Giuseppina Pilloni and Micaela Porta for their valuable support during the data acquisition process.

Conflicts of Interest: The authors declare no conflict of interest.

References

1. Bloem, B.R.; Okun, M.S.; Klein, C. Parkinson's disease. *Lancet* **2021**, *397*, 2284–2303. [CrossRef]
2. Miller-Patterson, C.; Buesa, R.; McLaughlin, N.; Jones, R.; Akbar, U.; Friedman, J.H. Motor asymmetry over time in Parkinson's disease. *J. Neurol. Sci.* **2018**, *393*, 14–17. [CrossRef] [PubMed]
3. Laihinen, A.; Ruottinen, H.; Rinne, J.O.; Haaparanta, M.; Bergman, J.; Solin, O.; Koskenvuo, M.; Marttila, R.; Rinne, U.K. Risk for Parkinson's disease: Twin studies for the detection of asymptomatic subjects using [18F]6-fluorodopa PET. *J. Neurol. Suppl.* **2000**, *247*, 110–113. [CrossRef]
4. Burn, D.J.; Mark, M.H.; Playford, E.D.; Maraganore, D.M.; Zimmerman, T.R.; Duvoisin, R.C.; Harding, A.E.; Marsden, C.D.; Brooks, D.J. Parkinson's disease in twins studied with 18–dopa and positron emission tomography. *Neurology* **1992**, *42*, 1894–1900. [CrossRef] [PubMed]
5. Sawle, G.V.; Wroe, S.J.; Lees, A.J.; Brooks, D.J.; Frackowiak, R.S.J. The identification of presymptomatic parkinsonism: Clinical and {^{18}F}dopa positron emission tomography studies in an irish kindred. *Ann. Neurol.* **1992**, *32*, 609–617. [CrossRef] [PubMed]
6. Djaldetti, R.; Ziv, I.; Melamed, E. The mystery of motor asymmetry in Parkinson's disease. *Lancet Neurol.* **2006**, *5*, 796–802. [CrossRef]
7. Hughes, A.J.; Ben-Shlomo, Y.; Daniel, S.E.; Lees, A.J. What features improve the accuracy of clinical diagnosis in Parkinson's disease: A clinicopathologic study. *Neurology* **1992**, *42*, 1142–1146. [CrossRef]
8. Lee, C.S.; Schulzer, M.; Mak, E.; Hammerstad, J.P.; Calne, S.; Calne, D.B. Patterns of asymmetry do not change over the course of idiopathic parkinsonism: Implications for pathogenesis. *Neurology* **1995**, *45*, 435–439. [CrossRef]
9. Gelb, D.J.; Oliver, E.; Gilman, S. Diagnostic criteria for Parkinson disease. *Arch. Neurol.* **1999**, *56*, 33–39. [CrossRef]
10. Toth, C.; Rajput, M.; Rajput, A.H. Anomalies of asymmetry of clinical signs in Parkinsonism. *Mov. Disord.* **2003**, *19*, 151–157. [CrossRef]
11. Elbaz, A.; Bower, J.H.; Peterson, B.J.; Maraganore, D.M.; McDonnell, S.K.; Ahlskog, J.E.; Schaid, D.J.; Rocca, W.A. Survival study of Parkinson disease in Olmsted County, Minnesota. *Arch. Neurol.* **2003**, *60*, 91–96. [CrossRef] [PubMed]
12. Uitti, R.J.; Baba, Y.; Wszolek, Z.K.; Putzke, D.J. Defining the Parkinson's disease phenotype: Initial symptoms and baseline characteristics in a clinical cohort. *Park. Relat. Disord.* **2005**, *11*, 139–145. [CrossRef] [PubMed]
13. Baltadjieva, R.; Giladi, N.; Gruendlinger, L.; Peretz, C.; Hausdorff, J.M. Marked alterations in the gait timing and rhythmicity of patients with de novo Parkinson's disease. *Eur. J. Neurosci.* **2006**, *24*, 1815–1820. [CrossRef]
14. Fling, B.W.; Curtze, C.; Horak, F.B. Gait asymmetry in people with Parkinson's disease is linked to reduced integrity of callosal sensorimotor regions. *Front. Neurol.* **2018**, *9*, 215. [CrossRef] [PubMed]
15. Nanhoe-Mahabier, W.; Snijders, A.; Delval, A.; Weerdesteyn, V.; Duysens, J.; Overeem, S.; Bloem, B. Walking patterns in Parkinson's disease with and without freezing of gait. *Neuroscience* **2011**, *182*, 217–224. [CrossRef] [PubMed]
16. Plotnik, M.; Giladi, N.; Balash, Y.; Peretz, C.; Hausdorff, J.M. Is freezing of gait in Parkinson's disease related to asymmetric motor function? *Ann. Neurol.* **2005**, *57*, 656–663. [CrossRef] [PubMed]

17. Yogev, G.; Plotnik, M.; Peretz, C.; Giladi, N.; Hausdorff, J.M. Gait asymmetry in patients with Parkinson's disease and elderly fallers: When does the bilateral coordination of gait require attention? *Exp. Brain Res.* **2006**, *177*, 336–346. [CrossRef]
18. Viteckova, S.; Kutilek, P.; Svoboda, Z.; Krupicka, R.; Kauler, J.; Szabo, Z. Gait symmetry measures: A review of current and prospective methods. *Biomed. Signal Process. Control* **2018**, *42*, 89–100. [CrossRef]
19. Zanardi, A.P.J.; da Silva, E.S.; Costa, R.R.; Passos-Monteiro, E.; dos Santos, I.O.; Kruel, L.F.M.; Peyré-Tartaruga, L.A. Gait parameters of Parkinson's disease compared with healthy controls: A systematic review and meta-analysis. *Sci. Rep.* **2021**, *11*, 752. [CrossRef]
20. Gibb, W.R.G.; Lees, A.J. The relevance of the Lewy body to the pathogenesis of idiopathic Parkinson's disease. *J. Neurol. Neurosurg. Psychiatry* **1988**, *51*, 745–752. [CrossRef]
21. Davis, R.B.; Õunpuu, S.; Tyburski, D.; Gage, J.R. A gait analysis data collection and reduction technique. *Hum. Mov. Sci.* **1991**, *10*, 575–587. [CrossRef]
22. Baker, R.; McGinley, J.; Schwartz, M.; Beynon, S.; Rozumalski, A.; Graham, H.K.; Tirosh, O. The Gait Profile Score and Movement Analysis Profile. *Gait Posture* **2009**, *30*, 265–269. [CrossRef] [PubMed]
23. Robinson, R.O.; Herzog, W.; Nigg, B.M. Use of force platform variables to quantify the effects of chiropractic manipulation on gait symmetry. *J. Manip. Physiol. Ther.* **1987**, *10*, 172–176.
24. Goswami, A. Kinematic quantification of gait symmetry based on bilateral cyclograms. In Proceedings of the International Society of Biomechanics XIXth Congress, Dunedin, New Zealand, 6–11 July 2003.
25. Hershler, C.; Milner, M. Angle-angle diagrams in the assessment of locomotion. *Am. J. Phys. Med.* **1980**, *59*, 109–125.
26. Goswami, A. A new gait parameterization technique by means of cyclogram moments: Application to human slope walking. *Gait Posture* **1998**, *8*, 15–36. [CrossRef]
27. Crenshaw, S.J.; Richards, J.G. A method for analyzing joint symmetry and normalcy, with an application to analyzing gait. *Gait Posture* **2006**, *24*, 515–521. [CrossRef]
28. Bruening, D.A.; Frimenko, R.E.; Goodyear, C.D.; Bowden, D.R.; Fullenkamp, A.M. Sex differences in whole body gait kinematics at preferred speeds. *Gait Posture* **2015**, *41*, 540–545. [CrossRef]
29. Corona, F.; Pau, M.; Guicciardi, M.; Murgia, M.; Pili, R.; Casula, C. Quantitative assessment of gait in elderly people affected by Parkinson's Disease. In Proceedings of the 2016 IEEE International Symposium on Medical Measurements and Applications (MeMeA), Benevento, Italy, 15–18 May 2016. [CrossRef]
30. Švehlík, M.; Zwick, E.B.; Steinwender, G.; Linhart, W.E.; Schwingenschuh, P.; Katschnig, P.; Ott, E.; Enzinger, C. Gait Analysis in Patients with Parkinson's Disease Off Dopaminergic Therapy. *Arch. Phys. Med. Rehabil.* **2009**, *90*, 1880–1886. [CrossRef]
31. Peppe, A.; Chiavalon, C.; Pasqualetti, P.; Crovato, D.; Caltagirone, C. Does gait analysis quantify motor rehabilitation efficacy in Parkinson's disease patients? *Gait Posture* **2007**, *26*, 452–462. [CrossRef]
32. Morris, M.E.; Huxham, F.; McGinley, J.; Dodd, K.; Iansek, R. The biomechanics and motor control of gait in Parkinson disease. *Clin. Biomech.* **2001**, *16*, 459–470. [CrossRef]
33. Morris, M.E.; Iansek, R.; Matyas, T.A.; Summers, J.J. Ability to modulate walking cadence remains intact in Parkinson's disease. *J. Neurol. Neurosurg. Psychiatry* **1994**, *57*, 1532–1534. [CrossRef] [PubMed]
34. Van Emmerik, R.E.A.; Wagenaar, R.C.; Winogrodzka, A.; Wolters, E.C. Identification of axial rigidity during locomotion in parkinson disease. *Arch. Phys. Med. Rehabil.* **1999**, *80*, 186–191. [CrossRef]
35. Sofuwa, O.; Nieuwboer, A.; Desloovere, K.; Willems, A.M.; Chavret, F.; Jonkers, I. Quantitative gait analysis in Parkinson's disease: Comparison with a healthy control group. *Arch. Phys. Med. Rehabil.* **2005**, *86*, 1007–1013. [CrossRef] [PubMed]
36. Inkster, L.M.; Eng, J.J.; MacIntyre, D.L.; Stoessl, A.J. Leg muscle strength is reduced in Parkinson's disease and relates to the ability to rise from a chair. *Mov. Disord.* **2002**, *18*, 157–162. [CrossRef]
37. Morris, M.E.; McGinley, J.; Huxham, F.; Collier, J.; Iansek, R. Constraints on the kinetic, kinematic and spatiotemporal parameters of gait in Parkinson's disease. *Hum. Mov. Sci.* **1999**, *18*, 461–483. [CrossRef]
38. Dietz, V.; Zijlstra, W.; Prokop, T.; Berger, W. Leg muscle activation during gait in Parkinson's disease: Adaptation and interlimb coordination. *Electroencephalogr. Clin. Neurophysiol. Mot. Control* **1995**, *97*, 408–415. [CrossRef]
39. DiPaola, M.; Pavan, E.E.; Cattaneo, A.; Frazzitta, G.; Pezzoli, G.; Cavallari, P.; Frigo, C.A.; Isaias, I.U. Mechanical energy recovery during walking in patients with Parkinson disease. *PLoS ONE* **2016**, *11*, e0156420. [CrossRef]
40. Porta, M.; Pau, M.; Leban, B.; Deidda, M.; Sorrentino, M.; Arippa, F.; Marongiu, G. Lower limb kinematics in individuals with hip osteoarthritis during gait: A focus on adaptative strategies and interlimb symmetry. *Bioengineering* **2021**, *8*, 47. [CrossRef]
41. Pau, M.; Leban, B.; Deidda, M.; Putzolu, F.; Porta, M.; Coghe, G.; Cocco, E. Kinematic analysis of lower limb joint asymmetry during gait in people with multiple sclerosis. *Symmetry* **2021**, *13*, 598. [CrossRef]
42. Sadeghi, H.; Allard, P.; Prince, F.; Labelle, H. Symmetry and limb dominance in able-bodied gait: A review. *Gait Posture* **2000**, *12*, 34–45. [CrossRef]
43. Barbieri, F.A.; Vitório, R. Parkinson's Disease and Gait Asymmetry. In *Locomotion and Posture in Older Adults: The Role of Aging and Movement Disorders*; Springer: Cham, Switzerland, 2017; pp. 161–175. [CrossRef]
44. Frazzitta, G.; Ferrazzoli, D.; Maestri, R.; Rovescala, R.; Guaglio, G.; Bera, R.; Volpe, D.; Pezzoli, G. Differences in muscle strength in Parkinsonian patients affected on the right and left side. *PLoS ONE* **2015**, *10*, e0121251. [CrossRef] [PubMed]

bioengineering

Article

Efficacy of Ciprofloxacin and Amoxicillin Removal and the Effect on the Biochemical Composition of *Chlorella vulgaris*

Rajamanickam Ricky [1], Fulvia Chiampo [2,*] and Subramaniam Shanthakumar [1]

[1] Department of Environmental and Water Resources Engineering, School of Civil Engineering, Vellore Institute of Technology (VIT), Vellore 632014, India; rickyraaaj@gmail.com (R.R.); shanthakumar.s@vit.ac.in (S.S.)
[2] Department of Applied Science and Technology, Politecnico di Torino, Corso Duca degli Abruzzi 24, 10129 Torino, Italy
* Correspondence: fulvia.chiampo@polito.it; Tel.: +39-011-090-4685

Abstract: Antibiotics are frequently detected in the aquatic environment due to their excessive usage and low-efficiency removal in wastewater treatment plants. This can provide the origin to the development of antibiotic-resistant genes in the microbial community, with considerable ecotoxicity to the environment. Among the antibiotics, the occurrence of ciprofloxacin (CIP) and amoxicillin (AMX) has been detected in various water matrices at different concentrations around the Earth. They are designated as emerging contaminants (ECs). Microalga *Chlorella vulgaris* (*C. vulgaris*) has been extensively employed in phycoremediation studies for its acclimatization property, non-target organisms for antibiotics, and the production of value-added bioproducts utilizing the nutrients from the wastewater. In this study, *C. vulgaris* medium was spiked with 5 mg/L of CIP and AMX, and investigated for its growth-stimulating effects, antibiotic removal capabilities, and its effects on the biochemical composition of algal cells compared to the control medium for 7 days. The results demonstrated that *C. vulgaris* adapted the antibiotic spiked medium and removed CIP (37 ± 2%) and AMX (25 ± 3%), respectively. The operating mechanisms were bioadsorption, followed by bioaccumulation, and biodegradation, with an increase in cell density up to 46 ± 3% (CIP) and 36 ± 4% (AMX), compared to the control medium. Further investigations revealed that, in the CIP stress-induced algal medium, an increase in major photosynthetic pigment chlorophyll-a (30%) and biochemical composition (lipids (50%), carbohydrates (32%), and proteins (65%)) was observed, respectively, compared to the control medium. In the AMX stress-induced algal medium, increases in chlorophyll-a (22%), lipids (46%), carbohydrates (45%), and proteins (49%) production were observed compared to the control medium. Comparing the two different stress conditions and considering that CIP is more toxic than AMX, this study provided insights on the photosynthetic activity and biochemical composition of *C. vulgaris* during the stress conditions and the response of algae towards the specific antibiotic stress. The current study confirmed the ability of *C. vulgaris* to adapt, bioadsorb, bioaccumulate, and biodegrade emerging contaminants. Moreover, the results showed that *C. vulgaris* is not only able to remove CIP and AMX from the medium but also can increase the production of valuable biomass usable in the production of various bioproducts.

Keywords: ciprofloxacin; amoxicillin; algae; *Chlorella vulgaris*; biomass; removal

Citation: Ricky, R.; Chiampo, F.; Shanthakumar, S. Efficacy of Ciprofloxacin and Amoxicillin Removal and the Effect on the Biochemical Composition of *Chlorella vulgaris*. *Bioengineering* **2022**, *9*, 134. https://doi.org/10.3390/bioengineering9040134

Academic Editor: Ali Zarrabi

Received: 22 February 2022
Accepted: 21 March 2022
Published: 24 March 2022

1. Introduction

Antibiotics are used in veterinary, human, and aquaculture targeting relative to the bacterial community to prevent or treat microbial diseases and infections. These antibiotics reach the aquatic environment persistently through various routes such as veterinary and human excretions, hospital wastewater, pharmaceutical wastewater, and sewage, reaching treatment plants in concentrations ranging from nanogram per liter to milligram per liter [1]. When properly untreated, the antibiotics reach the environment, causing chronic toxicity to some non-target organisms as they are designed to induce a biological response in living organisms [2]. The major problem associated with antibiotic polluted water is the development of antibiotic-resistant bacteria (ARB) and antibiotic-resistant genes (ARGs), which are responsible for 700,000 deaths per year. The treatment of ARB is highly problematic as they have developed resistance towards the antibiotics that are particularly prescribed for the treatment [3]. Ciprofloxacin (CIP) is an important antibiotic of the class of fluoroquinolones that targets Gram-positive and Gram-negative bacteria to treat severe infections, and its emission is remarkably found in municipal wastewater (58% of its total amount) and surface water (25% of its total emission) globally [4]. Amoxicillin (AMX) belongs to the β-lactam group, which accounts for more than 65% of the global antibiotic market and is recognized as one of the major threats to the environment as it is being frequently detected in surface waters [5,6]. The maximum detected worldwide concentration of CIP and AMX is 6.5 mg/L and 1.67 μg/L, respectively, with a median concentration range of 10–100 ng/L [7–9]. Their presence in the ecosystem is known to have effects on microorganism communities, such as bacteria, algae, invertebrates, and crustaceans. The effects of antibiotics in the aquatic ecosystem depend upon acute aquatic toxicity EC_{50} (concentration of pollutant that inhibits 50% of growth) present in water streams as they can inhibit the growth of beneficial microorganisms such as prokaryotes and eukaryotes [10].

Microalgae are eukaryotic organisms that play a crucial role in the production of oxygen in the aquatic ecosystem, as well as an important part of the food chain. Microalgae have drawn attention in the bioremediation research community for being non-target organisms for antibiotics and their ability to adapt and remove the antibiotics themselves from polluted water, producing valuable biomass [11–13]. *Chlorella vulgaris,* a unicellular microalga, is one of the most commonly investigated algae in the treatment of wastewater as it is commonly found in freshwater and soil habitats with fast growth rates and short production time when employed [14]. Antibiotic presence in the wastewater promotes dual responses on *C. vulgaris,* which either includes inhibitory effects or growth stimulation effects based upon the concentration [15]. Inhibitory effects depend on the EC_{50} value after 96 h of exposure. It has been reported that the 96 h EC_{50} value of *C. vulgaris* for CIP is 20.6 mg/L and 96 h EC_{50} for *P. subcapitata* when AMX is less than 50 mg/L [16,17]. *C. vulgaris* is known to adapt to antibiotics stress conditions by spontaneous physiological adaptation and can biodegrade pollutants, becoming a suitable candidate for antibiotic removal from wastewater [18]. Under stress and nutrient limitation, *C. vulgaris* is known to accumulate lipids and store energy in the form of proteins and carbohydrates.

In this study, the growth-stimulation effect caused by the presence of two different antibiotic stress conditions was investigated in comparison with the control medium. *C. vulgaris* medium was spiked with a tolerant concentration of antibiotics, considering EC_{50} value, world median concentration, maximum detected worldwide concentration, and instrument sensitivity for CIP and AMX detections during the analysis. The study lasted 7 days to investigate the mechanisms adapted by algae for the removal of CIP and AMX. The effects of these antibiotics on cell growth, photosynthetic activity (chlorophyll-a), and biochemical composition (lipids, carbohydrates, and proteins concentration) were monitored. The results were encouraging, even if the study needs further experimental runs to define the optimal operative conditions for a sound scale-up.

2. Materials and Methods

2.1. Reagents and Materials

Ciprofloxacin (CIP) (CAS No.: 85721-33-1) and amoxicillin (AMX) (CAS No.: 61336-70-7) (HPLC grade, >98% purity) stock solutions of 100 mg/L were prepared by dissolving reference standards in ultra-pure water and stored in a dark airtight Schott glass bottle at 4 °C for no longer than 10 days. The required concentration was prepared from a stock solution using serial dilution. All reagents and chemicals were of analytical grade.

2.2. Algae and Incubation Conditions

C. vulgaris (BDU GD003) was purchased from NRMC-F (National Repository for Microalgae and Cyanobacteria-Freshwater), Bharathidasan University, Tiruchirappalli, Tamil Nadu, India. The culture was maintained and sub-cultured in BG-11 medium at an orbital shaking incubator under a fluorescent light intensity of 50 μmol photon/m^2/s in 12 h light/12 h dark cycle at 30 °C until the stationary phase was attained.

2.3. Experimental Setup

All the experiments were conducted in batch culture using autoclaved 250 mL Erlenmeyer flasks containing 100 mL of BG-11 medium. A series of runs was conducted for each antibiotic, namely Batch 1 (CIP) and Batch 2 (AMX). In each series, four different experimental conditions (A, B, C, and D) were established for the determination of growth and removal mechanisms by *C. vulgaris*. Table 1 summarizes these conditions. Each run was performed in triplicates. In both batches, in runs A and B, algae cell pellets were inoculated with an average cell density of 5×10^6 cells/mL, and their growth profiles were monitored daily to study the impact of antibiotics.

The control (B) experiments were conducted with the same illumination conditions to elucidate the possible role of abiotic conditions in the removal of antibiotics and biomass production. In both batches, A, C, and D runs were spiked with 5 mg/L of antibiotic standard solution. The experimental runs A and B were incubated at 30 ± 1 °C under a fluorescent light intensity of 50 μmol photon/m^2/s in 12 h light/12 h dark cycles. These experimental conditions were kept for 7 days. In each run, monitoring was carried out by taking 5 mL aliquots of medium for the determination of biomass and antibiotic concentration.

Table 1. Experimental setup conditions and their respective abbreviations.

Batch 1—CIP	Experimental Condition	Culture	Antibiotic	Illumination
A	CIP Test	Algae	+	+
B	CIP Control	Algae	-	+
C	CIP Abiotic light	-	+	+
D	CIP Abiotic Dark	-	+	-
Batch 2—AMX	**Experimental Condition**	**Culture**	**Antibiotic**	**Illumination**
A	AMX Test	Algae	+	+
B	AMX Control	Algae	-	+
C	AMX Abiotic light	-	+	+
D	AMX Abiotic Dark	-	+	-

2.4. Determination of Algal Growth and Biochemical Composition

2.4.1. Algal Growth

Algal growth can be assessed by counting the algal cells using a hemocytometer, optical density (OD), and measuring chlorophyll-a content. The number of cells per mL of *C. vulgaris* was measured using a Neubauer improved hemocytometer under 40× magnification in a trinocular microscope, and the specific growth rate (μ) was calculated by using the following equation [19]:

$$\mu = (\ln N_2 - \ln N_0)/(t_2 - t_0) \quad (1)$$

where N_0 is the cell density at time t_0 (day 0), and N_2 is the cell density at time t_2 (day 7).

The maximum absorbance for *C. vulgaris* was inspected by scanning sample cultures between 500 and 800 nm using a UV-Vis spectrophotometer, and the maximum absorbance was found at 680 nm [20]. On this basis, the spectrophotometer was set to 680 nm wavelength to measure the OD values, as this parameter reflects the cell density in the medium by taking refraction into account.

Chlorophyll-a extraction and the concentration of the extract were calculated using the following equation [19]:

$$\text{Chlorophyll-a (mg/L)} = (9.90 \times OD_{660}) - (0.77 \times OD_{642}) \quad (2)$$

where OD_{660} and OD_{642} are the optical densities of the extracted chlorophyll pigment from the culture at 660 nm and 642 nm.

2.4.2. Biochemical Composition

Algae cells were harvested by centrifugation for the analysis of the biochemical composition. Harvested algae pellets were subjected to sonication for the extraction of lipids and quantified gravimetrically, as reported in [21,22]. Carbohydrates were extracted using the Anthrone method and quantified (in milligram per liter) by UV-Vis spectrophotometer using glucose standards [21]. Proteins were extracted using the Lowry method and quantified by a UV-Vis spectrophotometer using bovine serum albumin (BSA) standards [21]. Dry biomass was calculated gravimetrically by drying the harvested algae cell pellets.

Lipids, carbohydrates, and proteins concentrations were calculated as percentages by biomass weight by the following equation, respectively.

$$\%\text{Lipids} = \text{Lipid mass/Biomass weight} \times 100 \quad (3)$$

$$\%\text{Carbohydrates} = \text{Carbohydrate mass/Biomass weight} \times 100 \quad (4)$$

$$\%\text{Proteins} = \text{Protein mass/Biomass weight} \times 100 \quad (5)$$

2.5. Determination of Antibiotic Concentration

CIP and AMX concentrations were determined according to the USP 28-NF 23 s supplement [23,24], using 844 UV/VIS compact ion chromatography equipped with a Hichrom HPLC column (Alltima 5 µm C18 with dimensions 250 × 4.6 mm). The mobile phase consisted of 15% acetonitrile and 85% ultra-pure water (pH = 3.0) with a flow rate of 1 mL/min and 250 µL injection volume. The wavelength of the UV detector for CIP and AMX was 270 nm (retention time = 25.88 min) and 230 nm (retention time = 4.15 min), respectively. All samples were filtered through a 0.22 µm membrane filter before analysis. The concentration was estimated using IC net 2.3 software integrated with ion chromatography equipment. The overlay curves of the individual sample analysis were plotted using IC net software upon the completion of all analysis.

2.6. Determination of Antibiotic Removal Mechanisms in Algal Cells

Aliquots of 5 mL of microalgal suspension were withdrawn and separated by centrifugation at 2500 rpm for 10 min. The filtered supernatant was then used for analyzing the residual concentration (Cr) of antibiotics in the algal medium.

As a first information, total antibiotic removal can be calculated as follows:

$$\text{Total removal (\%)} = \frac{Ci - Cr}{Ci} \times 100 \quad (6)$$

where Ci is the initial antibiotic concentration in the medium, and Cr is the residual concentration.

Antibiotics tend to be adsorbed on algal cell walls due to the interaction between the pollutant and extra polymeric substances of microalgae. This amount of antibiotic can be desorbed from the cell wall by resuspending the harvested algal pellets with 5 mL of ultra-pure water and centrifuged again carefully by increasing the rotation speed from 2500 to 5000 rpm for 10 min without disrupting the cell wall [25,26]. The filtered supernatant was used for the determination of antibiotics adsorbed on the cell wall (Rad). The bioaccumulation (Rac) mechanism was determined by using the sonication method [27]. The centrifuged pellet was again suspended by adding 5 mL dichloromethane-methanol solution (concentration 1:2 by volume), sonicated for 30 min, and then centrifuged again for the analysis of antibiotics accumulated inside the algal cells.

As aforesaid, an amount of antibiotic can be removed in the abiotic condition (R_a) by photodegradation.

$$R_a\ (\%) = \frac{Cr(\text{abiotic dark}) - Cr(\text{abiotic light})}{Ci} \times 100 \quad (7)$$

At last, the biotic removal (R_b) of antibiotics in the *C. vulgaris* medium was calculated by the equation given by Xiong et al. [28], taking into account the adsorption, bioaccumulation, and abiotic removal.

$$R_b\ (\%) = \frac{(Ci - Cr - Rad - Rac - R_a)}{Ci} \times 100 \quad (8)$$

2.7. Statistical Analysis

All experiments were carried out in triplicates and the average results were reported. The data obtained from different experimental conditions were compared by ANOVA, having statistical significance at $p < 0.05$. All statistical analyses and graph plotting were carried out with JMP 16.2 software.2.1.1.

3. Results

3.1. Effect of Antibiotics on Algal Growth

C. vulgaris growth in the experimental conditions A and B for Batch 1 and Batch 2 was evaluated, and the results are shown in Figure 1. Previous toxicological studies have shown that a low concentration of antibiotics in the system can stimulate algal growth, whereas a concentration higher than EC_{50} will have toxic effects on the algal structures [15,29,30].

Figure 1a,b show the effect of CIP (a) and AMX (b), respectively, on the chlorophyll-a concentration, OD_{680}, and cell density in runs A and B for the tested batches.

In both batches, these parameters declined until day 3 in experimental condition A, compared to run B, indicating the stress caused by antibiotics upon the culture in the medium.

On day 5, in both batches, there was an increase in chlorophyll-a concentration, OD_{680}, and cell density for run A, indicating that the response mechanisms of *C. vulgaris* adapted and grew in the stress conditions induced by CIP and AMX.

The specific growth rate for the experimental condition B in Batch 1 was found to be $-0.03\ d^{-1}$ and $-0.029\ d^{-1}$ for Batch 2. A negative growth rate indicates the decline of algal cells due to the absence of nutrients in the medium, whereas for run A, it was found to be $0.04\ d^{-1}$ in Batch 1 and $0.038\ d^{-1}$ in Batch 2, respectively. A positive growth rate indicates the utilization of antibiotics as a carbon source for their growth.

To summarize, for both batches, the results revealed that induced stress caused by CIP and AMX stimulated algal growth. This is clearly evidenced by the comparison of the data achieved in runs A and B.

3.2. Effect of Antibiotics on Algal Biochemical Composition (Lipids, Carbohydrates, and Proteins)

Cell growth and biochemical composition accumulation are closely related to the adopted environmental conditions, such as light intensity, nutrient concentration in the medium, stress conditions, and CO_2 concentration [31–33]. During algal photosynthesis, chlorophyll transforms light energy into adenosine triphosphate (ATP) and nicotinamide adenine dinucleotide phosphate (NADPH), which converts the carbon source and CO_2 into glyceraldehydes-3-phosphate (G3P) during reaction in dark conditions [34]. G3P in the glycolytic pathway results in the biosynthesis of carbohydrates, while a part of G3P will be transformed into acetyl-CoA and pyruvate by the action of glycolysis and take part in the formation of protein. Acetyl-CoA catalyzed by acetyl-CoA carboxylase is converted into malonyl-CoA, which plays an important role in the synthesis of fatty acids.

The produced fatty acids becomes accumulated in the form of lipids in the chloroplast of the cell [35]. Chlorophyll-a, as a major light-harvesting pigment, releases protons during photosynthesis and enhances acetyl-CoA carboxylase activities inside the cell, thus increasing the production of biochemicals inside the cell.

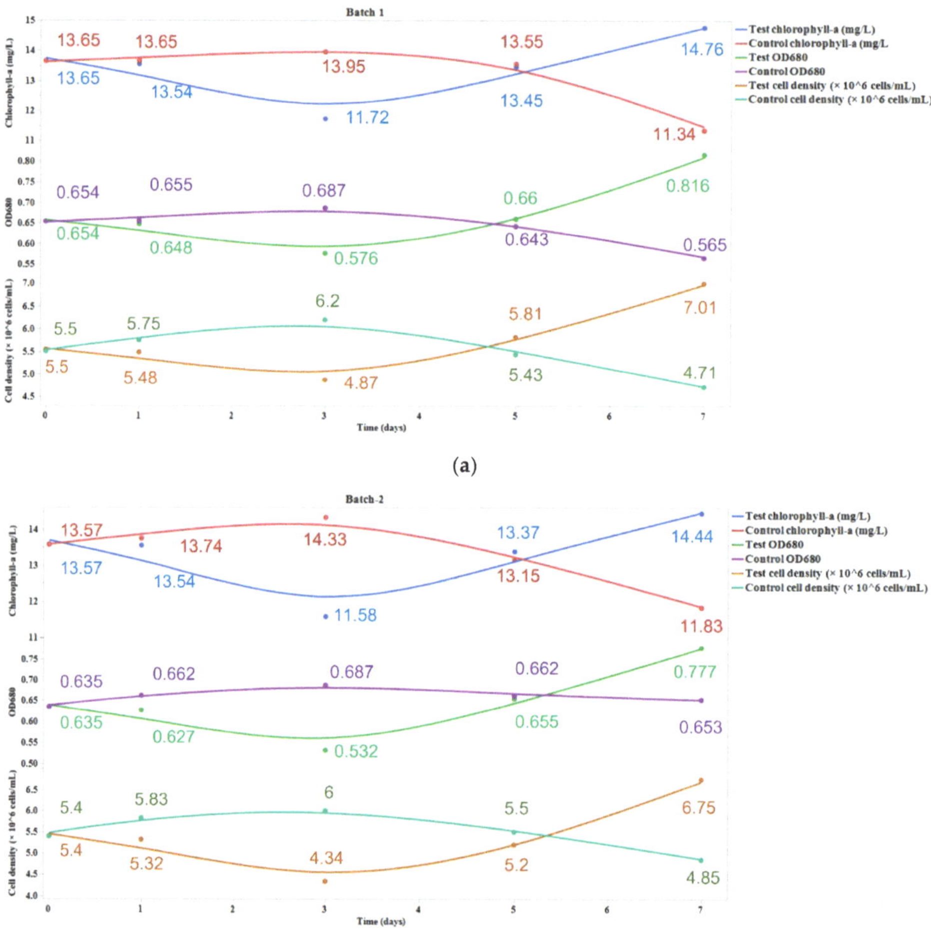

(b)

Figure 1. Effect of (**a**) CIP and (**b**) AMX on chlorophyll-a concentration, OD_{680}, and cell density.

Figure 2 reports the results achieved in the runs. The data refer to the lipid, carbohydrate, and protein concentration at the end of runs A and B after 7 days of incubation.

In this study, it was observed that lipids, carbohydrates, and protein increased in both the tested batches (1 and 2) compared to the control conditions, as shown in Figure 2, and the biomass dry weight for the experimental condition A of Batch 1 and Batch 2 is 72% and 41% higher than the control conditions. Furthermore, it can be noted that in run A of Batch 1, the lipid (464 mg/L), carbohydrate (39 mg/L), and protein (608 mg/L) accumulations were higher than the ones achieved in run A of Batch 2, where lipid, carbohydrate, and protein concentrations were 438 mg/L, 36 mg/L, and 580 mg/L, respectively (Figure 2). *C. vulgaris* adapted to CIP stress conditions and utilized this antibiotic better than with AMX. Previous studies have shown that the EC_{50} value for CIP is very low compared to AMX [16,17].

This indicates that, where CIP is more toxic to algae, it induces stress on *C. vulgaris* and because of this stress, there is an increase in chlorophyll-a, lipid, carbohydrate, and protein production in run A of Batch 1 compared to experimental condition A of Batch 2.

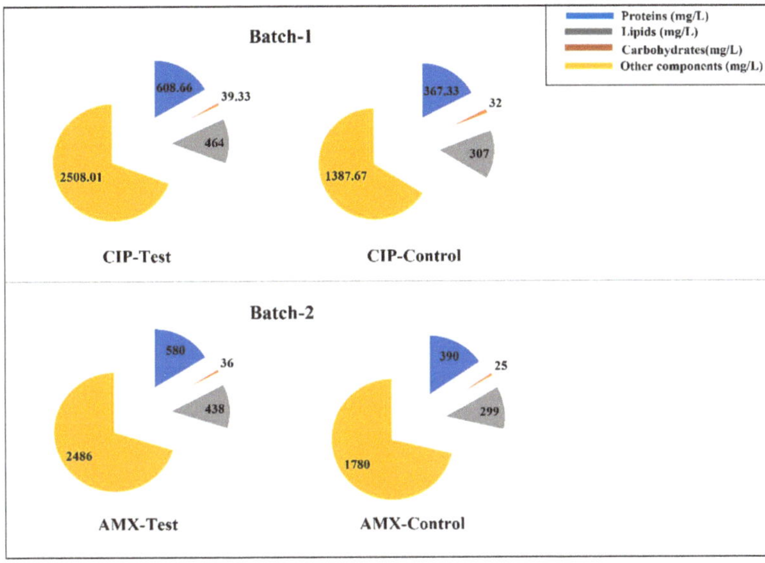

Figure 2. Effect of CIP and AMX upon lipid, carbohydrate, and protein accumulation in *Chlorella vulgaris* (incubation time = 7 days).

3.3. Antibiotics Removal Mechanisms Adopted by Algae

Removal mechanisms were determined from the experimental conditions given in Table 1. Bioadsorption, bioaccumulation, and biodegradation are major removal mechanisms adopted by algae to remove the organic contaminants [28]. The contribution of each mechanism to the total removal is shown in Table 2.

In this study, the removal of CIP and AMX via photodegradation (abiotic condition) is determined in runs C and D. This mechanism was negligible for CIP under the tested conditions, as shown by the value reported in Table 2.

Bioadsorption is an extracellular mechanism accomplished by the polymer assemblages (cellulose, hemicellulose, and proteins) and functional groups present on the cell wall [36]. Bioadsorption and bioaccumulation were not the major removal mechanisms in this study, and even other researchers have reported that they lead to the process for the biodegradation mechanism inside the cell.

In the current study, biodegradation accounts for the major removal mechanism, 76% for CIP and 46% for AMX, respectively. These values indicate the utilization and breakdown of antibiotics as a carbon source for their cellular growth. Figure 3 shows the process involved in the removal of antibiotics by algae and its effects on biochemical composition with the overlay peak curves of HPLC analysis performed to determine the contribution of each mechanism for the removal of antibiotics.

The increase in chlorophyll-a concentration, cell density, and biochemical composition (lipids, carbohydrates, and proteins) in the test conditions (run A), compared to the control medium (run B), indicates the response mechanism adopted by algae to utilize CIP and AMX after the depletion of nutrients in the medium.

Table 2. Antibiotic removal (%) and removal contribution (%) of mechanisms after 7 days.

Antibiotic	Total Removal	Photodegradation	Bioadsorption	Bioaccumulation	Biodegradation
CIP	36.9 ± 1.1	0.54 ± 0.02	15.4 ± 0.4	7.92 ± 0.06	76.09 ± 0.55
AMX	24. 7 ± 1.0	24.44 ± 9.71	18.48 ± 5.64	10.84 ± 9.56	46.23 ± 5.89

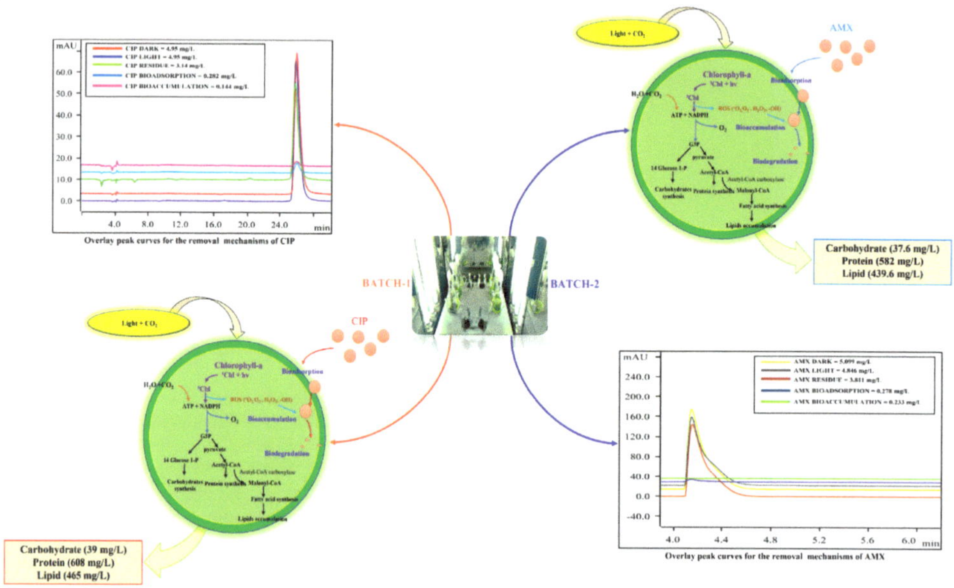

Figure 3. Removal mechanisms adopted by algae for their growth and their effects on lipids, carbohydrates, and proteins.

4. Discussion

Studies have reported that algae produce free radicals such as peroxyl radicals, single oxygen, and hydroxyl radicals during photolysis [37]. In photosynthesis, a single electron of chlorophyll molecule is excited to a higher energy state within the photosystem to form an excited triple-state chlorophyll molecule and produces free radicals in the medium. This helps to the breakdown of carbon sources for their growth [38] and at the same time for the production of lipids, carbohydrates, and proteins inside the algae themselves. However, the amounts of these biochemicals depend on the kind of algae and antibiotics.

About antibiotics, their removal depends strictly on their classes; that is to say that different antibiotics can be removed at different extents, even when they are present at the same concentrations.

The global average concentration of CIP and AMX in the surface freshwaters is in the range of 10–100 ng/L, whereas the maximum detected concentration of CIP in a lake is in the range of 2.5–6.5 mg/L [8]. In our study, we conducted the experimentation at a fixed concentration of 5 mg/L to provide insights on the impacts of these pollutants at selected concentrations. Antibiotics at lower concentrations have the potential to alter the community structure of algae in the surface waters, and it has been reported that altering the community structure can contribute to algal blooms [39]. The present study demonstrated that a concentration of 5 mg/L CIP and AMX not only stimulated growth but also removed antibiotics from the media by the mechanism of bioadsorption, bioaccumulation, and biodegradation.

The bioadsorption of compounds can be assessed by the octanol and water partition coefficient (log K_{ow}) value, which determines whether the compound is hydrophobic or lipophilic in nature [40]. The higher the value, the higher the adsorption of compounds onto the surface of the microorganism. The log K_{OW} value for AMX is 0.87, which is slightly higher than the one for CIP (0.28) [41]. The results demonstrate that the bioadsorption of AMX is higher than the CIP one. Bioaccumulation and bioadsorption are a continuous process, and adsorbed antibiotics become accumulated inside the cells through cell membrane diffusion [42]. AMX is more accumulated than CIP. However, being more toxic, CIP induces stress upon algae. To counteract this accumulated toxicity, *C. vulgaris* generates free radicals inside the cell by increasing the photosynthetic activity of the cell, and this helps in the process of biodegradation.

In this study, biodegradation was the major removal mechanism, followed by bioadsorption and bioaccumulation in the CIP and AMX test.

In line with this, CIP removal was higher than the AMX one, namely about 37% (CIP) against 25% (AMX). Taking these results into consideration, it is evident that *C. vulgaris* is more suitable to remove CIP than AMX. This can be justified by the chemistry of the tested antibiotics: CIP belongs to the fluoroquinolones, whereas AMX belongs to β-lactams, and their affinity to *C. vulgaris* is different.

It is interesting to note that, for both antibiotics, the chlorophyll-a concentration, cell density, and specific growth rates are very similar after 7 days.

About CIP, the achieved removal efficiency is lower than the one obtained by Hom-Diaz et al. [30], who studied the removal of this antibiotic by *C. sorokiniana*. Its efficiency was around 50% after 14 days, starting with a CIP concentration equal to 0.1 mg/L, which is much lower than the value used in the current study (5 mg/L). The longer process duration (more than double) could justify the higher removal efficiency, more than the different genus of *Chlorella*. Hom-Diaz et al. [30] found photodegradation as the main mechanism responsible for CIP removal. This result is in contrast with what was obtained in the current study, where photodegradation does not seem to be active in CIP removal (Table 2). The use of *C. vulgaris* instead of *C. sorokiniana* could be the reason. Moreover, the photodegradation process mainly depends on light intensity. Biodegradation showed to be the main removal mechanism also in the study of Xie et al. [43], where wastewater containing 5-mg/L of CIP was treated with *Chlamydomonas* sp. Tai-03 for 9 days.

Xiong et al. [28] used *C. vulgaris* to remove levofloxacin, a fluoroquinolone, from an initial concentration equal to 5 mg/L. After 7 days, about 15% of the antibiotic was removed, which is much lower than the amount achieved in the present study (37%). However, it must be evidenced also that these authors found biodegradation as the main mechanism able to remove the antibiotic and not activity by photodegradation, as was observed in this study. The reason could be the same algal kind.

For AMX, removal efficiency was 25%. Zhao et al. [44] studied the removal of AMX by *Chlorella regularis*. In their test, AMX was initially present at a concentration equal to 3 mg/L, and after 7 days, the concentration was reduced to 45%. Their cell density was 15×10^6 cells/mL, against about 7×10^6 cells/mL of the current study. Zhao et al. [44] also checked the concentration of lipids, carbohydrates, and proteins after 18 days, achieving values around 420 mg/L, 120 mg/L, and 120 mg/L, respectively. Notwithstanding the different removal efficiency, the lipid concentration is completely in line with the current one (440 mg/L), whereas carbohydrate concentration is higher (38 mg/L, in the current study), and the protein concentration is much lower (in this study, it is 580 mg/L). The comparison of these concentrations is not so easy due to the different processing times, which for Zhao et al. [44] was three times more.

A similar study was carried out by Zhang et al. [33] for the removal of AMX (starting concentration = 5 mg/L) by *C. regularis* for 5 days. At the run's end, about 90% of AMX was still present in the medium. At that time, cell density was around 3.5×10^6 cells/mL, and this value can confirm the low removal efficiency. The authors also analyzed the concentration of lipids, carbohydrates, and proteins at day 5, and the results were rather low at about 95 mg/L, 13 mg/L, and 13 mg/L, respectively. These values are much lower than the concentrations achieved in the current study. Both studies making use of

C. regularis show that the concentrations of lipids are always much higher than the other compounds (carbohydrates and proteins), which have the same concentration, even if at a different values: 120 mg/L in the study of Zhao et al. [44] and 13 mg/L in the study of Zhang et al. [33]. This does not hold for the current study, where protein production reached a very high concentration and was always over the lipid's concentration. At the moment, no hypothesis can be suggested and verified for this.

The need for further studies is evident to optimize the process in view of its scale-up.

This study provides new insights in utilizing microalgae for the treatment of CIP and AMX-polluted wastewater to remove these antibiotics. Moreover, the results demonstrated a second advantage linked to the removal and represented by valuable biomass production, containing lipids, carbohydrates, and proteins.

Author Contributions: Conceptualization, S.S.; methodology, S.S.; software, R.R.; validation, S.S.; formal analysis, R.R.; investigation, R.R.; data curation, F.C.; writing—original draft preparation, R.R.; writing—review and editing, F.C.; visualization, R.R.; supervision, S.S.; project administration, S.S.; funding acquisition, S.S. All authors have read and agreed to the published version of the manuscript.

Funding: Vellore Institute of Technology financially supported the research by a seed grant.

Institutional Review Board Statement: Not applicable.

Informed Consent Statement: Not applicable.

Conflicts of Interest: The authors declare no conflict of interest.

References

1. Danner, M.; Robertson, A.; Behrends, V.; Reiss, J. Science of the Total Environment Antibiotic pollution in surface fresh waters: Occurrence and effects. *Sci. Total Environ.* **2019**, *664*, 793–804. [CrossRef] [PubMed]
2. Liu, L.; Wu, W.; Zhang, J.; Lv, P.; Xu, L.; Yan, Y. Progress of research on the toxicology of antibiotic pollution in aquatic organisms. *Acta Ecol. Sin.* **2018**, *38*, 36–41. [CrossRef]
3. Editors, P.M. Antimicrobial Resistance: Is the World UNprepared? *PLoS Med.* **2016**, *13*, e1002130. [CrossRef]
4. Kelly, K.R.; Brooks, B.W. Global Aquatic Hazard Assessment of Ciprofloxacin: Exceedances of Antibiotic Resistance Development and Ecotoxicological Thresholds. In *Progress in Molecular Biology and Translational Science*; Teplow, D., Ed.; Academic Press: Cambridge, MA, USA, 2018; Volume 159, pp. 59–77. ISBN 9780128162354.
5. Freitas-Silva, J.; Silva-Oliveira, T.; Muricy, G.; Laport, M.S. Bacillus Strains Associated to Homoscleromorpha Sponges are Highly Active Against Multidrug Resistant Bacteria. *Curr. Microbiol.* **2020**, *77*, 807–815. [CrossRef] [PubMed]
6. Szymańska, U.; Wiergowski, M.; Sołtyszewski, I.; Kuzemko, J.; Wiergowska, G.; Woźniak, M.K. Presence of antibiotics in the aquatic environment in Europe and their analytical monitoring: Recent trends and perspectives. *Microchem. J.* **2019**, *147*, 729–740. [CrossRef]
7. Hughes, S.R.; Kay, P.; Brown, L.E. Global synthesis and critical evaluation of pharmaceutical data sets collected from river systems. *Environ. Sci. Technol.* **2013**, *47*, 661–677. [CrossRef]
8. Kovalakova, P.; Cizmas, L.; McDonald, T.J.; Marsalek, B.; Feng, M.; Sharma, V.K. Occurrence and toxicity of antibiotics in the aquatic environment: A review. *Chemosphere* **2020**, *251*, 126351. [CrossRef]
9. Lee, S.; Kim, C.; Liu, X.; Lee, S.; Kho, Y.; Kim, W.K.; Kim, P.; Choi, K. Ecological risk assessment of amoxicillin, enrofloxacin, and neomycin: Are their current levels in the freshwater environment safe? *Toxics* **2021**, *9*, 196. [CrossRef]
10. Välitalo, P.; Kruglova, A.; Mikola, A.; Vahala, R. Toxicological impacts of antibiotics on aquatic micro-organisms: A mini-review. *Int. J. Hyg. Environ. Health* **2017**, *220*, 558–569. [CrossRef]
11. Leng, L.; Wei, L.; Xiong, Q.; Xu, S.; Li, W.; Lv, S.; Lu, Q.; Wan, L.; Wen, Z.; Zhou, W. Use of microalgae based technology for the removal of antibiotics from wastewater: A review. *Chemosphere* **2020**, *238*, 124680. [CrossRef]
12. Li, S.; Show, P.L.; Ngo, H.H.; Ho, S.-H. Algae-mediated antibiotic wastewater treatment: A critical review. *Environ. Sci. Ecotechnol.* **2022**, *9*, 100145. [CrossRef]
13. Wollmann, F.; Dietze, S.; Ackermann, J.U.; Bley, T.; Walther, T.; Steingroewer, J.; Krujatz, F. Microalgae wastewater treatment: Biological and technological approaches. *Eng. Life Sci.* **2019**, *19*, 860–871. [CrossRef] [PubMed]
14. Wirth, R.; Pap, B.; Böjti, T.; Shetty, P.; Lakatos, G.; Bagi, Z.; Kovács, K.L.; Maróti, G. *Chlorella vulgaris* and Its Phycosphere in Wastewater: Microalgae-Bacteria Interactions During Nutrient Removal. *Front. Bioeng. Biotechnol.* **2020**, *8*, 1108. [CrossRef] [PubMed]
15. Mao, Y.; Yu, Y.; Ma, Z.; Li, H.; Yu, W.; Cao, L.; He, Q. Azithromycin induces dual effects on microalgae: Roles of photosynthetic damage and oxidative stress. *Ecotoxicol. Environ. Saf.* **2021**, *222*, 112496. [CrossRef]
16. Nie, X.; Wang, X.; Chen, J.; Zitko, V.; An, T. Response of the freshwater alga *Chlorella vulgaris* to trichloroisocyanuric acid and ciprofloxacin. *Environ. Toxicol. Chem.* **2008**, *27*, 168–173. [CrossRef]

17. Andreozzi, R.; Caprio, V.; Ciniglia, C.; De Champdoré, M.; Lo Giudice, R.; Marotta, R.; Zuccato, E. Antibiotics in the environment: Occurrence in Italian STPs, fate, and preliminary assessment on algal toxicity of amoxicillin. *Environ. Sci. Technol.* **2004**, *38*, 6832–6838. [CrossRef]
18. Hena, S.; Gutierrez, L.; Croué, J.P. Removal of metronidazole from aqueous media by *C. vulgaris*. *J. Hazard. Mater.* **2020**, *384*, 121400. [CrossRef]
19. Turner, M. *Microalgae—Biotechnology and Microbiology*; Cambridge University Press: Cambridge, UK, 1994; Volume 183, ISBN 0521350204.
20. Santos-Ballardo, D.U.; Rossi, S.; Hernández, V.; Gómez, R.V.; del Carmen Rendón-Unceta, M.; Caro-Corrales, J.; Valdez-Ortiz, A. A simple spectrophotometric method for biomass measurement of important microalgae species in aquaculture. *Aquaculture* **2015**, *448*, 87–92. [CrossRef]
21. Bligh, E.G.; Dyer, W.J. A rapid method of total lipid extraction and purification. *Can. J. Biochem. Physiol.* **1959**, *37*, 911–917. [CrossRef]
22. Chen, Y.; Vaidyanathan, S. Simultaneous assay of pigments, carbohydrates, proteins and lipids in microalgae. *Anal. Chim. Acta* **2013**, *776*, 31–40. [CrossRef]
23. Raju, C.B.V.N.; Sharma, H.K.; Rao, S.C.; Rao, G. RP-HPLC method for analysis of related substances in amoxicillin drug substance. *Acta Chromatogr.* **2009**, *21*, 57–70. [CrossRef]
24. Pistos, C.; Tsantili-Kakoulidou, A.; Koupparis, M. Investigation of the retention/pH profile of zwitterionic fluoroquinolones in reversed-phase and ion-interaction high performance liquid chromatography. *J. Pharm. Biomed. Anal.* **2005**, *39*, 438–443. [CrossRef] [PubMed]
25. Bai, X.; Acharya, K. Removal of trimethoprim, sulfamethoxazole, and triclosan by the green alga *Nannochloris* sp. *J. Hazard. Mater.* **2016**, *315*, 70–75. [CrossRef] [PubMed]
26. Xiong, J.-Q.; Kurade, M.B.; Abou-Shanab, R.A.I.; Ji, M.-K.; Choi, J.; Kim, J.O.; Jeon, B.-H. Biodegradation of carbamazepine using freshwater microalgae Chlamydomonas mexicana and Scenedesmus obliquus and the determination of its metabolic fate. *Bioresour. Technol.* **2016**, *205*, 183–190. [CrossRef] [PubMed]
27. Ding, T.; Yang, M.; Zhang, J.; Yang, B.; Lin, K.; Li, J.; Gan, J. Toxicity, degradation and metabolic fate of ibuprofen on freshwater diatom *Navicula* sp. *J. Hazard. Mater.* **2017**, *330*, 127–134. [CrossRef]
28. Xiong, J.Q.; Kurade, M.B.; Jeon, B.H. Biodegradation of levofloxacin by an acclimated freshwater microalga, *Chlorella vulgaris*. *Chem. Eng. J.* **2017**, *313*, 1251–1257. [CrossRef]
29. Song, C.; Wei, Y.; Qiu, Y.; Qi, Y.; Li, Y.; Kitamura, Y. Biodegradability and mechanism of florfenicol via *Chlorella* sp. UTEX1602 and L38: Experimental study. *Bioresour. Technol.* **2019**, *272*, 529–534. [CrossRef]
30. Hom-Diaz, A.; Jaén-Gil, A.; Rodríguez-Mozaz, S.; Barceló, D.; Vicent, T.; Blánquez, P. Insights into removal of antibiotics by selected microalgae (*Chlamydomonas reinhardtii*, *Chlorella sorokiniana*, *Dunaliella tertiolecta* and *Pseudokirchneriella subcapitata*). *Algal Res.* **2022**, *61*, 102560. [CrossRef]
31. Sathasivam, R.; Radhakrishnan, R.; Hashem, A.; Abd_Allah, E.F. Microalgae metabolites: A rich source for food and medicine. *Saudi J. Biol. Sci.* **2019**, *26*, 709–722. [CrossRef]
32. Lv, J.M.; Cheng, L.H.; Xu, X.H.; Zhang, L.; Chen, H.L. Enhanced lipid production of *Chlorella vulgaris* by adjustment of cultivation conditions. *Bioresour. Technol.* **2010**, *101*, 6797–6804. [CrossRef]
33. Zhang, C.; Zhang, Q.; Dong, S.; Zhou, D. Could co-substrate sodium acetate simultaneously promote Chlorella to degrade amoxicillin and produce bioresources? *J. Hazard. Mater.* **2021**, *417*, 126147. [CrossRef] [PubMed]
34. Debnath, C.; Bandyopadhyay, T.K.; Bhunia, B.; Mishra, U.; Narayanasamy, S.; Muthuraj, M. Microalgae: Sustainable resource of carbohydrates in third-generation biofuel production. *Renew. Sustain. Energy Rev.* **2021**, *150*, 111464. [CrossRef]
35. Alishah Aratboni, H.; Rafiei, N.; Garcia-Granados, R.; Alemzadeh, A.; Morones-Ramírez, J.R. Biomass and lipid induction strategies in microalgae for biofuel production and other applications. *Microb. Cell Fact.* **2019**, *18*, 178. [CrossRef] [PubMed]
36. Xiong, Q.; Hu, L.X.; Liu, Y.S.; Zhao, J.L.; He, L.Y.; Ying, G.G. Microalgae-based technology for antibiotics removal: From mechanisms to application of innovational hybrid systems. *Environ. Int.* **2021**, *155*, 106594. [CrossRef] [PubMed]
37. Bai, X.; Liang, W.; Sun, J.; Zhao, C.; Wang, P.; Zhang, Y. Enhanced production of microalgae-originated photosensitizer by integrating photosynthetic electrons extraction and antibiotic induction towards photocatalytic degradation of antibiotic: A novel complementary treatment process for antibiotic removal from effluent of conventional biological wastewater treatment. *J. Environ. Manag.* **2022**, *308*, 114527. [CrossRef]
38. Petsas, A.S.; Vagi, M.C. Effects on the Photosynthetic Activity of Algae after Exposure to Various Organic and Inorganic Pollutants: Review. In *Chlorophyll*; Jacob-Lopez, E., Queiroz Zepka, L., Queiroz, M.I., Eds.; IntechOpen: Rijeka, Croatia, 2017; Chapter 4, ISBN 978-953-51-3108-3.
39. Ma, J.; Chen, J.; Wang, P.; Tong, S. Comparative sensitivity of eight freshwater phytoplankton species to isoprocarb, propargite, flumetralin and propiconazol. *Pol. J. Environ. Stud.* **2008**, *17*, 525–529.
40. Rogers, H.R. Sources, behaviour and fate of organic contaminants during sewage treatment and in sewage sludges. *Sci. Total Environ.* **1996**, *185*, 3–26. [CrossRef]
41. Anjali, R.; Shanthakumar, S. Insights on the current status of occurrence and removal of antibiotics in wastewater by advanced oxidation processes. *J. Environ. Manag.* **2019**, *246*, 51–62. [CrossRef]

42. Yu, Y.; Zhou, Y.; Wang, Z.; Torres, O.L.; Guo, R.; Chen, J. Investigation of the removal mechanism of antibiotic ceftazidime by green algae and subsequent microbic impact assessment. *Sci. Rep.* **2017**, *7*, 4168. [CrossRef]
43. Xie, P.; Chen, C.; Zhang, C.; Su, G.; Ren, N.; Ho, S.H. Revealing the role of adsorption in ciprofloxacin and sulfadiazine elimination routes in microalgae. *Water Res.* **2020**, *172*, 115475. [CrossRef]
44. Zhao, Z.; Xue, R.; Fu, L.; Chen, C.; Ndayisenga, F.; Zhou, D. Carbon dots enhance the recovery of microalgae bioresources from wastewater containing amoxicillin. *Bioresour. Technol.* **2021**, *335*, 125258. [CrossRef] [PubMed]

MDPI AG
Grosspeteranlage 5
4052 Basel
Switzerland
Tel.: +41 61 683 77 34

Bioengineering Editorial Office
E-mail: bioengineering@mdpi.com
www.mdpi.com/journal/bioengineering

www.ingramcontent.com/pod-product-compliance
Lightning Source LLC
Chambersburg PA
CBHW041410210926
51850CB00013B/3228
* 9 7 8 3 7 2 5 8 2 9 3 2 3 *